Praise for *Visions of Technology*

"An impressionistic image revealing the edgy American ambivalence toward the technologies that, mostly, our nation pioneered. . . . Intrigued readers are bound to want more."

—Lynn Yarris, *San Jose Mercury News*

"Fascinating. . . . Rhodes has done a wonderful job of selecting the works and placing them in an order that keeps the reader delightfully jumping along from one to the next."

—Tom Danehy, *Tucson Weekly*

"An interesting collection of what well-known people thought would happen as we moved into the twentieth century with all this newfangled technology. . . . All in all, a fascinating book."

—G. William Gray, *The Tampa Tribune*

"A provocative collection dedicated to the twentieth century's passion for technology. Voices from 1900 to the present speak for themselves, including visionaries from Robert Frost to Senator John Glenn and Henry Ford. This book is significant because of its point of view: different thinkers draw vastly different conclusions, some leaving hope for the future, others despairing of technology's negative impact. All explicate the potential and prospects of technology with wit and profundity. A chronological time line leads the reader through the evolution of technology. Highly recommended."

—*Library Journal*

"For those who like to study predictions, *Visions of Technology* offers plenty—everything from the foretelling of airplanes to sex with robots. The reader will be amazed by what the prognosticators got right, and perhaps smile at what they got wrong."

—Jeff Minerd, *The Futurist*

"A fascinating compendium of scientific writing, criticism, and satire. . . . *Visions of Technology* has a new surprise lurking in every page crease."

—*Flaunt* magazine

"An eclectic collection of 200 brief excerpts providing a series of chronologically arranged perspectives on the technological advances of the century just ending. The tale is extraordinary."

—Karl Giberson, *Books & Culture*

OTHER BOOKS BY RICHARD RHODES

Visions of

Technology

A Century of Vital Debate About Machines,
Systems and the Human World

EDITED BY

RICHARD RHODES

A TOUCHSTONE BOOK
PUBLISHED BY SIMON & SCHUSTER
NEW YORK LONDON TORONTO SYDNEY SINGAPORE

For Elting E. Morison
1909–1995

TOUCHSTONE
Rockefeller Center
1230 Avenue of the Americas
New York, NY 10020

Designed by Ruth Lee
Photo research by Natalie Goldstein
Research associates to Mr. Rhodes: Stephen Kim and Jeff Wheelwright

Manufactured in the United States of America

10 9 8 7 6 5 4 3 2 1

The Library of Congress has cataloged the Simon & Schuster edition as follows:

Visions of technology : a century of vital debate about machines, systems, and the
   human world / edited by Richard Rhodes.
      p.   cm.   —(The Sloan technology series)
   Includes bibliographical references and index.
      1. Technology—History—20th century. 2. Technology—United States—
   History—20th century. I. Rhodes, Richard. II. Series.
   T20.V57   1999
   609'.04—DC21   98-37209
                                          CIP

ISBN 0-684-83903-2
      0-684-86311-1 (Pbk)
The author gratefully acknowledges permission from the following sources to reprint
material in their control:

   Edward Abbey Estate and Don Congdon Associates, Inc., for an excerpt from
Desert Solitaire by Edward Abbey © 1968 by Edward Abbey, renewal © 1996 by
Clarke Abbey.
   Academic Press for an excerpt from *Laser Pioneers* (Revised Edition) by Jeff Hecht
© 1991 by Jeff Hecht.

(continued on page 396)

Contents

II. DEPRESSION AND WAR: 1932-1945
111

III. POSTWAR BOOM: 1945-1970
169

IV. YESTERDAY, TODAY AND TOMORROW: 1970-
297

Preface to the Sloan Technology Series

Technology is the application of science, engineering and industrial organization to create a human-built world. It has led, in developed nations, to a standard of living inconceivable a hundred years ago. The process, however, is not free of stress; by its very nature, technology brings change in society and undermines convention. It affects virtually every aspect of human endeavor: private and public institutions, economic systems, communications networks, political structures, international affiliations, the organization of societies and the condition of human lives. The effects are not one-way; just as technology changes society, so too do societal structures, attitudes and mores affect technology. But perhaps because technology is so rapidly and completely assimilated, the profound interplay of technology and other social endeavors in modern history has not been sufficiently recognized.

The Sloan Foundation has had a long-standing interest in deepening public understanding about modern technology, its origins and its impact on our lives. The Sloan Technology Series, of which the present volume is a part, seeks to present to the general reader the stories of the development of critical twentieth-century technologies. The aim of the series is to convey both the technical and human dimensions of the subject: the invention and effort entailed in devising the technologies and the comforts and stresses they have introduced into contemporary life. As the century draws to an end, it is hoped that the series will disclose a past that might provide perspective on the present and inform the future.

The Foundation has been guided in its development of the Sloan Technology Series by a distinguished advisory committee. We express deep gratitude to John Armstrong, Simon Michael Bessie, Samuel Y. Gibbon, Thomas P. Hughes, Victor McElheny, Robert K. Merton, Elting E. Morison (deceased) and Richard Rhodes. The Foundation has been represented on the committee by Ralph E. Gomory, Arthur L. Singer Jr., Hirsh G. Cohen and Doron Weber.

ALFRED P. SLOAN FOUNDATION

See page 395 for a list of Sloan Technology Series titles.

Introduction

Richard Rhodes

The Western world has argued passionately about technology—what it is, where it's going, whether it's good or bad for us—throughout the twentieth century, even while inventing it at a ferocious and accelerating rate. This anthology samples that vital debate, drawing primarily on American sources. It's an impressionistic sampling. It had to be, given the sheer volume of statements, articles, books and documents generated across a hundred years. I sorted for variety, for felicity and succinctness of expression, for range not only of subject but also of mood. I looked for humor to balance solemnity, prediction to balance explication, recollection to balance abstraction. I included a share of the canonical texts all of us have heard (or, sometimes, misheard)—H. G. Wells's prediction of atomic bombs, Arthur C. Clarke's vision of geosynchrony, Murphy's Law, Moore's Law, the silent spring of Rachel Carson. I left out most commentary on medicine, which is regularly attended because of its mortal impact on our lives. Something you think should be here is probably missing; but I hope you will also be surprised by what you find. If my witching methods work, drinking from this particular Pierian spring may at least leave you thirsty to explore the original texts, a mighty river of discourse. Those texts are referenced in the bibliography that begins on page 381.

I could pretend innocence of America's environmental and cultural wars and say that technology is human making. At first inspection, it is that—from lemon pie to computer chips, from plowshares to gene sequencers. Along with language, it's what distinguishes us from the other species with which we share the planet. People used to speak of craft or "practical arts"; in that guise, technology has been around for a good two million years. The Pleistocene spearpoint flaked from pink flint that I display on my coffee table was the high technology of its day, as sophisticated and effective as a samurai sword or a fighter jet.

But for many of us, "technology" means something more specific (and problematic) than craft or practical art; hearing the word applied to cooking or gardening might surprise us, at least initially. In this more recent sense, the term came into use only about 150 years ago, adapted from a classical Greek noun meaning a systematic treatment, as of grammar or philosophy. Ask a friend today to define technology and you might hear words like "machines," "engineering," "science." Most of us aren't even sure where science leaves off and technology begins. Neither are the experts. We also usually think of technology as hardware rather than software, although many organized systems serve technological ends no less than machines; your computer isn't complete without its programs. Arguably the greatest technological triumph of the century has been the public-health system, which is sophisticated preventive and investigative medicine organized around mostly low- and medium-tech equipment; as two demographers report late in this book, fully half of us are alive today because of its improvements.

This barely acknowledged distinction between old, low technology and new, high technology probably reflects our ambivalence toward machines and machinelike systems, especially when they're recent and unfamiliar, and particularly when they're large and not in our control. We swim in technology as fish swim in the sea, depend on it from birth to final hours, but many of us trust it only in its older and more familiar guises. Recent technology is more often seen as a threat—to our jobs, our health, our values—than a blessing. Even public health takes its lumps; I've had otherwise decent people, people who donate a share of their worldly goods to feed the poor, tell me that saving all those lives just crowds the planet. Your life too, I argue, and they nod guiltily, embarrassed but unwilling to concede the point.

Technological wariness is an enduring disturbance, with roots in religion. Prometheus stealing fire from the gods and giving it to humans carries the sense of it; so does the serpent persuading Eve to taste the knowledgeable apple, and the Jewish myth of the Golem, a Frankenstein's monster animated by incorporations of holy words. (Stanislaw Ulam, the Polish mathematician who conceived the design breakthrough that led to the development of the U.S. hydrogen bomb, once told MIT polymath Norbert Wiener that the Golem had been in his family, since he was a descendant of the rabbi supposed to have constructed it; Wiener, thinking of the Bomb, responded, "It still is.") Technology competes with the gods at miracleworking and the gods take revenge: no wonder we're nervous about it. At its most fundamental, our distress probably reflects angst at automata, at organized systems without souls, like nature itself in its destructive and predatory forms—isn't that why we argue whether computers can think?

C. P. Snow, the English physicist and novelist, identified more specific hostility toward technology among intellectuals, particularly literary intellectuals, in his well-known 1959 lectures on "The Two Cultures." Such hostility becomes obvious when you survey the literature; it's obvious in this book, and not because I biased the sample. To the contrary, appreciation of technology among intellectuals not technologically trained was hard to find. Since many intellectuals are concerned with social justice and not devoid of ordinary compassion, it's surprising that they don't value technology; by any fair assessment, it has reduced suffering and improved welfare across the past hundred years. Why doesn't this net balance of benevolence inspire at least grudging enthusiasm for technology among intellectuals?

Snow traced the conflict to class differences that widened with the progress of the industrial revolution. I've included an excerpt from that discussion at its appropriate place in this book. The landed classes resisted the revolution, Snow notes, since it threatened their predominately agricultural interests. The new industrialists and engineers emerged from the craft and working classes. The landed classes neglected technical education, taking refuge in classical studies; as late as 1930, for example, long after Ernest Rutherford at Cambridge had discovered the atomic nucleus and begun transmuting elements, the physics laboratory at Oxford University had not yet been wired for electricity. Intellectuals neglect technical education to this day. Since most intellectuals aren't from upper-class backgrounds, Snow seems to be implying that their hostility to technology results from aping their betters—not a very generous assessment.

Given the pervasiveness of the intellectual bias against technology, technologists are probably justified in concluding that it derives in some measure from technical and scientific illiteracy as well as jealousy and competition for influence. But such conclusions risk trivializing the debate—much as do intellectuals' tiresome accusations that greed is the technologist's primary motivation for enterprise. If the record of technological innovation in this century is, on balance, clearly positive, it's also true that technologists have been prodigal at excusing themselves from moral responsibility for weapons of mass destruction, pollution and other well-known horrors.

They do so in part by refusing to acknowledge the extent to which belief systems intrude into their operations. Claims that a "technical imperative" drives technological change, for example, much as an invisible hand is supposed to drive the capitalist marketplace, fall into this category. The recent literary-intellectual assault on science as an arbitrary construction no more anchored in the real world than any other religious or social institution is an extreme but predictable consequence of such denial. The assault has found support precisely because the scientific and technological community has chosen to deny knowledge of its own complicity in installing and maintaining structural violence. Structural violence—violence such as racial discrimination that is built into the structure of societies—remains the largest-scale and most intractable form of violence left in a world where knowledge of how to release nuclear energy has foreclosed world war. As methodologies, science and technology are demonstrably objective and effective; but they're unquestionably bound up with power relations as social systems.

All this is to anticipate the vital and continuing debate I've sampled in this book. I'm reluctant to generalize from my sample. It's designed to be an animated performance of itself, its four parts anchored in the major events that set its terms. Enthusiasm for technology grew among technologists in the first quarter of the century as the expanding mass production of consumer goods, particularly automobiles, created great wealth. But critics attacked the application of technology to industrial production even before the First World War showed how technology could mass-produce slaughter (one theorist described the machine gun, the basic killing tool of the war, as "concentrated essence of infantry"). The Great Depression shifted the debate from industrial to social transformation, borrowing metaphors and solutions from technology even as technology was challenged. By the end of the Second World War the shift from an agricultural to a technological society was essentially complete. The second half of the century filled in the spaces while a new transformation to an electronically based information technology began—to reach its maturity, presumably, in the twenty-first century now opening.

These tidal highs and lows hardly obscure the persistent, continuing enlargement of the influence of science and technology on human affairs. By whatever measure you choose, science and technology came to dominate the human project in the twentieth century. Public health more than doubled the average lifespan. The discovery of how to release nuclear energy made world-scale war suicidal. Birth control subdued the Malthusian multiplication of human population. Agriculture fed the multitudes. Electronics wired the world and put human communication beyond the reach of tyranny. Manned vessels of discovery cast off beyond the earth; automated voyagers—notes in high-tech bottles—even escaped the solar system. At the same time, human activities drove a catastrophic decline in species diversity and began global warming; from a wild place the earth became a garden, well tended in some districts, ruthlessly exploited in others. The evolutionary neural enlargement that spun out technology (which imitates evolution culturally, propagating in memes rather than in genes) is not only open ended; it's also myopic, which makes invention and application acts of faith. The deep truth about the debate that fills this book is that it's a debate among the orthodox, a debate about speed limits and barricades rather than the necessity of the quest. No one, not even the Unabomber, has proposed a return to the Hobbesian garden of the primates.

Visions of Technology originated in discussions among the members of the Alfred P. Sloan Foundation Technology Book Series advisory committee. In the midst of commissioning histories of major twentieth-century technologies, we realized that there must also be a thick vein of debate *about* technology to be mined into a book, and that such a book might serve as a meta-history of the effect of technological change on the twentieth-century human world. Fools walk in where angels fear to tread: I volunteered to assemble an anthology provided a professional historian could be found to join me in the work. Elting E. Morison agreed to undertake that partnership. Unfortunately, his final illness intervened before he could contribute beyond reading and approving the initial proposal I drafted. His participation would have broadened the range of selections and enriched the running commentary. It wasn't to be. I wish it had been.

I planned at the outset to arrange selections by theme within their roughly quarter-century periods. That plan foundered on the breadth of issues many contributors explore. Finally, chronology alone seemed adequate and appropriate; I try to sketch themes and connections in my introductory comments. Chronology—usually of publication, occasionally of subject matter—reveals characteristic preoccupations and repetitions with minimal anachronism. It exposes, for example, the crisis of confidence in technology that arose with the Great Depression, the challenge to technology the environmental movement of the 1960s and 1970s launched, the recurring testing of links between innovation and job loss. The result is a species of textual archeology, levels exposed from the earliest to the most recent in turn. The index cross-references them; the bibliography points to the original source.

Thanks to Paul Kennedy, who recommended my editorial associate Stephen Kim. Stephen spent two summers sorting through the first half of the century in the stacks and archives of Sterling Memorial Library at Yale University. Jeff Wheelwright then contributed from his own extensive experience and archives as a science writer and from library investigations into the second half of the century. I consulted the distinguished members of the advisory committee and queried Technology Book Series contributors, but minimized overlap with their books. I've been reading about technology since early childhood and reporting and writing about it for more than thirty years, and obviously drew on that knowledge and experience as well.

Here then is a chronological and topical range of twentieth-century assessments of what technology is, who does it, how it works and what values it sustains.

GLADE

MAY 1993–JUNE 1998

I. THE NEW TECHNOLOGY: 1900–1933

AMERICA IN 1900

MARK SULLIVAN

Journalist Mark Sullivan's six-volume compendium Our Times *chronicles the first decades of the new century from the perspective of the late 1920s. Early in the first volume, Sullivan catalogs what Americans had not yet experienced of technology and social mores at the turn of the century.*

In his newspapers of January 1, 1900, the American found no such word as radio,* for that was yet twenty years from coming; nor "movie," for that too was still mainly of the future; nor chauffeur, for the automobile was only just emerging and had been called "horseless carriage" when treated seriously, but rather more frequently, "devil-wagon," and the driver, the "engineer." There was no such word as aviator—all that that word implies was still a part of the Arabian Nights. Nor was there any mention of income tax or surtax, no annual warning of the approach of March 15 [*sic:* the income-tax-filing deadline was later moved to April]—all that was yet thirteen years from coming. In 1900 doctors had not yet heard of 606 [i.e., biochemist Paul Ehrlich's "magic bullet" for syphilis, reported in 1910] or of insulin; science had not heard of relativity or the quantum theory. Farmers had not heard of tractors, nor bankers of the Federal Reserve System. Merchants had not heard of chain-stores nor "self-service"; nor seamen of oil-burning engines. Modernism had not been added to the common vocabulary of theology, nor futurist and "cubist" to that of art. . . .

The newspapers of 1900 contained no menton of smoking by women,† nor of "bobbing," nor "permanent wave," nor vamp, nor flapper, nor jazz, nor feminism, nor birth-control. There was no such word as rum-runner,

---

*The number of new words added to the dictionary between 1900 and 1925, and new uses of old words, is a reflection of the expanding of man's intelligence and the increasing complexity of civilization. Nearly a thousand were used to adjust the radio to the language. Hundreds were required for the automobile, its parts and associations; yet more hundreds for the popular and technical terminology of aviation. [M.S.]

†Indeed there was a distinct movement against smoking by men. Three important railroads put in effect, on January 1, 1900, a rule against smoking what the Washington, Pa., *Reporter,* in recording the edict, called "the nasty cigarette." [M.S.]

nor hijacker, nor bolshevism, fundamentalism, behaviorism, Nordic, Freudian, complexes, ectoplasm, brain-storm, Rotary, Kiwanis, blue-sky law, cafeteria, automat, sundae; nor mah-jong, nor crossword puzzle. Not even military men had heard of camouflage; neither that nor "propaganda" had come into the vocabulary of the average man. "Over the top," "zero hour," "no man's land" meant nothing to him. "Drive" meant only an agreeable experience with a horse. The newspapers of 1900 had not yet come to the lavishness of photographic illustration that was to be theirs by the end of the quarter-century. There were no rotogravure sections. If there had been, they would not have pictured boy scouts, nor State constabularies, nor traffic cops, nor Ku Klux Klan parades; nor women riding astride, nor the nudities of the Follies, nor one-piece bathing suits, nor advertisements of lipsticks, nor motion-picture actresses, for there were no such things.

In 1900, "short-haired woman" was a phrase of jibing; women doctors were looked on partly with ridicule, partly with suspicion. Of prohibition and votes for women, the most conspicuous function was to provide material for newspaper jokes. Men who bought and sold lots were still real-estate agents, not "realtors." Undertakers were undertakers, not having yet attained the frilled euphemism of "mortician." There were "star-routes" yet—rural free delivery had only just made a faint beginning; the parcel post was yet to wait thirteen years. In 1900, "bobbing" meant sliding down a snow-covered hill; women had not yet gone to the barber-shop. For the deforestation of the male countenance, the razor of our grandfathers was the exclusive means; men still knew the art of honing. The hairpin, as well as the bicycle, the horseshoe, and the buggy were the bases of established and, so far as anyone could foresee, permanent businesses. Ox-teams could still be seen on country roads; horse-drawn street-cars in the cities. Horses or mules for trucks were practically universal;* livery-stables were everywhere. The blacksmith beneath the spreading chestnut-tree was a reality; neither the garage mechanic nor the chestnut blight had come to retire that scene to poetry. The hitching-post had not been supplanted by the parking problem. Croquet had not yet given way to golf. "Boys in blue" had not yet passed into song. Army blue was not merely a sentimental memory, had not yet succumbed to the invasion of utilitarianism in olive green. G. A. R. [Grand Army of the Republic, an organization established by Civil War veterans of the Union army and navy] were still potent letters. . . .

RICHARD RHODES

*A very few motor-trucks of European make had come to America just before 1900. The first American-made gasoline truck was built and sold in 1900. Practically all the transformation of city streets and country roads that attended gasoline motive power came after 1900. Even for street-cars, horses were still much in use in 1900. That a time should come when horses would be a rare sight on city streets seemed, in 1900, one of the least credible of prophecies. [M.S.]

Only the Eastern seaboard had the appearance of civilization having really established itself and attained permanence. From the Alleghenies to the Pacific Coast, the picture was mainly of a country still frontier and of a people still in flux: the Allegheny mountainsides scarred by the axe, cluttered with the rubbish of improvident lumbering, blackened with fire; mountain valleys disfigured with ugly coal-breakers, furnaces, and smokestacks; western Pennsylvania and eastern Ohio an eruption of ungainly wooden oil-derricks; rivers muddied by the erosion from lands cleared of trees but not yet brought to grass, soiled with the sewage of raw new towns and factories; prairies furrowed with the first breaking of sod. Nineteen hundred was in the floodtide of railroad-building: long fingers of fresh dirt pushing up and down the prairies, steam-shovels digging into virgin land, rock-blasting on the mountainsides. On the prairie farms, sod houses were not unusual. Frequently there were no barns, or, if any, mere sheds. Straw was not even stacked, but rotted in sodden piles. Villages were just past the early picturesqueness of two long lines of saloons and stores, but not yet arrived at the orderliness of established communities; houses were almost wholly frame, usually of one story, with a false top, and generally of a flimsy construction that suggested transiency; larger towns with a marble Carnegie Library at Second Street, and Indian tepees at Tenth. Even as to most of the cities, including the Eastern ones, their outer edges were a kind of frontier, unfinished streets pushing out to the fields; sidewalks, where there were any, either of brick that loosened with the first thaw, or wood that rotted quickly; rapid growth leading to rapid change. At the gates of the country, great masses of human raw materials were being dumped from immigrant ships. Slovenly immigrant trains tracked westward. Bands of unattached men, floating labor, moved about from the logging camps of the winter woods to harvest in the fields, or to railroad-construction camps. Restless "sooners" wandered hungrily about to grab the last opportunities for free land.

One whole quarter of the country, which had been the seat of its most ornate civilization, the South, though it had spots of melancholy beauty, presented chiefly the impression of the weedy ruins of thirty-five years after the Civil War, and comparatively few years after Reconstruction—ironic word. . . .

In 1900 the United States was a nation of just under 76,000,000 people.

MESSAGES WITHOUT WIRES

GUGLIELMO MARCONI, 1901

The twentieth century opened to major invention. Here a faint signal in Morse code inaugurates a new era in world communication.

RICHARD RHODES

Shortly before midday [December 12, 1901, in a hut on the cliffs at St. John's, Newfoundland] I placed the single earphone to my ear and started listening. The receiver on the table before me was very crude—a few coils and condensers and a coherer—no [vacuum tubes], no amplifiers, not even a crystal. But I was at last on the point of putting the correctness of all my beliefs to test. The answer came at 12:30 when I heard, faintly but distinctly, *pip-pip-pip*. I handed the phone to Kemp: "Can you hear anything?" I asked. "Yes," he said, "the letter S"—he could hear it. I knew then that all my anticipations had been justified. The electric waves sent out into space from Poldhu [Cornwall] had traversed the Atlantic—the distance, enormous as it seemed then, of 1,700 miles—unimpeded by the curvature of the earth. The result meant much more to me than the mere successful realization of an experiment. As Sir Oliver Lodge has stated, it was an epoch in history. I now felt for the first time absolutely certain that the day would come when mankind would be able to send messages without wires not only across the Atlantic but between the farthermost ends of the earth.

A VERY LOUD ELECTROMAGNETIC VOICE

P. T. McGRATH, 1902

A writer in Century Magazine *imagines how Marconi's invention might work and descries in the distance the cellular phone.*

If a person wanted to call to a friend he knew not where, he would call in a very loud electromagnetic voice, heard by him who had the electromagnetic ear, silent to him who had it not. "Where are you?" he would say. A small reply would come "I am at the bottom of a coal mine, or crossing the Andes, or in the middle of the Atlantic." Or, perhaps in spite of all the calling, no reply would come, and the person would then know that his friend was dead. Think of what this would mean, of the calling which goes on every day from room to room of a house, and then think of that calling extending from pole to pole, not a noisy babble, but a call audible to him who wants to hear, and absolutely silent to all others. It would be almost like dreamland and ghostland, not the ghostland cultivated by a heated imagination, but a real communication from a distance based on true physical laws.

WILBUR WRIGHT'S AFFLICTION

WILBUR WRIGHT, 1900

When Wilbur Wright wrote engineer and aviation pioneer Octave Chanute for the first time, on May 13, 1900, asking advice, he and his brother Orville had already begun flying aircraft-scale kites and were three years away from powered flight.

For some years I have been afflicted with the belief that flight is possible to man. My disease has increased in severity and I feel that it will soon cost me an increased amount of money if not my life. I have been trying to arrange my affairs in such a way that I can devote my entire time for a few months to experiment in this field.

My general ideas of the subject are similar to those held by most practical experimenters, to wit: that what is chiefly needed is skill rather than machinery. The flight of the buzzard and similar sailers is a convincing

▶ First sustained flight at Kitty Hawk, 1903. Orville Wright: "I got on the machine at 10:35 for the first trial. . . . Time about 12 seconds."

demonstration of the value of skill, and the partial needlessness of motors. It is possible to fly without motors, but not without knowledge & skill. This I conceive to be fortunate, for man, by reason of his greater intellect, can more reasonably hope to equal birds in knowledge, than to equal nature in the perfection of her machinery.

Assuming then that Lilienthal was correct in his ideas of the principles on which man should proceed, I conceive that his failure was due chiefly to the inadequacy of his method, and of his apparatus. As to his method, the fact that in five years' time he spent only about five hours, altogether, in actual flight is sufficient to show that his method was inadequate. Even the simplest intellectual or acrobatic feats could never be learned with so short practice, and even Methuselah could never have become an expert stenographer with one hour per year for practice. I also conceive Lilienthal's apparatus to be inadequate not only from the fact that he failed, but my observations of birds convince me that birds use more positive and energetic methods of regaining equilibrium than that of shifting the center of gravity.

With this general statement of my principles and belief I will proceed to describe the plan and apparatus it is my intention to test. In explaining these, my object is to learn to what extent similar plans have been tested and found to be failures, and also to obtain such suggestions as your great knowledge and experience might enable you to give me. I make no secret of my plans for the reason that I believe no financial profit will accrue to the inventor of the first flying machine, and that only those who are willing to give as well as to receive suggestions can hope to link their names with the honor of its discovery. The problem is too great for one man alone and unaided to solve in secret.

A HORSELESS CARRIAGE WAS A COMMON IDEA

HENRY FORD

Farm labor is hard. It was even harder before the advent of power machinery. In his autobiography, My Life and Work, *Henry Ford recalled that he began the work that led to the automobile intending to invent a better tractor.*

A horseless carriage was a common idea. People had been talking about carriages without horses for many years back—in fact, ever since the steam engine was invented—but the idea of the carriage at first did not seem so practical to me as the idea of an engine to do the harder farm work, and of all the work on the farm ploughing was the hardest. . . .

▶ Henry Ford in his first automobile, 1896.

It was not difficult for me to build a steam wagon or tractor. In the building of it came the idea that perhaps it might be made for road use. I felt perfectly certain that horses, considering all the bother of attending them and the expense of feeding, did not earn their keep. The obvious thing to do was to design and build a steam engine that would be light enough to run an ordinary wagon or to pull a plow. I thought it most important first to develop the tractor. To lift farm drudgery off flesh and blood and lay it on steel and motors has been my most constant ambition. It was circumstances that took me first into the actual manufacture of motor cars. I found eventually that people were more interested in something that would travel on the road than in something that would do the work on the farms.

HENRY ADAMS, 1900

Not everyone admired the new technologies, much less understood them, and even scientists and engineers found some of them uncomfortable. Henry Adams, for one, the historian and social observer, born in 1838 to a family that "for three generations," as an heir wrote, ". . . had power to direct the destiny of nations," thought the Dynamo as incomprehensible as the Virgin; his friend Samuel Pierpoint Langley, the physicist, Smithsonian secretary and aeronautical pioneer, was queasy about the new discovery of radioactivity. Writing of himself in the third person, Adams in his Education *defines the conflict of the two cultures that C. P. Snow would find still dissonant half a century later.*

U ntil the Great Exposition of 1900 closed its doors in November, Adams haunted it, aching to absorb knowledge, and helpless to find it. He would have liked to know how much of it could have been grasped by the best-informed man in the world. While he was thus meditating chaos, Langley came by, and showed it to him. At Langley's behest, the Exhibition dropped its superfluous rags and stripped itself to the skin, for Langley knew what to study, and why, and how; while Adams might as well have stood outside in the night, staring at the Milky Way. Yet Langley said nothing new, and taught nothing that one might not have learned from Lord Bacon, three hundred years before; but though one should have known the "Advancement of Science" as well as one knew the "Comedy of Errors," the literary knowledge counted for nothing until some teacher should show how to apply it. Bacon took a vast deal of trouble in teaching King James I and his subjects, American or other, towards the year 1620, that true science was the development or economy of forces; yet an elderly American in 1900 knew neither the formula nor the forces; or even so much as to say to himself that his historical business in the Exposition concerned only the economies or developments of force since 1893, when he began the study at Chicago.

Nothing in education is so astonishing as the amount of ignorance it accumulates in the form of inert facts. Adams had looked at most of the accumulations of art in the storehouses called Art Museums; yet he did not know how to look at the art exhibits of 1900. He had studied Karl Marx and his doctrines of history with profound attention, yet he could not apply them at Paris. Langley, with the ease of a great master of experiment, threw out of the field every exhibit that did not reveal a new application of force, and naturally threw out, to begin with, almost the whole art exhibit. Equally, he ignored almost the whole industrial exhibit. He led his pupil directly to the forces. His chief interest was in new motors to make his airship feasible, and he taught Adams the astonishing complexities of the new Daimler motor, and of the automobile, which, since 1893, had become a nightmare at a hundred kilometers an hour, almost as destructive as the electric tram which was only ten years older; and threatening to become as terrible as the locomotive steam-engine itself, which was almost exactly Adams's own age.

Then he showed his scholar the great hall of dynamos, and explained how little he knew about electricity or force of any kind, even of his own special sun, which spouted heat in inconceivable volume, but which, as far as he knew, might spout less or more, at any time, for all the certainty he felt in it. To him, the dynamo itself was but an ingenious channel for conveying somewhere the heat latent in a few tons of poor coal hidden in a dirty engine-house carefully kept out of sight; but to Adams the dynamo became a symbol of infinity. As he grew accustomed to the great gallery of machines, he began to feel the forty-foot dynamos as a moral force, much as the early Christians felt the Cross. The planet itself seemed less impressive, in its old-fashioned, deliberate, annual or daily revolution, than this huge wheel, revolving within an arm's-length at some vertiginous speed, and barely murmuring—scarcely humming an audible warning to stand a hair's-breadth further for respect of power—while it would not wake the baby lying close against its frame. Before the end, one began to pray to it; inherited instinct taught the natural expression of many before silent and infinite force. Among the thousand symbols of ultimate energy, the dynamo was not so human as some, but it was the most expressive.

Yet the dynamo, next to the steam-engine, was the most familiar of objects. For Adams's objects its value lay chiefly in its occult mechanism. Between the dynamo in the gallery of machines and the engine-house outside, the break of continuity amounted to abysmal fracture for a historian's objects. No more relation could he discover between the steam and the electric current than between the Cross and the cathedral. The forces were interchangeable if not reversible, but he could see only an absolute *fiat* in electricity as in faith. Langley could not help him. Indeed, Langley seemed to be worried by the same trouble, for he constantly repeated that the new forces were anarchical, and specially that he was not responsible for the new rays, that were little short of parricidal in their wicked spirit towards science. His own rays, with which he had doubled the solar spectrum, were altogether harmless and beneficent; but Radium denied its God—or, what was to Langley the same thing, denied the truths of his Science. The force was wholly new.

THE AMERICAN WILL NOT
LIVE NEAR HIS WORK

CHARLES M. SKINNER, 1902

The "electric tram" that Henry Adams thought "destructive" was creating new suburbs around American cities at the turn of the century, this Atlantic Monthly *essayist notes, and changing much else besides.*

W hat are the conditions [the electric trolley car] has made? Quicker transit, with cleaner, larger cars, heated and lighted by electricity. . . .

The dismissal of the horse from car service, to the cheapening of that animal, the saving and cleanness of our streets, and the sparing of no end of feelings.

Increased scope in service, for not only are the usual closed and open cars operated in the cities, but postal cars, parlor cars, express cars, repair cars, coal cars, freight cars, even, and there is a report that one line, with an important rural extension, is to have dining and sleeping cars!

A much up-building of suburbs and the emergence on the map of a thousand Daisy Knolls, Sparrow Parks, and Maplehursts.

An increase in the size and number of melancholy institutions called pleasure resorts, within reach of the cities; therefore, the vexation of hitherto tranquil regions by rowdies and picnic parties.

The hurt to faraway hotels and boarding places, through this diversion of holiday makers to beaches and beer gardens near home.

The disfigurement of streets and injury to roads worked by the erection of poles, the stringing of wires, the cutting of pavements, the lopping of shade trees, and the blight of vegetation due to escaping currents.

A multiplication of the dangers and bothers of street traffic through increased speed in the cars, blown-out fuses, broken wires, and charged rails.

An immense increase in the capital invested in local transportation; hence, an increase in corporate and public wealth through dividends and taxes.

The promise of a wide extension of electric power to other vehicles and other industries.

The lowering of our standard of public manners, due to the overcrowding of cars.

Of these conditions, or elements in a condition, that is happiest which tends to deplete the city and persuade the people into roomier, healthier districts, where factories and slums are not; where flowers and trees are many. . . .

On one point the American is determined: he will not live near his work. You shall see him in the morning, one of sixty people in a car built for twenty-four, reading his paper, clinging to a strap, trodden, jostled, smirched, thrown into harrowing relations with men who drink whiskey, chew tobacco, eat raw onions, and incontinently breathe; and after thirty minutes of this contact, with the roar of the streets in his ears, with languid clerks and pinguid market women leaning against him, he arrives at his office. The problems of his homeward journey in the evening will be still more difficult, because, in addition to the workers, the cars must carry the multitude of demoiselles who shop and go to matinees. To many men and women of business a seat is an undreamed luxury. Yet, they would be insulted if one were to ask why they did not live over their shops, as Frenchmen do, or back of them, like Englishmen. It is this uneasy instinct of Americans, this desire of their families to separate industrial and social life, that makes the use of the trolley car imperative, and the street railway in this manner widens the life and dominion of the people; it enables them to distribute themselves over wider spaces and unwittingly to symbolize the expansiveness of the nation.

THE PROFESSION
OF ENGINEERING

HERBERT HOOVER

Engineering was just becoming a profession at the turn of the century, one of its most eminent practitioners reports in his Memoirs.

It was the American universities that took engineering away from rule-of-thumb surveyors, mechanics, and Cornish foremen and lifted it into the realm of application of science. . . . The European universities did not acknowledge engineering as a profession until long after America had done so. I took part in one of the debates at Oxford as to whether engineering should be included in its instruction. . . . I cited the fact that while various special technical colleges had been existent in England for a long time, yet there were more than a thousand American engineers of all breeds in the British Empire, occupying top positions.

Soon after the Oxford discussions, I returned to America. At my ship's table sat an English lady of great cultivation and a happy mind, who contributed much to the evanescent conversation on government, national customs, literature, art, industry, and whatnot. We were coming up New York harbor at the final farewell breakfast, when she turned to me and said:

"I hope you will forgive my dreadful curiosity, but I should like awfully to know—what is your profession?"
I replied that I was an engineer. She emitted an involuntary exclamation, and "Why, I thought you were a gentleman!"

A RESERVOIR OF
SUFFERING HUMANITY

RICHARD RHODES

Innovations in medicine at places like the newly founded Mayo Clinic of Rochester, Minnesota, met a backlog of accumulated human need in the first decades of the new century. I visited the Mayo Clinic to write about it for a new edition of my first book, The Inland Ground.

There was a reservoir of suffering humanity in the Upper Middle West then—at the close of the 19th century—as everywhere else in the world. The dam that confined it was medical ignorance, ignorance that was just giving way. Anesthesia first, then antiseptic technique and finally the superior gloves and gowns and sterilization of aseptic surgery began to make healing possible.

The Mayo brothers didn't often originate the new surgical procedures. What they usually did was refine them, perfect them, perform them in incredible numbers and with observant care and reduce their mortality until finally they became routine. Ten gallbladder operations in 1895 became 75 in 1900 and 324 in 1905. In 1905 the Mayos performed 2,157 abdominal operations in all. They were boyish-looking men when they were young, dressed in country suitings. When they showed up at Eastern medical meetings with their careful reports, claiming to have performed hundreds of operations in a town of six thousand souls (in 1904 they reported jointly on 1,000 gallbladder procedures when many surgeons had not yet performed ten), they were sometimes taken for charlatans. They evolved a stock answer for doubters: "Come and see." Eventually the doubters did, and went away convinced.

Long before patients visited them from everywhere in the world, the Mayos healed the sick around them: the chronic gallbladders and infected appendixes misdiagnosed as "colic" and "stomach disease" and "dyspepsia"; the tens of thousands of goiters in those regions without iodine in their soil (the Mayo Clinic treated 37,228 cases of goiter between 1892 and 1934); the ovarian cysts that grew so large, filling with fluid, that women sometimes wore special harnesses their husbands made for them to hold their abdomens up (the largest ever removed at Mayo, in 1920, weighed 139.5

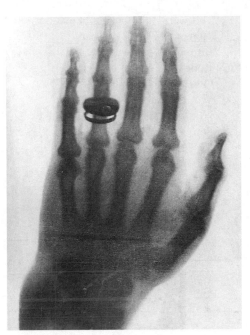

▶ Wilhelm Roentgen's first X ray, 1898.

pounds) but that no one had dared to extract except in extremity. These the Mayos worked night and day to heal, and the word went out that, almost alone in those parts at first, they could. "As I look back over those early years," Dr. Will said later, "I am impressed with the fact that much of our success, if not most of it, was due to the time at which we entered medicine."

THE CHARGE OF THE FOUR HUNDRED

ANONYMOUS, 1904

At the turn of the century, automobiles were an ostentatious luxury of the rich, as this parody of "The Charge of the Light Brigade," from the old Life *magazine, implies.*

Half a block, half a block,
Half a block onward,
All in their motobiles
Rode the Four Hundred.

"Forward!" the owners shout,
"Racing-car!" "Runabout!"
Into Fifth Avenue
Rode the Four Hundred.
"Forward!" the owners said.
Was there a man dismay'd?
Not, though the chauffeurs knew
Someone had blundered.
Theirs not to make reply,
Theirs not to reason why,
Theirs but to kill or die.
Into Fifth Avenue
Rode the Four Hundred.

THE DISCIPLINE OF
THE MACHINE

THORSTEIN VEBLEN, 1904

As well as revealing the striving for social status that he believed underlay economic decision, Veblen explored the effects on values of what he called "the machine process" in his The Discipline of the Machine.

The machine process pervades the modern life and dominates it in a mechanical sense. Its dominance is seen in the enforcement of precise mechanical measurements and adjustment and the reduction of all manner of things, purposes and acts, necessities, conveniences, and amenities of life, to standard units. . . .

Wherever the machine process extends, it sets the pace for the workmen, great and small. . . . Mechanically speaking, the machine is not his to do with as his fancy may suggest. His place is to take thought of the machine and its work in terms given him by the process that is going forward. . . .

The machine process compels a more or less unremitting attention to phenomena of an impersonal character and to sequences and correlations not created by habit and custom. The machine throws out anthropomorphic habits of thought. . . . [It] gives no insight into questions of good and evil, merit and demerit, except in point of material causation, nor into the foundations or the constraining force of law and order, except such mechanically enforced law and order as may be stated in terms of pressure, temperature, velocity, tensile strength, etc. The machine technology takes no cognizance of conventionally established rules of precedence; it knows neither manners nor breeding and can make no use of any of the attributes of worth. Its scheme of knowledge and of inference is based on the laws of material causation, not on those of immemorial custom, authenticity, or authoritative enactment. Its metaphysical basis is the law of cause and effect. . . .

The ubiquitous presence of the machine, with its spiritual concomitant—workday ideals and skepticism of what is only conventionally valid— is the unequivocal mark of the western culture of today as contrasted with the culture of other times and places.

A GOOD REASON FOR
EVERYTHING THAT HE TRIES

RICHARD C. MACLAURIN

Thomas Alva Edison's extraordinary record of invention in the last quarter of the nineteenth century and the first quarter of the twentieth made him famous throughout the world. He cultivated the common touch, an observer comments in the journal Science; *it obscured the debt he owed to scientific method.*

43

Edison more than any one else in this country has taught men to see something of what science can do. . . . With such an embarrassment of riches, it is scarcely practicable even to single out a few of his great accomplishments. Many of you are familiar with what he did in the early days by way of improving the duplex and quadruplex systems of telegraphy, you know of his invention of the contact transmitter and his development of the loud-speaking telephone, of his marvelous invention of the phonograph (Edison being the first to make a record that would *reproduce* sound), you think of his wonderful work in 1878 and later years in developing the incandescent lamp, and you realize that he practically made the *whole* incandescent system, not only inventing the lamp, but turning his attention to all its adjuncts, improving the dynamos for such work and providing the necessary means for the distribution of power over large areas. You recognize that he laid the foundations for the design of central power stations and that his Pearl Street Station was a landmark in the history of science. His work in this field is truly phenomenal, the three-wire distribution, the system of feeders entering the network of mains at different points, the underground conductor system, the bus system in stations, the innumerable accessories of switches, fuses, meters, etc., that he provided are each achievements that would make the fame of any individual. You appreciate the remarkable character of his later work in developing the apparatus of moving pictures and you agree that what he has done still more recently in perfecting the alkaline storage cell is a splendid example of energy and persistence in attacking a difficult problem. Thinking of all these things, you can not fail to be impressed with two things—the enormous range of his activities and the wonderful simplicity of many of his devices. After all, simplicity of device is always the sign of the master, whether in science or in art. In studying Edison you have something of the same impression as in studying Newton— you are surprised how easy are the steps. Someone asked Lord Kelvin why no one before Edison had invented so *simple* a thing as the feeder system. "The only reason I can think of," he said, "is that no one else was Edison." As to the range of his activities, he has been associated in some way with so many of the great modern developments that people sometimes speak as if he had invented *everything*, even electricity itself, or if they do not go to this length, they find it necessary to explain why he did not invent this or that. The fact that his name is not intimately associated with one of the great modern achievements—the development of the aeroplane—has called forth numerous ingenious explanations. . . . Seeing that he has done so much, we need not spend much time in wondering why he has not done more. Nor need we attempt the impossible in the effort to measure the debt that mankind owes to him. Such statements as have been made to the effect that his inventions have given rise to industries that employ nearly a million of men and thousands of millions of capital really give no adequate sense of the value of his achievements, although they may be of some use as a very rough indication of the scale of his activities.

Not only has he shown his faith in science by great achievements, but he has proved himself a great force in education by giving so brilliant an exhibition of the *method* of science, the method of experimentation. When we get to the root of the matter we see that nearly all great advances are made by improvements in method. There is no evidence that men are abler in the twentieth century than they were in the Middle Ages, but they have learned a new method. "It was in Boston," said Edison, "that I bought Faraday's works, and appreciated that he was the master experimenter." It is interesting to think what Edison's appreciation of this fact has meant for the world. . . . He is no slave to theory; he is ready, as every scientific man is ready, to try anything that seems reasonable, but practically always he has what seems to him a good reason for everything that he tries. In the rare case where he has tried blindly, it has been because there was absolutely no light.

> One man, the Archbishop of Canterbury, when asked [in 1900] what was the chief danger threatening the coming century, replied: "I have not the slightest idea."
>
> MARK SULLIVAN

It is our confident claim that applied science, if carried out according to our program, will succeed in achieving for humanity, above all for the city industrial worker, results even surpassing in value those today in effect on the farm.

> THEODORE ROOSEVELT TO WILLIAM
> HOWARD TAFT, 1908

In the past the man has been first; in the future the system must be first.

> FREDERICK WINSLOW TAYLOR, 1911

Electricity, carrier of light and power; devourer of time and space; bearer of human speech over land and sea; greatest servant of man—yet itself unknown.

> CHARLES W. ELIOT, INSCRIPTION FOR
> UNION DEPOT, WASHINGTON, D.C.

My main interest is in the aeroplane as a real promoter of civilization.

> ORVILLE WRIGHT, 1917

On the day when two armies will be able to annihilate each other in one second all civilized nations will recoil from war in horror and disband their forces.

> ALFRED NOBEL, 1890

Pioneering don't pay.

> ANDREW CARNEGIE, C. 1910

It is like writing history with lightning.

> WOODROW WILSON, 1915 (after seeing D. W.
> Griffith's THE BIRTH OF A NATION, the first
> motion picture ever shown in the White House)

FORTY DOSES OF
CHEMICALS AND COLORS

MARK SULLIVAN

The purity of food was a burning issue in 1903 just as it is today.

At the annual meeting of the National Association of State Dairy and Food Departments at St. Paul in 1903, a report on "The Use of Coloring Matter and Antiseptics in Food Products" was read by the State Analyst of South Dakota, James H. Shepard, to whom training in chemistry had given a bent for austere truth, as well as a convincing way of setting out proof. That extraneous substances appeared in most of the articles in common use for food, he took for granted. . . . What he was intent on showing was the aggregate of such substances absorbed in a day by the average person, and the effect of such an aggregate on the human body. This was his answer to some food manufacturers who, unable to conceal their use of chemicals and other adulterants, claimed the amount in the portion for a single meal was so small as to be harmless.

"In order," said Professor Shepard, "to bring this matter out more forcibly, I have prepared a menu for one day such as any family in the United States might possibly use, and I am not sure but the working man in our cities would be quite likely to use it." Professor Shepard's menu, omitting the few articles not commonly adulterated, such as potatoes, was:

BREAKFAST

Sausage, coal-tar dye and borax.
Bread, alum.
Butter, coal-tar dye.
Canned cherries, coal-tar dye and salicylic acid.
Pancakes, alum.
Syrup, sodium sulphite.

This gives eight doses of chemicals and dyes for breakfast.

DINNER

Tomato soup, coal-tar dye and benzoic acid.
Cabbage and corned beef, saltpetre.
Canned scallops, sulphurous acid and formaldehyde.
Canned peas, salicylic acid.
Catsup, coal-tar dye and benzoic acid.
Vinegar, coal-tar dye.
Bread and butter, alum and coal-tar dye.
Mince pie, boracic acid.
Pickles, copperas, sodium sulphite, and salicylic acid.
Lemon ice cream, methyl alcohol.

This gives sixteen doses for dinner.

SUPPER

Bread and butter, alum and coal-tar dye.
Canned beef, borax.

Canned peaches, sodium sulphite, coal-tar dye and salicylic acid.
Pickles, copperas, sodium sulphite, and formaldehyde.
Catsup, coal-tar dye and benzoic acid.
Lemon cake, alum.
Baked pork and beans, formaldehyde.
Vinegar, coal-tar dye.
Currant jelly, coal-tar dye and salicylic acid.
Cheese, coal-tar dye.

This gives sixteen doses for supper.

Congressman Augustus O. Stanley of Kentucky defended the blending, but not the adulteration, of whiskey in a rollicking 1906 speech in the U.S. House of Representatives.

I want to say this, that I have no objection to a man blending two kinds of whiskey, but I do object to his making any kind of whiskey "while you wait." Here is a quart of alcohol [holding it up]. It will eat the intestines out of a coyote. It will make a howling dervish out of an anchorite. It will make a rabbit spit in a bulldog's face. It is pure alcohol, and under the skill of the rectifier he will put in a little coloring matter and then a little bead oil [illustrating]. I drop that in it. Then I get a little essence of Bourbon whiskey, and there is no connoisseur in this House who can tell that hellish concoction from the genuine article; and that is what I denounce. [Applause.] I say that the coloring matter is not harmful; I say that the caramels are not harmful; but I say that the body, the stock, of the whiskey I made is rank alcohol, and when it gets into a man it is pure hell. [Applause.]

THE WONDERFUL CULEBRA CUT

DAVID MCCULLOUGH

The Panama Canal, which the United States built in the years 1904–1914, captured the imagination of the world, the Pulitzer Prize–winning historian reports.

The "special wonder of the canal" was Culebra Cut. It was the great focus of attention, regardless of whatever else was happening at Panama. The building of Gatun Dam or the construction of the locks, projects of colossal scale and expense, were always of secondary interest so long as the battle raged in that nine-mile stretch between Bas Obispo and Pedro

▶ Digging the Panama Canal: Culebra Cut.

Miguel. The struggle lasted seven years, from 1907 through 1913, when the rest of the world was still at peace, and in the dry seasons, the tourists came by the hundreds, by the thousands as time went on, to stand and watch from grassy vantage points hundreds of feet above it all. . . . "He who did not see the Culebra Cut during the mighty work of excavation," declared an author of the day, "missed one of the great spectacles of the ages—a sight that no other time, or place was, or will be, given to man to see." Lord Bryce called it the greatest liberty ever taken with nature.

A spellbound public read of cracks opening in the ground, of heart-breaking landslides, of the bottom of the canal mysteriously rising. Whole sides of mountains were being brought down with thunderous blasts of dynamite. A visiting reporter engaged in conversation at a tea party felt his chair jump half an inch and spilled a bit of scalding tea on himself.

To Joseph Bucklin Bishop, writing of "The Wonderful Culebra Cut," the most miraculous element was the prevailing sense of organization one felt. "It was organization reduced to a science—the endless-chain system of activity in perfect operation."

On either side were the grim, forbidding, perpendicular walls of rock, and in the steadily widening and deepening chasm between—the first man-made canyon in the world—a swarming mass of men and rushing railway trains, monster-like machines, all working with ceaseless activity, all animated seemingly by human intelligence, without confusion or conflict anywhere. . . . The rock walls gave place here and there to ragged sloping banks of rock and earth left by the great slides, covering many acres and reaching far back into the hills, but the ceaseless human activity prevailed everywhere. Everybody knew what he was to do and was doing it, apparently without verbal orders and without getting in the way of anybody else. . . .

Generally, the more the observer knew of engineering and construction work, the higher and warmer was his appreciation.

Panoramic photographs made at the height of the work gave an idea of how tremendous that canyon had become. But the actual spectacle, of course, was in vibrant color. The columns of coal smoke that towered above the shovels and locomotives—"a veritable Pittsburgh of smoke"—were blue-black turning to warm gray; exposed clays were pale ocher, yellow, bright orange, slate blue, or a crimson like that of the soil of Virginia; and the vibrant green of the near hills was broken by cloud shadow into great patchworks of sea blue and lavender.

The noise level was beyond belief. On a typical day there would be more than three hundred rock drills in use and their racket alone—apart from the steam shovels, the trains, the blasting—could be heard for miles. In the crevice between Gold Hill and Contractors Hill, where the walls were chiefly rock, the uproar, reverberating from wall to wall, was horrible, head-splitting. . . .

Construction of the canal would consume more than 61,000,000 pounds of dynamite, a greater amount of explosive energy than had been expended in all the nation's wars until that time. A single dynamite ship arriving at Colón carried as much as 1,000,000 pounds—20,000 fifty-pound boxes of dynamite in one shipload—all of which had to be unloaded by hand, put aboard special trains, and moved to large concrete magazines built at various points back from the congested areas.

At least half the labor force was employed in some phase of dynamite work. Those relatively few visitors permitted to walk about down in the Cut saw long lines of black men march by with boxes of dynamite on their heads, gangs of men on the rock drills, more men doing nothing but loading sticks of dynamite into the holes that had been drilled. The aggregate depth of the dynamite holes drilled in an average month in Culebra Cut (another of those statistics that defy the imagination) was 345,223 feet, or more than sixty-five miles. In the same average month more than 400,000 pounds of dynamite were exploded, which meant that all together more than 800,000 dynamite sticks with their brown paper wrappings, each eight inches long and weighing half a pound, had been placed in those sixty-five miles of drill holes, and again all by hand. . . .

Premature explosions occurred all too often as the pace of work increased. "We are having too many accidents with blasts," [chief engineer George W.] Goethals noted in June 1907. "One killed 9 men on Thursday at Pedro Miguel. The foreman blown all to pieces." Several fatal accidents were caused when shovels struck the cap of an unexploded charge. Another time a twelve-ton charge went off prematurely when hit by a bolt of lightning, killing seven men. Looking back years later, one West Indian remembered, "The flesh of men flew in the air like birds many days."

DIGNITY VERSUS INTOXICATION

ROGER BURLINGAME

In his vigorous study Engines of Democracy, *this historian of technology explores the emotional differences between railroads and the automobile in the latter's earliest days.*

America of the pre-war years was laid out, socially, according to railroad geography. Nowhere in the world had the railroad developed to such a point and everywhere were the marks of its culture. The vast system had been simplified and coordinated to facilitate long "through" journeys, eliminate stopovers and changes for passengers, route freight efficiently and promote speed. This had necessitated great changes in financial control, close interrelation of managements, standardization of freight rates. The whole industrial system depended on the effective handling of the immense bulk of freight. Without it, mass production on any considerable scale could not exist. Every factory must be on a spur. The perfect interplay of this elaborate system with all its mechanical aids, switches, signals, yards; the essential simplification of the labyrinthic complexity of rail—such things were the greatest achievements of the American genius in collectivity to date. The railroads might well be proud. And the people, always awed by the dignity of the giant rhythmic locomotive—colossal, sonorous, deep breathing, yet supple and flexible as a young athlete—the people were proud of their railroads.

Along the steel lines lay the great congested cities. Like magnets they had drawn the people from the wide pioneer dispersion. Had we, in the late nineteenth century, been able to see the whole land from the high air, we should have watched these little bodies, like iron filings moving in jerks over the lines of the railroad fields toward the magnet centers. Every day the trains disgorged their multitudes of bewildered boys and girls, men and women, sick in their souls of cows and corn and spaces, into the iron "deepos" [i.e., depots, train stations] from which a trolley car, harassed by its curves, would carry them into a new life, for better or worse.

Along thousands of spurs, smaller, jerkier trains moved through tens of thousands of small towns and villages. At each of their "deepos" stood a line of sleepy horses, hitched to patched buggies, muddy surreys, hotel stages. Leaning against every post and wall stood the bicycles. Occasionally, in 1910, there would be a strange, high, ugly contraption which the horses eyed aslant, uncertain whether it or the train were more alarming. Yet the train, the horses and the men knew, must stick to the tracks and there was no knowing what the newfangled thing might do. It was surrounded always by a crowd of curious, prophetic little boys waiting for the moment when a hand would seize its crank and startle it into violent explosion. It had neither dignity nor beauty; it was high, hideous and instinct with doubt.

If, in 1910, we should get into one of the horse buggies, we might drive for an hour over mud and ruts and find, at the end, a moribund hamlet with a decayed church, a few sleeping houses, a square brick school, a general store, a post office containing, perhaps, a telegraph key. There were thousands of these dead groupings, forgotten by the railroad world. In them, as in the isolated farms round about, people lived a life so remote and so monotonous that even the notorious assembly line would have been a relief to them. Indeed, from such places, many men were in constant migration toward the machines of production.

Should we take, instead of the buggy, the "contraption" we should be in for a period of high excitement. Whether "she" would get us home was always a live subject of speculation, though if our driver were a good mechanic there was reasonable expectation of arrival. As long as she moved, the sense of superiority to other men was acute beyond anything the human mind had yet experienced. The first railroad train bound by its track could have been nothing to it. The thrill of power and freedom in the driver, feeling the vibration and violence under his guiding hand, was intoxicating beyond his own belief.

Predictions:
The Flying-Machine
of the Future

Waldemar Kaempffert, 1911

The science editor of The New York Times *gets it partly right in this* Harper's *speculation:*

What will [the] flying-machine of the future be like? He would be a wise man indeed who could predict with any degree of accuracy its exact form and dimensions. The dreams of the old-time imaginative novelist seem almost to be realized now. Mr. R. W. A. Brewer, an English authority, sees a larger and a heavier machine than we have at present, a kind of air-yacht, weighing at least three tons, and built with a boat body. The craft of his fancy will be decked in. It will carry several persons conveniently, and will be provided with living and sleeping accommodations. He prophesies that it will fly at speeds of 150 to 200 miles an hour, for the reason that high speeds in flying mean less expenditure of power than lower speeds. Mr. F. W. Lancaster, another authority, entertains similar views on the necessity of high speed. He argues that the aeroplane speed must be twice that of the maximum wind in which the machine is to be driven. A certain amount of automatic stability is thus obtained; for a machine travelling at a hundred miles an hour is practically uninfluenced by gusts and eddies that might prove disastrous at thirty-five miles an hour. A modern *Lusitania* plunges undaunted through waves that would be perilous to a schooner. If it is ever possible for an aeroplane to travel at such terrific velocities, the United States will become the playground of the Chicago aviator. Daily trips of one thousand miles would not be extraordinary.

It seems certain that special starting and alighting grounds will be ultimately provided throughout the world. If street-cars must have their stables and their yards, it is not unreasonable to demand the provision of suitable aeroplane stations. Depots or towers will be erected for the storage of fuel and oil—garages on stilts, in a word. The aviator in need of supplies may some day signal his wants, lower a trailing line, and pick up gasoline by some such device as we now employ to catch mail-sacks on express trains. . . .

Compared with the flying-machine of the future, the motor-car will seem as tame and dull as a cart drawn by a weary nag on a dusty country road. Confined to no route in particular, free as a bird, an adventurous pilot can satisfy his craving for speed in the high-powered monoplane of the future. Even the most leisurely of air-touring machines will travel at velocities

RICHARD RHODES

52

that only a racing-automobile now attains, while the air racer will flit over us, a mere blur to the eye and a buzz to the ear. In an hour or two a whole province will be traversed; in a day half a continent. Swifter than any storm will be the flight of its pilot. If the black, whirling maelstrom of a cyclone looms up before him, he can make a detour or even outspeed it; for the velocity of his machine will be greater than that of the fiercest of howling, wintry blasts. At a gale which now drives every aviator timorously to cover, he snaps a contemptuous finger, plunges through it in a breathless dash, and emerges again in the sunshine, as indifferent to his experience as a locomotive engineer after running through a drizzling rain.

A RAPID SUCCESSION OF IMPROVEMENTS

GEORGE EASTMAN, 1912

George Eastman, who invented the film camera and founded Eastman Kodak, was one of many early twentieth-century technologists who recognized the value of industrial research.

I have come to think that the maintenance of a lead in the apparatus trade will depend greatly upon a rapid succession of changes and improvements, and with that aim in view, I propose to organize the Experimental Department in the Camera Works and raise it to a high degree of efficiency. If we can get out improved goods every year nobody will be able to get out original goods the same as we do.

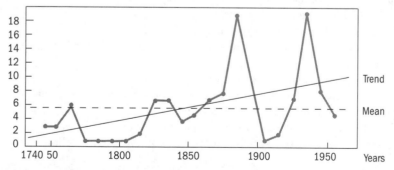

▶ Frequency of basic innovations (new industries), 1740–1960.

Every Woman an Engineer

Ellen Swallow Richards, 1910

Ellen Swallow Richards, the first woman to earn a Bachelor of Science degree at MIT (in 1873), founded the home economics movement.

The work of homemaking in this scientific age must be worked out on engineering principles and with the cooperation of trained men and trained women. The mechanical setting of life becomes an important factor, and this new impulse is showing itself so clearly today for the modified construction and operation of the family home is the final crown or seal of the conquest of the last stronghold of conservatism, the homekeeper. Tomorrow, if not today, the woman who is to be really mistress of her house must be an engineer, so far as to be able to understand the use of machines.

A High-Priced Man

Frederick Winslow Taylor, 1911

Frederick Winslow Taylor defined a new movement of industrial efficiency based on observation and measurement in his book The Principles of Scientific Management.

The task before us . . . narrowed itself down to getting Schmidt to handle 47 tons of pig iron per day and making him glad to do it. This was done as follows. Schmidt was called out from among the gang of pig-iron handlers and talked to somewhat in this way:

"Schmidt, are you a high-priced man?"

"Vell, I don't know vat you mean."

"Oh yes, you do. What I want to know is whether you are a high-priced man or not."

"Vell, I don't know vat you mean."

"Oh, come now, you answer my questions. What I want to find out is whether you are a high-priced man or one of these cheap fellows here. What I want to find out is whether you want to earn $1.85 a day or whether you are satisfied with $1.15, just the same as all those cheap fellows are getting."

"Did I vant $1.85 a day? Vas dot a high-priced man? Vell, yes, I vas a high-priced man."

"Oh, you're aggravating me. Of course you want $1.85 a day—every

RICHARD RHODES

one wants it! You know perfectly well that that has very little to do with your being a high-priced man. For goodness' sake answer my questions and don't waste any more of my time. Now come over here. You see that pile of pig-iron?"

"Yes."

"You see that car?"

"Yes."

"Well, if you are a high-priced man, you will load that pig-iron on that car tomorrow for $1.85. Now do wake up and answer my question. Tell me whether you are a high-priced man or not."

"Vell—did I got $1.85 for loading dot pig iron on dot car tomorrow?"

"Yes, of course you do, and you get $1.85 for loading a pile like that every day right through the year. That is what a high-priced man does, and you know it just as well as I do."

"Vell, dot's all right. I could load dot pig iron on the car tomorrow for $1.85, and I get it every day, don't I?"

"Certainly you do—certainly you do."

"Vell, den, I vas a high-priced man."

"Now, hold on, hold on. You know just as well as I do that a high-priced man has to do exactly as he's told from morning till night. You have seen this man here before, haven't you?"

"No, I never saw him."

"Well, if you are a high-priced man, you will do exactly as this man tells you tomorrow, from morning till night. When he tells you to pick up a pig and walk, you pick it up and you walk, and when he tells you to sit down and rest, you sit down. You do that right straight through the day. And what's more, no back talk. . . ."

Schmidt started to work, and all day long, and at regular intervals, was told by the man who stood over him with a watch, "Now pick up a pig and walk. Now sit down and rest. Now walk—now rest," etc. He worked when he was told to work, and rested when he was told to rest, and at half past five in the afternoon had his 47 tons loaded on the car. And he practically never failed to work at this pace and do the task that was set him during the three years that the writer was at Bethlehem [Steel]. And throughout this time he averaged a little more than $1.85 per day, whereas before he had never received over $1.15 per day, which was the ruling rate of wages at that time in Bethlehem. That is, he received 60 percent higher wages than were paid to other men who were not working on task work. One man after another was picked out and trained to handle pig iron at the rate of 47 tons per day until all of the pig iron was handled at this rate, and the men were receiving 60 percent more wages than other workmen around them.

. . . The essence of scientific management [is] first, the careful selection of the workman, and, second and third, the method of first inducing and then training and helping the workman to work according to the scientific method.

PRODUCING WEALTH
BUT GRINDING MAN

SAMUEL GOMPERS, 1911

Industrial efficiency had a price, Samuel Gompers, president of the American Federation of Labor, testified in 1911 before a committee of the U.S. House of Representatives investigating Taylorism. The new methodology might increase the productivity of common laborers, but it reduced the wages of skilled craftsmen. Worse, Gompers argued, it was inhumane.

I grant you that if this Taylor system is put into operation, as we see it and as we understand it, it will mean great production in goods and things, but in so far as man is concerned it means destruction. So long as the supply of human labor is not exhausted the system can go on. I mean by human labor not only that now in the United States and not only that on the American Continent, but I mean labor from wherever the supply can be bought. It is producing wealth but grinding man, and, while I think we all agree that production is one of the essentials of life and that while greater productions must go on in order to satisfy our growing needs, there are other considerations of a primary and more important character, and that is that the intelligence, that the physique, that the spirit, the mind, hopes, and aspirations of man shall be also cultivated and given an opportunity for higher achievements.

Why should employers want a system that preys upon the independence, the development and the character of the worker? This last statement by Mr. Taylor, in reply to the query, discloses at once the idea. It is that four or five foremen shall stand over a worker and by a species of industrial "third degree" compel obedience. . . . Workers in the United States in every industry of which we know anything, or, at least, of which I know anything, work harder, quicker, and more effectively, so far as results are concerned, than the workers in any other country. When workers from foreign countries come to the United States for the first few months they are simply dazed with the rapid motion and the great production of the American worker, and it is only after months and sometimes a year before they can acquire the same movement. Sometimes they never attain the rapidity of movement and effectiveness of result possessed and demonstrated by the American worker. If we had a dearth of workers, if our production lagged, if

RICHARD RHODES

56

we needed as essential to our well-being all these things, they might be justified for a time. But there is no need; we produce and can produce with economy applied all around and under a humane system of management all that is required. I wish to say this for the men of labor—and I speak not only for the men of labor, because I am a worker myself—I worked at my trade for 26 years and I have a right to speak as a worker—that there are some limits beyond which we will not allow you to go with your domination as captains of industry. You are our employers, but you are not our masters.

A BABE STILL IN THE CRADLE

ARTHUR D. LITTLE, 1913

The babe still in the cradle was industrial research; note this pioneer's emphasis on minimizing waste.

The country of Franklin, Morse and Rumford; of McCormick, Howe and Whitney; of Edison, Thomson, Westinghouse and Bell; and of Wilbur and Orville Wright, is obviously a country not wholly hostile to industrial research or unable to apply it to good purpose. It is, however, not surprising that with vast areas of virgin soil of which a share might be had for the asking; with interminable stretches of stately forest; with coal and oil and gas, the ores of metals and countless other gifts of nature scattered broadcast by her lavish hand, our people entered upon this rich inheritance with the spirit of the spendthrift, and gave little heed to refinements in methods of production and less to minimizing waste. That day and generation is gone. Today, their children, partly through better recognition of potential values, but mainly by the pressure of greatly increased population and the stress of competition among themselves and in the markets of the world, are rapidly acquiring the knowledge that efficiency of production is a sounder basis for prosperity than mere volume of product, however great. Many of them have already learned that the most profitable output of their plant is that resulting from the catalysis of raw materials by brains. A far larger number are still ignorant of these fundamental truths, and so it happens that most of our industrial effort still proceeds under the guidance of empiricism with a happy disregard of basic principles. A native ingenuity often brings it to a surprising success and seems to support the aphorism "Where ignorance is profitable, 'tis folly to be wise." Whatever may be said, therefore, of industrial research in America at this time is said of a babe still in the cradle but which has nevertheless, like the infant Hercules, already destroyed its serpents and given promise of its performance at man's estate.

A City Built by Experts

Frederic C. Howe, 1913

The new vision of expert analysis and application extended to the American city as well.

City planning is the art of building cities as men build homes, as engineers project railroad systems, as landscape artists lay out garden cities, as manufacturing corporations build factory towns like Gary, Indiana, or Pullman, Illinois. City planning treats the city as a unit, as an organic whole. It lays out the land on which a city is built as an individual plans a private estate. It locates public buildings so as to secure the highest architectural effects, and anticipates the future with the farsightedness of an army commander, so as to secure the orderly, harmonious, and symmetrical development of the community.

City planning makes provision for people as well as for industry. It coordinates play with work, beauty with utility. It lays out parks, boulevards, and playgrounds, and links up water, rail, and street traffic so as to reduce the wastes of production to a minimum. . . .

In a big way, city planning is the first conscious recognition of the unity of society. It involves a socializing of art and beauty and the control of the unrestrained license of the individual. It enlarges the power of the State to include the things men own as well as the men themselves, and widens the idea of sovereignty so as to protect the community from him who abuses the rights of property, as it now protects the community from him who abuses his personal freedom.

City planning involves a new vision of the city. It means a city built by experts, by experts in architecture, in landscape gardening, in engineering, and housing; by students of health, sanitation, transportation, water, gas, and electricity supply; by a new type of municipal officials who visualize the complex life of a million people as the builders of an earlier age visualized an individual home.

PREDICTIONS:
ATOMIC BOMBS

H. G. WELLS, 1914

From a 1909 book by British radiochemist Frederick Soddy, the popular science-fiction visionary borrowed the idea of an atomic explosive for his 1914 novel The World Set Free. *"The man who put his hand on the lever by which a parsimonious nature regulates so jealously the output of this store of energy [i.e., radioactivity] would possess a weapon by which he could destroy the earth if he chose," Soddy had written. Wells imagined an atomic war in 1958 which destroys the major cities of the world and leads the survivors to world government. His bombs are hocus-pocus: "continuing explosives" made of an artificial element he calls "Carolinum." Notice that he fails to update his aircraft; the bombardiers lift the bombs from open cockpits and toss them over the side.*

Never before in the history of warfare had there been a continuing explosive; indeed, up to the middle of the twentieth century the only explosives known were combustibles whose explosiveness was due entirely to their instantaneousness; and these atomic bombs which science burst upon the world that night were strange even to the men who used them. Those used by the Allies were lumps of pure Carolinum, painted on the outside with unoxidized cydonator inducive enclosed hermetically in a case of membranium. A little celluloid stud between the handles by which the bomb was lifted was arranged so as to be easily torn off and admit air to the inducive, which at once became active and set up radioactivity in the outer layer of the Carolinum sphere. This liberated fresh inducive, and so in a few minutes the whole bomb was blazing continual explosion. The Central European bombs were the same, except that they were larger and had a more complicated arrangement for animating the inducive.

Always before in the development of warfare the shells and rockets fired had been but momentarily explosive, they had gone off in an instant once for all, and if there was nothing living or valuable within reach of the concussion and the flying fragments, then they were spent and over. But Carolinum, which belonged to the β-Group of Hyslop's so-called "suspended degenerator" elements, once its degenerative process had been induced, continued a furious radiation of energy, and nothing could arrest it. Of all Hyslop's artificial elements, Carolinum was the most heavily stored with energy and the most dangerous to make and handle. To this day it remains the most potent degenerator known. What the earlier twentieth-century chemists called its half period was seventeen days; that is to say, it poured out half of the huge store of energy in its great molecules in the space of seventeen days, the next seventeen days' emission was a half of that first period's outpouring, and so on. As with all radioactive substances, this Carolinum, though every seventeen days its power is halved, though constantly it diminishes toward the imperceptible, is never entirely exhausted, and to this day the battlefields and bomb-fields of that frantic time in human history are sprinkled with radiant matter and so centers of inconvenient rays. . . .

What happened then when the celluloid stud was opened was that the inducive oxidized and became active. Then the surface of the Carolinum began to degenerate. This degeneration passed only slowly into the substance of the bomb. A moment or so after its explosion began it was still mainly an inert sphere exploding superficially, a big, inanimate nucleus wrapped in flame and thunder. Those that were thrown from aeroplanes fell in this state; they reached the ground still mainly solid and, melting soil and rock in their progress, bored into the earth. There, as more and more of the Carolinum became active, the bomb spread itself out into a monstrous cavern of fiery energy at the base of what became very speedily a miniature active volcano. The Carolinum, unable to disperse freely, drove into and mixed up with a boiling confusion of molten soil and superheated steam, and so remained, spinning furiously and maintaining an eruption that lasted for years or months or weeks according to the size of the bomb employed and the chances of its dispersal. Once launched, the bomb was absolutely unapproachable and uncontrollable until its forces were nearly exhausted, and from the crater that burst open above it, puffs of heavy incandescent vapor and fragments of viciously punitive rock and mud, saturated with Carolinum, and each a center of scorching and blistering energy, were flung high and far.

Such was the crowning triumph of military science, the ultimate explosive, that was to give the "decisive touch" to war. . . .

A recent historical writer has described the world of that time as one that "believed in established words and was invincibly blind to the obvious in things." Certainly it seems now that nothing could have been more obvious to the people of the early twentieth century than the rapidity with which war was becoming impossible. And as certainly they did not see it. They did not see it until the atomic bombs burst in their fumbling hands.

PAVING PROPAGANDA

MORRIS LLEWELLYN COOKE, 1915

The power of public opinion began to be recognized as a force for technological change, a mechanical engineer observes.

The development of some varieties of municipal engineering is absolutely dependent upon the development of public opinion and must proceed with it. The matter of street cleaning is largely a question of an improved public taste in the matter of street paving. Unless streets are well paved they cannot be well cleaned except at a prohibitive cost. To jump from one degree of cleanliness in this respect, to another, without a supporting public opinion, may be enough to wreck an administration and to set the tide of civic improvement running in the opposite direction.

The newspaper is the great educator in these matters today. But we are already using in Philadelphia moving pictures, parades and exhibitions. The possibilities of these and other means of publicity are not yet fully understood.

MAKING HISTORY

Henry Ford, 1916

Henry Ford was referring to the problem of international disarmament when he minimized the importance of history in his comments to a reporter, but his American impatience with tradition enshrined his remarks in myth.

History is more or less bunk. It's tradition. We don't want tradition. We want to live in the present and the only history that is worth a tinker's dam is the history we make today.

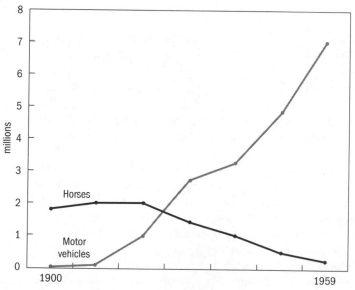

▶ U.S. motor vehicles vs. horses on farms, 1900–1959.

THE LINE-GANG

ROBERT FROST, 1916

The arrival of a new technology engenders awe, often tinged with suspicion. From 1953 in the Midwest, where I grew up, I remember newspaper stories heralding with bold headlines ("The Cable is Coming!") the progressive approach of coaxial cable and microwave towers that would bring us national television. Four decades earlier, the New Hampshire poet observed the city encroaching on the countryside in the form of telephone lines.

Here come the line-gang pioneering by.
They throw a forest down less cut than broken.
They plant dead trees for living, and the dead
They string together with a living thread.
They string an instrument against the sky
Wherein words whether beaten out or spoken
Will run as hushed as when they were a thought.
But in no hush they string it: they go past
With shouts afar to pull the cable taut,
To hold it hard until they make it fast,
To ease away—they have it. With a laugh,
An oath of towns that set the wild at naught,
They bring the telephone and the telegraph.

GASOLINE DRIPPING FROM THE TREES

MAX W. BALL, 1916

Concern for the environment did not begin with Rachel Carson: here an engineer for the U.S. Bureau of Mines bemoans a shortsighted profligacy.

Are we practicing conservation? Within the last few weeks I have seen millions of cubic feet of natural gas wasting into the air—gas so rich in gasoline that it dripped from the trees like an April shower. I have seen wells capable of yielding 40,000,000 cu. ft. of gas each being deliberately drowned out by pumping water into the gas sands. Reckless drilling, defective casing, careless plugging are flooding great areas with water and losing forever enormous quantities of oil. It has been testified before the Corporation Commission of Oklahoma that ordinary methods leave from 25 to 85% of the oil in the ground, and this estimate is concurred in by careful engineers and practical oil men.

Nor are these underground losses the only ones. When the oil is brought to the surface before transportation and market are ready for it, it must go into storage. Indeed, in many fields oil has been produced before storage was available, and millions of gallons have gone down the streams or seeped away from earthen reservoirs. Even when the best steel tankage has been provided, evaporation losses still go on. Cushing crude stored in steel tanks for a few months lost approximately a fifth of its gasoline content. The State Mineralogist's office of California has estimated that even with the heavy oils of that state the loss by evaporation represents perhaps 25% of the total value of the production at the well. An official of one of the largest companies in the Midcontinent field recently told me that last year fire destroyed 6% of his company's production.

Just consider these examples: 25 to 85% left underground; 20 to 50% of the value of oil produced lost through evaporation in storage; 6% of stored oil lost by fire! These losses are staggering and are not exceptional! What a small percentage of this wonderful natural resource is saved to run your machine or to deliver goods at your door or to plow the fields from which your food must come!

If we feel these losses in the high prices of the present day, how much more will we feel them five, ten, twenty or thirty years from now? Is it not time we considered them seriously and tried to determine upon some remedy?

BILLY MITCHELL TAKES TO THE AIR

BRIGADIER GENERAL WILLIAM
MITCHELL, 1917

Bombing from the air, the U.S. Army's Billy Mitchell believed, would revolutionize warfare. For his insubordination in promoting aerial bombing he was court-martialed and forced to resign from the army in 1925. He observed a revolution when he toured the French lines during the Great War.

We left the ground at about 5:30 P.M. The pilot, Lieutenant François Lafont, took the machine off the ground in fine style. He jumped her up pretty well, then straightened out to get speed and climbed up in long spirals, with the two Hispanos [engines] buzzing regularly. The fuselage was so constructed that most of the wind was blown over one's head. The pilot and gunner wore no glasses. I took some as I was used to wearing them, but soon pulled them off. There was very little vibration and I found I could see very easily with field glasses. . . .

While we were over the lines, several units of French pursuit machines came by us, for my benefit I believe. I was impressed with the compact formation that the flights of five were keeping. The machines were not more than one hundred yards from each other, sometimes not over fifty, and passed very close to me, both under and over.

We could look down the Marne river and see Reims plainly at a distance of ten miles, in spite of its being in the direction of the setting sun. All about were villages of which there was nothing left except crumbling walls. The enemy destroyed them to prevent their being used for shelter or points of supply and concealment. Villages were shot up at distances of ten miles or more behind the lines, particularly places where roads or railways joined.

At 7:00 P.M. we started back. . . . I asked the pilot how the big machine handled when he threw her around a little, and he said that looping had been done with one, but he could not do it then because the gunner was not tied in. He made several vertical banks, nose dives and extremely tight spirals, however, and the machine handled beautifully, with little vibration. He said it was almost impossible to get into a tailspin with it, and anyway if it did, he could stop it quickly by applying opposite control.

We came down at very slow speed. He used his motor until within three hundred feet of the ground, gave her a kick just before we landed; and we did not roll more than one hundred feet. It was a beautiful landing. . . .

We had dinner with the squadron officers in their mess, a small frame building by a clear little stream. On the walls were hung pieces of a couple of German machines that had been destroyed by the large observation machines.

The conversation was all about the tactical use of airplanes. The Commanding Officer considered that the observation part was well worked out, that is, the map, radio and seeing end of it. He wanted more performance from his machine, more armor, and above all, armament. He wanted a gun of ten to twenty barrels, each one of which could fire as fast as a Lewis or Vickers. He only fought defensively and needed all the fire he could get.

A few days before, one of his machines had been engaged by five enemy pursuits; they attacked both from above and underneath. Those underneath would stick their machines straight up in a stall and let go their guns, while those from above attacked in the usual manner. It was impossible for the pilot to keep all of them under the fire of his guns. Soon the observer was killed, the gunner shot through both arms and the pilot's left arm broken; the wires to both ailerons were shot away, and the gas tanks of both motors punctured. He glided for the ground and when making a landing, turned the machine upside down. The pilot yelled to the gunner to come and let him out, as he could not release his belt with one hand; but the gunner was wedged in the cockpit and, on account of both his arms being broken, could not get out. Some infantry were near them, and came over and released them.

This was a normal happening in the war, something that could be expected to happen every day. Everyone knew when he started on a flight that there was sure to be a fight if he was looking for one, or if his mission was such that he had to carry out some special thing regardless.

One flight over the lines gave me a much clearer impression of how the armies were laid out than any amount of traveling around on the ground. A very significant thing to me was that we could cross the lines of these contending armies in a few minutes in our airplane, whereas the armies had been locked in the struggle, immovable, powerless to advance, for three years. To even stick one's head over the top of a trench invited death. This whole area over which the Germans and French battled was not more than sixty miles across. It was as though they kept knocking their heads against a stone wall, until their brains were dashed out. They got nowhere, as far as ending the war was concerned.

It looked as though the war would keep up indefinitely until either the airplanes brought an end to the war or the contending nations dropped from sheer exhaustion.

THE MECHANIZATION OF WAR: I

WILLIAM L. SIBERT, 1919

The Great War, the first of two world wars the twentieth century would suffer, saw extensive mechanization, an engineer reports.

In the beginning of the hostilities the first things that appeared out of the ordinary were the extended use of machine guns and the use of longer range, larger calibered guns in the field. This extensive use of machine guns, firing from hastily prepared positions, covered in front by barbed wire, was so destructive to life that it soon became apparent that a position so defended could not be taken by direct infantry attack with the ordinary artillery support. The machine gun has done more to determine the trend of changes in military appliances than any other thing in this war.

THE MECHANIZATION
OF WAR: II

ORVILLE WRIGHT, 1917

RICHARD RHODES

I really believe that the aeroplane will help peace in more ways than one—in particular I think it will have a tendency to make war impossible. Indeed, it is my conviction that, had the European governments foreseen the part which the aeroplane was to play [in the Great War], especially in reducing all their strategical plans to a devastating deadlock, they would never had entered upon the war. . . . This illustrates the mistaken notions which were entertained concerning the practical uses of the aeroplane in warfare. Most of us saw its use for scouting purposes, but few foresaw that it would usher in an entirely new form of warfare. As a result of its activities, every opposing general knows precisely the strength of his enemy and precisely what he is going to do. Thus surprise attacks, which for thousands of years have determined the event of wars, are no longer possible, and thus all future wars, between forces which stand anywhere near an equality, will settle down to tedious deadlocks. Civilized countries, knowing this in advance, will hesitate before taking up arms—a fact which makes me believe that the aeroplane, far more than Hague conferences and Leagues to enforce peace, will exert a powerful influence in putting an end to war.

66

BREAKING CRUST

THOMAS P. HUGHES

In his masterful study Networks of Power, *about the electrification of Europe and North America from 1880 to 1930, the historian finds lessons for prediction in the significant changes that followed the Great War.*

Because technology is often manifested in material form—machines, processing equipment, structures, and tools—its lasting effects are easily observed. Ideas and events, historians argue, have effects far beyond the time in which they occur, but their effects are less visible. In the case of technology, even the casual observer knows that he is surrounded by things that were made in the past, things that were often made under substantially different circumstances. In a sense, therefore, surviving technology brings to the present the character of the past, a past that imposed its characteristics on the technology when it was first invented, developed, and introduced

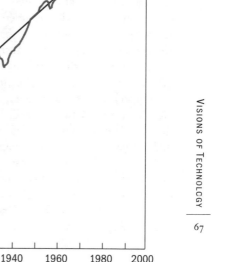

▶ World energy consumption in millions of tons of coal equivalent, 1860–2000. Trend line has 2 percent slope.

into use. The technology of electric power systems that was introduced during World War I not only caused perturbations in trends but also carried into peacetime certain aspects of the wartime environment. The extremely large electric generating systems that were built to fill the pressing and unusual needs for electric power during World War I survived the war and became, in a sense, a solution in search of a problem. Another, less obvious case is the large number of interconnections of electric light and power systems that were made during the emergencies of wartime and carried over into peacetime.

. . . War brought the governments of Germany and the United States to fund the development of power plants of unprecedented size. . . . The series of wartime interconnections in Great Britain and the United States . . . resulted from the unusual, even radical, demands that were made on formerly reluctant utilities by central governments concerned about survival. . . . The three governments [made efforts] to maintain in peacetime the momentum of their wartime planning and nurturing of technology. The efforts failed in the three cases considered here. The reason for the failure was in part a conservative reaction comparable to that which is often encountered in the political realm after radical changes have been introduced. Once the disruptive force—in this case, war—is removed, the prewar context again prevails. There were other causes, however, and they differed from country to country.

The fact that the exigencies of war caused the accelerated development of certain technological characteristics—in this instance, large size and interconnection—shows again that the rate and direction of technological change can be shaped by nontechnological factors. In other words, technology is not necessarily a simple extrapolation of its past, or a working out of inherent technological implications. The cases considered suggest that war did not so much stimulate the invention and development of new technologies as clear away the political, economic, and other nontechnological factors that prevented or retarded the utilization of existing technologies. The imperatives of war did not reverse the direction of technological evolution nor did they cause mutations; rather, they broke a conservative crust that had restrained adjustments in course and velocity. The wartime history of electric power supply suggests that a society under the influence of even more pressing demands—a moral equivalent of war—could alter the shape of technology even more drastically.

BIRTH CONTROL

MARGARET SANGER, 1920

Long before the invention of the Pill, feminists like Margaret Sanger identified birth control as the key to liberating women from what Sanger called "sex servitude."

Woman has, through her reproductive ability, founded and perpetuated the tyrannies of the Earth. Whether it was the tyranny of a monarchy, an oligarchy or a republic, the one indispensable factor of its existence was, as it is now, hordes of human beings—human beings so plentiful as to be cheap, and so cheap that ignorance was their natural lot. Upon the rock of an unenlightened, submissive maternity have these been founded; upon the product of such a maternity have they flourished.

No despot ever flung forth his legions to die in foreign conquest, no privilege-ruled nation ever erupted across its borders, to lock in death embrace with another, but behind them loomed the driving power of a population too large for its boundaries and its natural resources.

No period of low wages or of idleness with their want among the workers, no peonage or sweatshop, no child-labor factory, ever came into being, save from the same source. Nor have famine and plague been as much "acts of God" as acts of too prolific mothers. They, also, as all students know, have their basic causes in over-population.

The creators of over-population are the women, who, while wringing their hands over each fresh horror, submit anew to their task of producing the multitudes who will bring about the *next* tragedy of civilization.

While unknowingly laying the foundations of tyrannies and providing the human tinder for racial conflagrations, woman was also unknowingly creating slums, filling asylums with insane, and institutions with other defectives. She was replenishing the ranks of the prostitutes, furnishing grist for the criminal courts and inmates for prisons. Had she planned deliberately to achieve this tragic total of human waste and misery, she could hardly have done it more effectively.

Woman's passivity under the burden of her disastrous task was almost altogether that of ignorant resignation. She knew virtually nothing about her reproductive nature and less about the consequences of her excessive childbearing. It is true that, obeying the inner urge of their natures, *some* women revolted. They went even to the extreme of infanticide and abortion. Usually their revolts were not general enough. They fought as individuals, not as a mass. In the mass they sank back into blind and hopeless subjection. They went on breeding with staggering rapidity those numberless, undesired children who become the clogs and the destroyers of civilizations.

Today, however, woman is rising in fundamental revolt. Ever her efforts at mere reform are . . . steps in that direction. Underneath each of them is the feminine urge to complete freedom. Millions of women are asserting their right to voluntary motherhood. They are determined to decide for themselves whether they shall become mothers, under what conditions and when. This is the fundamental revolt referred to. It is for woman the key to the temple of liberty.

Even as birth control is the means by which woman attains basic freedom, so it is the means by which she must and will uproot the evil she has wrought through her submission. As she has unconsciously and ignorantly brought about social disaster, so must and will she consciously and intelligently *undo* that disaster and create a new and a better order. . . .

War, famine, poverty and oppression of the workers will continue while woman makes life cheap. They will cease only when she limits her reproductivity and human life is no longer a thing to be wasted.

The Radio Age: I

Clayton R. Koppes

Commercial radio in the 1920s evoked hopes even more extravagant than those expressed by Internet optimists in the 1990s, a historian reports.

The radio age in America was inaugurated almost inaudibly on election night, November 2, 1920, when station KDKA broadcast the news of a return to normalcy from a two-story garage in East Pittsburgh. By 1922 radio had become the equal of mahjongg as a national mania—and proved to be considerably more lasting. . . .

As radio established itself as more than a fad, many writers began trying to ascertain what effect the invention would have on American life. . . . In general the early twenties in almost all quarters, even the liberal enclaves, hailed the radio as the means of automatically achieving most of mankind's golden dreams. . . .

Writers of the twenties held four main aspirations for radio. The first, reiterated by persons in diverse quarters, was that radio would be a powerful force for world peace. . . . Why? M. H. Aylesworth, president of the National Broadcasting Company, explained, "People in all countries of the civilized world, hearing the same programs—music, speeches, sermons and so on—cannot fail to have a more friendly feeling for each other."

. . . The people of the twenties also believed that the radio would save democracy, a goal frequently linked with world peace. Political observers hastened to predict the death of the political demagogue. In his place would rise the man who spoke the cool voice of reason. . . .

A third tenet of the faith in radio was that it would introduce an unparalleled advance in education and mass culture. . . .

Religion formed the fourth quarter of the faith in radio. Some churchmen viewed the radio as a tool that would enable them to outstrip in decades what it had taken the church centuries to accomplish. . . . The final design was clear. Radio alone, unaided, could solve the pressing problems of the day. A mechanical device was virtually to usher in the millennium.

THE INSIDIOUS DANGERS OF
RADIO ADVERTISING

PRINTER'S INK, 1923

It is a matter of general advertising interest to record that the American Telephone and Telegraph Company is trying to establish a new advertising medium. Through its station WEAF, New York, it is permitting advertisers to broadcast messages. So far the company's venture is only in the experimental stage. As a tryout, it has placed a nominal charge of $100 on a ten-minute talk. During this time about 750 words can be delivered.

The fact that several advertisers have already availed themselves of this service would seem to indicate that there is a demand for it. Just the same, it is our advice to the American Telephone and Telegraph Company to "stop, look and listen" before extending this new branch of its business. The plan is loaded with insidious dangers. The company, itself, evidently recognizes this, as it is proceeding cautiously in this advertising broadcasting experiment. For one thing, it is restricting the number of times a product may be mentioned during the course of a talk. It feels that the radio audience may regard the advertising message as an unwarranted imposition on its time. For this reason, it is insisted that the advertiser make his announcement subtle. No bald statements are permitted.

But regardless of how carefully censored the messages may be, the objection to this form of advertising still stands. Station WEAF has built up its reputation on the fine quality of its programs. Radio fans who tune in on this station are accustomed to get high-class entertainment. If they are obliged to listen to some advertiser exploit his wares, they will very properly resent it. . . . An audience that has been wheedled into listening to a selfish message will naturally be offended. Its ill-will would be directed not only against the company that delivered the story, but also against the advertiser who chose to talk shop at such an inopportune time.

There are several objections to the sending out of advertising messages through radio broadcasting stations, but we are opposed to the scheme principally because it is against good public policy. We are opposed to it for the same reason that we object to sky writing. People should not be forced to read advertising unless they are so inclined.

IN THE HANDS OF
THE TECHNICIANS

THORSTEIN VEBLEN, 1921

With the Russian Revolution spreading fear among capitalists, Veblen contrasted the new professionals of industry—the technicians—with the old entrepreneurial class in his book The Engineers and the Price System. *The technicians had become indispensable, he thought, but they were too compliant to foment revolt.*

Anything like a Soviet of Technicians is at the most a remote contingency in America. . . . By settled habit the technicians, the engineers and industrial experts, are a harmless and docile sort, well fed on the whole, and somewhat placidly content with the "full dinner-pail" which the lieutenants of the Vested Interests habitually allow them. It is true, they constitute the indispensable General Staff of that industrial system which feeds the Vested Interests; but hitherto at least, they have had nothing to say in the planning and direction of this industrial system, except as employees in the pay of the financiers. . . .

But it remains true that they and their dear-bought knowledge of ways and means—dear-bought on the part of the underlying community—are the pillars of that house of industry in which the Vested Interests continue to live. Without their continued and unremitting supervision and direction the industrial system would cease to be a working system at all; whereas it is not easy to see how the elimination of the existing businesslike control could bring anything but relief and heightened efficiency to this working system. The technicians are indispensable to productive industry of this mechanical sort; the Vested Interests and their absentee owners are not. . . .

It follows that the material welfare of all the advanced industrial peoples rests in the hands of these technicians, if they will only see it that way, take counsel together, constitute themselves the self-directing General Staff of the country's industry, and dispense with the interference of the lieutenants of the absentee owners. . . .

But there is assuredly no present promise of the technicians' turning their insight and common sense to such a use. There need be no present apprehension. The technicians are a "safe and sane" lot, on the whole. . . . And herein lies the present security of the Vested Interests, as well as the fatuity of any present alarm about Bolshevism and the like.

Rube Goldberg™ and © Rube Goldberg Inc. Distributed by United Media.

▶ Rube Goldberg's popular cartoons satirized and domesticated technological complexity.

ROSSUM'S
UNIVERSAL ROBOTS

KAREL CAPEK, 1922

73

The Bohemian dramatist coined the word "robot" in his "fantastic melodrama" R.U.R. to characterize technological marvels: factory-produced "living machines" that perform the work formerly assigned to slaves, workers and beasts of burden. Helena Glory, "the daughter of our president," visits the central office of the factory of Rossum's Universal Robots intent on liberating them. Eventually, in love with her, Dr. Gall, the head of the Physiological and Experimental Department, contrives to supply the robots with something like souls, and they revolt and destroy humanity. Before then, we learn the benevolent vision of the creators.

HELENA

Excuse me, gentlemen, for—for—. Have I done something dreadful?

ALQUIST

Not at all, Miss Glory. Please sit down.

HELENA

I'm a stupid girl. Send me back by the first ship.

DR. GALL

Not for anything in the world, Miss Glory. Why should we send you back?

HELENA

Because you know I've come to disturb your Robots for you.

DOMIN

My dear Miss Glory, we've had close upon a hundred saviors and prophets here. Every ship brings us some. Missionaries, anarchists, Salvation Army, all sorts. It's astonishing what a number of churches and idiots there are in the world.

HELENA

And you let them speak to the Robots?

DOMIN

So far we've let them all, why not? The Robots remember everything, but that's all. They don't even laugh at what the people say. Really, it is quite incredible. If it would amuse you, Miss Glory, I'll take you over to the Robot warehouse. It holds about three hundred thousand of them.

BUSMAN

Three hundred and forty-seven thousand.

DOMIN

Good! And you can say whatever you like to them. You can read the Bible, recite the multiplication table, whatever you please. You can even preach to them about human rights.

HELENA

Oh, I think that if you were to show them a little love—.

FABRY

Impossible, Miss Glory. Nothing is harder to like than a Robot.

HELENA

What do you make them for, then?

BUSMAN

Ha, ha, ha, that's good! What are Robots made for?

FABRY

For work, Miss Glory! One Robot can replace two and a half workmen. The human machine, Miss Glory, was terribly imperfect. It had to be removed sooner or later.

BUSMAN

It was too expensive.

FABRY

It was not effective. It no longer answers the requirements of modern engineering. Nature has no idea of keeping pace with modern labor. For example: from a technical point of view, the whole of childhood is a sheer absurdity. So much time lost. And then again—.

HELENA

Oh, no! No!

FABRY

Pardon me. But kindly tell me what is the real aim of your League—the . . . the Humanity League.

HELENA

Its real purpose is to—to protect the Robots—and—and ensure good treatment for them.

FABRY

Not a bad object, either. A machine has to be treated properly. Upon my soul, I approve of that. I don't like damaged articles. Please, Miss Glory, enroll us all as contributing, or regular, or foundation members of your League.

HELENA

No, you don't understand me. What we really want is to—to liberate the Robots.

HALLEMEIER

How do you propose to do that?

HELENA

They are to be—to be dealt with like human beings.

HALLEMEIER

Aha. I suppose they're to vote? To drink beer? To order us about?

HELENA

Why shouldn't they drink beer?

HALLEMEIER

Perhaps they're even to receive wages?

HELENA

Of course they are.

HALLEMEIER

Fancy that, now! And what would they do with their wages, pray?

HELENA

They would buy—what they need . . . what pleases them.

HALLEMEIER

That would be very nice, Miss Glory, only there's nothing that does please the Robots. Good heavens, what are they to buy? You can feed them on pineapples, straw, whatever you like. It's all the same to them, they've no appetite at all. They've no interest in anything, Miss Glory. Why, hang it all, nobody's ever yet seen a Robot smile.

HELENA

Why . . . why don't you make them happier?

HALLEMEIER

That wouldn't do, Miss Glory. They are only workmen.

HELENA

Oh, but they're so intelligent.

HALLEMEIER

Confoundedly so, but they're nothing else. They've no will of their own. No passion. No soul.

HELENA

No love?

HALLEMEIER

Love? Rather not. Robots don't love. Not even themselves.

HELENA

Nor defiance?

HALLEMEIER

Defiance? I don't know. Only rarely, from time to time.

HELENA

What?

HALLEMEIER

Nothing particular. Occasionally they seem to go off their heads. Something like epilepsy, you know. It's called Robot's cramp. They'll suddenly sling down everything they're holding, stand still, gnash their teeth—and then they have to go to the stamping mill. It's evidently some breakdown in the mechanism.

DOMIN

A flaw in the works that has to be removed.

HELENA

No, no, that's the soul.

FABRY

Do you think that the soul first shows itself by a gnashing of teeth?

HELENA

Perhaps it's a sort of revolt. Perhaps it's just a sign that there's a struggle within. Oh, if you could infuse them with it!

DOMIN

That'll be remedied, Miss Glory. Dr. Gall is just making some experiments——

DR. GALL

Not with regard to that, Domin. At present I am making pain-nerves.

HELENA

Pain-nerves?

DR. GALL

Yes, the Robots feel practically no bodily pain. You see, young Rossum provided them with too limited a nervous system. We must introduce suffering.

HELENA

Why do you want to cause them pain?

DR. GALL

For industrial reasons, Miss Glory. Sometimes a Robot does damage to himself because it doesn't hurt him. He puts his hand into the machine, breaks his finger, smashes his head, it's all the same to him. We must provide them with pain. That's an automatic protection against damage.

HELENA

Will they be happier when they feel pain?

DR. GALL

On the contrary; but they will be more perfect from a technical point of view.

HELENA

Why don't you create a soul for them?

DR. GALL

That's not in our power.

FABRY

That's not in our interest.

BUSMAN

That would increase the cost of production. Hang it all, my dear young lady, we turn them out at such a cheap rate. A hundred and fifty dollars each fully dressed, and fifteen years ago they cost ten thousand. Five years ago we used to buy the clothes for them. Today we have our own weaving mill, and now we even export cloth five times cheaper than other factories. What do you pay a yard for cloth, Miss Glory?

HELENA

I don't know really, I've forgotten.

BUSMAN

Good gracious, and you want to found a Humanity League? It only costs a third now, Miss Glory. All prices are today a third of what they were and they'll fall still lower, lower, lower, like that.

HELENA

I don't understand.

BUSMAN

Why, bless you, Miss Glory, it means that the cost of labor has fallen. A Robot, food and all, costs three quarters of a cent per hour. That's mighty important, you know. All factories will go pop like chestnuts if they don't at once buy Robots to lower the cost of production.

HELENA

And get rid of their workmen?

BUSMAN

Of course. But in the meantime, we've dumped five hundred thousand tropical Robots down on the Argentine pampas to grow corn. Would you mind telling me how much you pay a pound for bread?

HELENA

I've no idea.

BUSMAN

Well, I'll tell you. It now costs two cents in good old Europe. A pound of bread for two cents, and the Humanity League knows nothing about it. Miss Glory, you don't realize that even that's too expensive. Why, in five years' time I'll wager———

HELENA

What?

BUSMAN

That the cost of everything won't be a tenth of what it is now. Why, in five years we'll be up to our ears in corn and everything else.

ALQUIST

Yes, and all the workers throughout the world will be unemployed.

DOMIN

Yes, Alquist, they will. Yes, Miss Glory, they will. But in ten years Rossum's Universal Robots will produce so much corn, so much cloth, so much everything, that things will be practically without price. There will be no poverty. All work will be done by living machines. Everybody will be free from worry and liberated from the degradation of labor. Everybody will live only to perfect himself.

HELENA

Will he?

DOMIN

Of course. It's bound to happen. But then the servitude of man to man and the enslavement of man to man will cease. Of course, terrible things may happen at first, but that simply can't be avoided. Nobody will get bread at the price of life and hatred. The Robots will wash the feet of the beggar and prepare a bed for him in his house.

ALQUIST

Domin, Domin. What you say sounds too much like Paradise. There was something good in service and something great in humility. There was some kind of virtue in toil and weariness.

DOMIN

Perhaps. But we cannot reckon with what is lost when we start out to transform the world. Man shall be free and supreme; he shall have no other aim, no other labor, no other care than to perfect himself. He shall serve neither matter nor man. He will not be a machine and a device for production. He will be Lord of creation.

BUSMAN

Amen.

ON THE LINE

HENRY FORD, 1922

More than simply the automobile, Henry Ford perfected the method of assembly-line production. Thomas Edison called it a "radical innovation" and saw that it was "open to all in nearly every line of business." As Ford had dreamed of lifting "farm drudgery," so he envisioned his assembly line lifting the drudgery of labor.

Every piece of work in the shop moves. It may move on hooks or overhead chains going to assembly in the exact order in which the parts are required; it may travel on a moving platform, or it may go by gravity, but the point is that there is no lifting or trucking of anything other than materials.

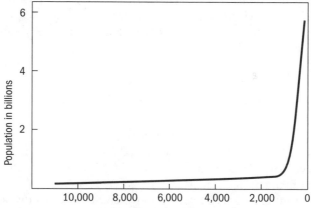

▶ Growth of world population since 10,000 B.P. (before the present).

YOU MUST GRIND
YOUR BEARINGS

ALFRED P. SLOAN, JR.

Mass production depended on interchangeable parts, a lesson General Motors pioneer Alfred P. Sloan, Jr., remembered learning in 1930, when he was general manager for the Hyatt Roller Bearing Company.

The white beard of Henry M. Leland seemed to wag at me, he spoke with such long-faced emphasis. He was the general manager of Cadillac. "Mr. Sloan, Cadillacs are made to run, not just to sell."

On his desk were some of our roller bearings, like culprits before a judge. We did not ship bearings to Cadillac; we shipped them to Weston-Mott, who were making 500 axles for Cadillac, an emergency trial order. Under Mr. Leland's brown hand with its broad thumb was a micrometer. He had measured the diameters of several specimen bearings. Then he had drawn lines and written down the variations from the agreed tolerances. I listened humbly as he went on talking.

"Your Mr. Steenstrup told me these bearings would be accurate, one like another, to within one thousandth of an inch. But look here!" I heard the click of his ridged fingernail as he tapped it against a guilty bearing. "There is nothing like that uniformity. . . . You must grind your bearings. Even though you make thousands, the first and the last should be precisely alike." We discussed interchangeability of parts. A genuine conception of what mass production should mean really grew in me with that conversation.

I was an engineer and a manufacturer, and I considered myself conscientious. But after I had said goodby to Mr. Leland, I began to see things differently. I was determined to be as fanatical as he in obtaining precision in our work. An entirely different standard had been established for Hyatt Roller Bearings.

Incidentally, a few years later Mr. Leland's rigid standards were given a dramatic test. Three Cadillac cars, taken from the dockside by Royal Automobile Club officials in London, were dismembered and jumbled into a pile of parts. Thereafter, Cadillac mechanics, with wrenches, screw drivers, hammers and pliers, swiftly assembled three Cadillacs, which were forthwith started on a track test—for 500 miles. All cars finished with a perfect score. Nothing, I think, ever did more to establish the reputation of American cars.

This may seem inconsequential, but truly it is of great importance, because the ability to produce large quantities of parts, each one just like the other or sufficiently alike, is the foundation of mass production, as we understand the term today. This conception has been the basis of American predominance in such methods of production. It is definitely an American approach, and even today it is still limited in its application elsewhere as compared with that in American industry.

THE RAILROADS HAVE
REACHED CAPACITY

CHARLES PIERCE BURTON, 1923

A Harper's essayist marks the moment when the railroads reached their maximum point of expansion—the moment as well, though he could not know it at the time, when their influence began to wane.

That the railroads of the United States have reached their capacity and find it difficult or impossible to raise the vast sums needed for expansion, upon which the continued prosperity of the country depends, is a serious matter. Thirty-six thousand carloads of embargoed freight stood on the rails of the New York, New Haven & Hartford System on April 5th of this year. On the same day the Delaware, Lackawanna & Western was holding 4,000 carloads of freight, to be turned over to the New Haven road; and, presumably, other Eastern lines correspondingly were congested. Such a situation, both directly and in its significance, endangered the growing prosperity of the country. Were all railroads similarly situated it would spell disaster. Fortunately, the New Haven road is not typical of America's transportation systems. A multitude of industries in the congested territory which it serves originate much of the freight of the country. In a less degree this is true of all New England railroads. It is notorious that the carriers of New England are unable to handle the maximum traffic of that great center of intensive industry. In a general way, it can be truthfully said that the railroads of America are unable to distribute properly the maximum production of the country. It is serious. What is the answer?

In terms of money, the answer is billions in expenditure. . . . In six and one-half years, from June 30, 1914, to December 31, 1920, the money expended by all the railroads of the United States for new construction amounted only to about $1,700,000,000, approximately $260,000,000 a year. Based on a tentative valuation of the railroads at $20,000,000,000, this is one and three-tenths per cent. During the same period three times as much money was spent on our highways, and even that great expenditure was inadequate. During the same six and one-half years the railroads spent $1,250,000,000 on rolling stock, less than $200,000,000 a year, not quite 1 percent based on a physical valuation of $20,000,000,000. . . . It is an astounding and alarming condition of affairs when the greatest business in the country—a business on whose shoulders falls largely the burden of distribution—whose functioning is vital to the prosperity and safety of the nation, has been able to spend less than two and one-half percent a year for needed expansion.

A BRIDGE TOO LOW

ROBERT CARO

Technology can intensify as well as reduce structural violence—violence that is built into the structure of a society. Beginning in 1924, planner and power broker Robert Moses built the elegant parkways that grace New York and Long Island, but as his biographer discovered, he deliberately scaled them to limit access by public transportation, thus effectively excluding lower-income citizens. "Many of his monumental structures of concrete and steel embody a systematic social inequality," writes technologist Langdon Winner, "a way of engineering relationships among people that, after a time, became just another part of the landscape."

"The building of the bridges is an example of his foresight and vision," Sid Shapiro says in his quiet way. "I've often been astonished myself that he was so right in those days, and not only so right but so indispensably right. Mr. Moses had an instinctive feeling that someday politicians would try to put buses on the parkways, and that would break down the whole parkway concept—and he used to say to us fellows, 'Let's design the bridges so the clearance is all right for passenger cars but not for anything else.' All the original bridges were designed with nine feet of clearance at the curb. Later we went up to eleven feet, but that has the same effect. Well, yes, buses *could* use the center lane, but that's an impractical thing. No bus would do that. Mr. Moses did this because he knew that something might happen after he was dead and gone. He wrote legislation [e.g., clauses prohibiting the use of parkways by "buses or other commercial vehicles"] but he knew you could change the legislation. You can't change a bridge after it's up. And the result of this is that a bus from New York couldn't use the parkways if we wanted it to." A quiet smile broke across Shapiro's seamed face, and he almost laughed as a pleasant recollection crossed his mind. "You know," he said, "we've had cases where buses mistakenly got on a parkway—buses from a foreign state, I suppose, and the first bridge stopped them dead. One had its roof rolled up like the top of a sardine can."

"Foresight and vision." Apt, if the vision was one man's private vision. Building parkway bridges low was indeed an example of Moses' foresight in trying to keep intact his original concept, the bright and shining dream he had dreamed in 1924—of roads that would not be just roads but works of art, that would be "ribbon parks" for "pleasure driving." Building the parkway bridges low was indeed an example of Moses' foresight in trying to keep intact his original concept of the area through which those roads ran—lovely Long Island—as a serene and sparsely populated suburban setting, a home for the relatively small number of people wealthy enough to live there, a playground for the larger but still restricted number of people wealthy enough to drive there and play in the parks he had built there.

But the reality of Moses's vision was grimmer, Caro demonstrates earlier in his biography.

Underlying Moses' strikingly strict policing for cleanliness in his parks was, Frances Perkins realized with "shock," deep distaste for the public that was using them. "He doesn't love the people," she was to say. "It used to shock me because he was doing all these things for the welfare of the people.... He'd denounce the common people terribly. To him they were lousy, dirty people, throwing bottles all over Jones Beach. 'I'll get them! I'll teach them!' He loves the public, but not as people. The public is just *the* public. It's a great amorphous mass to him; it needs to be bathed, it needs to be aired, it needs recreation, but not for personal reasons—just to make it a better public." Now he began taking measures to limit use of his parks. He had restricted the use of state parks by poor and lower-middle-class families in the first place, by limiting access to the parks by rapid transit; he had vetoed the Long Island Rail Road's proposed construction of a branch spur to Jones Beach for this reason. Now he began to limit access by buses; he instructed Shapiro to build the bridges across his new parkways low—too low for buses to pass. Bus trips therefore had to be made on local roads, making the trips discouragingly long and arduous.

> Could not explosives even of the existing type be guided automatically in
> flying machines by wireless or other rays, without a human pilot, in
> ceaseless procession upon a hostile city, arsenal, camp, or dockyard?
>   WINSTON CHURCHILL, 1925

LARGE CITIES HAVE
COME TO STAY

ROBERT RIDGWAY, 1925

The myth of the Jeffersonian yeoman died hard. Here an engineer wrestles with his ambivalence between admiring the "wonderful growth" of cities and fearing a loss of individuality and independence. His concerns anticipate the conservative distaste for social welfare programs of half a century later.

Perhaps the most notable effect of the application of those laws of Nature which have been brought to light by the patient investigations of the scientist during the past century and a half is evidenced in the wonderful growth of cities everywhere. . . .

The change to urban life has affected deeply the customs, the habits, and the thoughts of the people. In the simple days during the early life of the Republic, when men lived generally distant from one another, they were largely individualists, believing with Thomas Jefferson that government was a necessary evil and that they should get along with as little of it as possible, depending on their own efforts for success. Modern concentration of vast numbers of people in cities has led to the organization of industrial corporations with hundreds and frequently thousands of employees working under the same roof. Diversities of custom are disappearing. Organizations based on class consciousness are formed with the idea of bettering the material condition of particular classes of workers. More government is demanded. There is a drift toward paternalism and socialism; a tendency to lean on the State or National Government for help out of all difficulties; to lose that Anglo-Saxon spirit of independence which formerly prevailed and which was the foundation on which the structure of our national life was built. The fear exists that these tendencies are going too far; that the worker is being made into a machine and that so much effort is given to developing his material side, that the moral and the spiritual sides are forgotten. While city workers have escaped the hardness of the farm life of the old days, they are paying for the ease of life in a loss of that touch of fundamental things which is so necessary for true happiness and for the full rounding out and proper balance of humanity. . . .

The price we have paid for the many advantages which life in the great cities has given us is a large one. The opportunities and advantages of urban life have not yet come to compensate humanity for the restful quiet of the open country, for the simple pleasures it affords, and for the spirit of introspection which it fosters. Their comforts and luxuries are to be enjoyed only at the expense of a certain softening of character. The joy of doing is marred by the prevalent feeling of unrest, and the ever mounting cost of living in cities is a cause of anxiety to those of moderate means. . . .

Large cities have come to stay and will doubtless continue to exist as long as industrial conditions and human nature remain as they are. . . . Recognizing this, the efforts of every good citizen should be directed to making cities what they should be, by each giving of his best thought to improving the quality rather than the number of inhabitants.

UNSTABLE AGES

ALFRED NORTH WHITEHEAD, 1925

Philosopher and mathematician Alfred North Whitehead found the pace of modern change challenging—a theme that recurs.

Modern science has imposed on humanity the necessity for wandering. Its progressive thought and its progressive technology make the transition through time, from generation to generation, a true migration into uncharted seas of adventure. The very benefit of wandering is that it is dangerous and needs skill to avert evils. We must expect, therefore, that the future will disclose dangers. It is the business of the future to be dangerous; and it is among the merits of science that it equips the future for its duties. The prosperous middle classes, who ruled the nineteenth century, placed an excessive value upon the placidity of existence. They refused to face the necessities for social reform imposed by the new industrial system, and they are now refusing to face the necessities for intellectual reform imposed by the new knowledge. The middle class pessimism over the future of the world comes from a confusion between civilization and security. In the immediate future there will be less security than in the immediate past, less stability. It must be admitted that there is a degree of instability which is inconsistent with civilization. But, on the whole, the great ages have been unstable ages.

THE MEANING OF POWER

HENRY FORD, 1926

The source of material civilization is developed power. If one has this developed power at hand, then a use for it will easily be found. One way to use the power is through a machine, and just as we often think of the automobile as a thing of itself instead of a way of using power, so also do we think of the machine as something of itself instead of as a method of making power effective. We speak of a "machine age." What we are entering is a power age, and the importance of the power age lies in its ability, rightly used with the wage motive behind it, to increase and cheapen production so that all of us may have more of this world's goods. The way to liberty, the way to equality of opportunity, the way from empty phrases to actualities, lies through power; the machine is only an incident.

The function of the machine is to liberate man from brute burdens, and release his energies to the building of his intellectual and spiritual powers for conquests in the fields of thought and higher action. The machine is the symbol of man's mastery of his environment.

One has only to go to other lands to see that the only slave left on earth is man minus the machine. We see men and women hauling wood and stone and water on their backs. We see artisans clumsily spending long hours and incredible toil for a paltry result. We see the tragic disproportion between laborious hand culture of the soil and the meager fruits thereof. We meet with unbelievably narrow horizons, low standards of life, poverty always on the edge of disaster—these are the conditions where men have not learned the secrets of power and method—the secrets of the machine.

To release himself to more human duties, man has trained beasts to carry burdens. The ox team and camel represent man's mind plus brute strength. The sail is man's release from the slavery of the oar. The use of the swift horse was man's dim sensing that time had value for himself and his concerns.

Did man thus increase his slavery, or did he increase his liberty?

It is true that the machine has sometimes been used by those who owned it, not to liberate men, but to exploit them. This was never accepted by society as right. It has always been challenged, and as the use of the machine became more widespread, it effectually checked the misuse that had been made of it. The right and serviceable use of the machine always makes unprofitable and at last impossible the abuse of it.

IMMENSE DECREASE IN
THE DEATH RATE

MARK SULLIVAN, 1926

In 1901 the average expectation of life in America, as determined for certain registered areas, was 49 24/100 years. In 1920 it was 54 9/100 years. These figures are from the official Bureau of Public Health at Washington. . . .

In 1900, out of every 100,000 persons living in the then registered area of the United States, 1755 died. In 1921, the death-rate per hundred thousand had been reduced to 1163.9, a decrease of 34 percent.

The saving in human lives was at the rate of about 6 for each 1000 individuals. With a total population of approximately 110,000,000, this means that at the end of a single year, 1924 for example, a total of 660,000 Americans were alive who would have died during the year had the conditions affecting health and longevity in 1900 remained unchanged. This addition to the security and length of tenure of human life, when worked out mathematically, turns out to be many times greater than the number of lives lost by Americans in the Great War. . . .

The agencies that have enriched man in this respect, that have increased his immunity to disease, the security of his life, his tenure on continued existence, have been, chiefly: advances in understanding the causes of disease, advances in sanitation and other agencies for the prevention of dis-

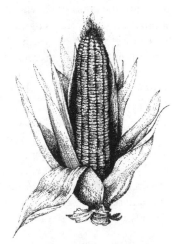

▶ Hybrid corn, introduced in the 1920s, transformed American agriculture.

ease, the discovery of specific cures or specific preventives for some diseases, advances in surgery, professional nursing, greater watchfulness over water supply, meat and milk inspection, better housing, better conditions of living and labor. Of all these the greatest single cause has been the discovery of means for safeguarding babies from the intestinal diseases that attack them up to the age of two years. In 1900 the death rate from this cause was 108.8 per hundred thousand. By 1921 it had decreased to 41.9. . . .

The least tolerable of all causes of death, the one which a civilized society ought to be able to control most surely, murder and the other forms of homicide, has increased to exactly four times what it was, from a rate of 2.1 in 1900 to 8.4 in 1922. Suicide has remained about stationary: 111.5 in 1900; 111.9 in 1922. The one appalling increase in the number of deaths, from a cause that is among the least excusable, has been in automobile accidents. The number of deaths from this cause was so small in 1900 as not to be reported. . . . In 1923 the number of deaths from automobiles reached the unforgivable total of 22,600. At this rate, automobiles in the United States were killing almost as many Americans in two years as were killed in the Great War; and in three years as many as the number of Union soldiers killed in battle in the Civil War. As civilization expressed itself in the United States in the year 1925, the tolerance of this was probably its least lovely aspect.

COMFORT

ALDOUS HUXLEY, 1927

Before Brave New World, *and anticipating it, the thirty-three-year-old British novelist and social satirist examined an unexamined expectation, providing another perspective on Mark Sullivan's observation the same year that "luxuries became necessities."*

Comfort is a thing of recent growth, younger than steam, a child when telegraphy was born, only a generation older than radio. The invention of the means of being comfortable and the pursuit of comfort as a desirable end—one of the most desirable that human beings can propose to themselves—are modern phenomena, unparalleled in history since the time of the Romans. Like all phenomena with which we are extremely familiar, we take them for granted, as a fish takes the water in which it lives, not realizing the oddity and novelty of them, not bothering to consider their significance. The padded chair, the well-sprung bed, the sofa, central heating, and the regular hot bath—these and a host of other comforts enter into the daily

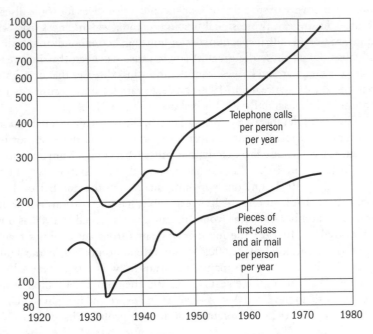

▶ U.S. telephone calls vs. mail, 1920–1980.

lives of even the most moderately prosperous of the Anglo-Saxon bourgeoisie. Three hundred years ago they were unknown to the greatest kings. This is a curious fact which deserves to be examined and analyzed.

The first thing that strikes one about the discomfort in which our ancestors lived is that it was mainly voluntary. Some of the apparatus of modern comfort is of purely modern invention; people could not put rubber tires on their carriages before the discovery of South America and the rubber plant. But for the most part there is nothing new about the material basis of our comfort. Men could have made sofas and smoking-room chairs, could have installed bathrooms and central heating and sanitary plumbing any time during the last three or four thousand years. . . . If the men of the Middle Ages and early modern epoch lived in filth and discomfort, it was not for any lack of ability to change their mode of life; it was because they chose to live in this way, because filth and discomfort fitted in with their principles and prejudices, political, moral and religious.

What have comfort and cleanliness to do with politics, morals and religion? At first glance one would say that there was and could be no causal connection between armchairs and democracies, sofas and the relaxation of the family system, hot baths and the decay of Christian orthodoxy. But look more closely and you will discover that there exists the closest connection between the recent growth of comfort and the recent history of ideas. . . .

Old furniture reflects the physical habits of the hierarchical society for which it was made. It was in the power of medieval and renaissance craftsmen to create armchairs and sofas that might have rivaled in comfort those of today. But society being what, in fact, it was, they did nothing of the kind. It was not, indeed, until the sixteenth century that chairs became at all common. Before that time a chair was a symbol of authority. . . . In the Middle Ages only the great had chairs. When a great man travelled, he took his chair with him, so that he might never be seen detached from the outward and visible sign of his authority. To this day the Throne no less than the Crown is the symbol of royalty. . . .

It is . . . to the doctors that the bath-lovers owe their greatest debt. The discovery of microbic infection has put a premium on cleanliness. We wash now with religious fervor, like the Hindus. Our baths have becoming something like magic rites to protect us from the powers of evil, embodied in the dirt-loving germ. We may venture to prophesy that this medical religion will go still further in undermining the Christian ascetic tradition. Since the discovery of the beneficial effects of sunlight, too much clothing has become, medically speaking, a sin. Immodesty is now a virtue. It is quite likely that the doctors, whose prestige among us is almost equal to that of the medicine men among their savages, will have us stark naked before long. That will be the last stage in the process of making clothes more comfortable. . . .

Made possible by changes in the traditional philosophy of life, comfort is now one of the causes of its own further spread. For comfort has now become a physical habit, a fashion, an ideal to be pursued for its own sake. The more comfort is brought into the world, the more it is likely to be valued. To those who have known comfort, discomfort is a real torture. And the fashion which now decrees the worship of comfort is quite as imperious as any other fashion. Moreover, enormous material interests are bound up with the supply of the means of comfort. The manufacturers of furniture, of heating apparatus, of plumbing fixtures, cannot afford to let the love of comfort die. In modern advertisement they have means for compelling it to live and grow. . . .

I am inclined to think that our present passion for comfort is a little exaggerated. Though I personally enjoy comfort, I have lived very happily in houses devoid of almost everything that Anglo-Saxons deem indispensable. Orientals and even South Europeans, who know not comfort and live very much as our ancestors lived centuries ago, seem to get on very well without our elaborate and costly apparatus of padded luxury. I am old-fashioned enough to believe in higher and lower things, and can see no point in material progress except in so far as it subserves thought. I like labor-saving devices, because they economize time and energy which may be devoted to mental labor. (But then I enjoy mental labor; there are plenty of people who detest it, and who feel as much enthusiasm for thought-saving devices as for automatic dishwashers and sewing machines.) I like rapid and easy transport, because by enlarging the world in which men can live it enlarges their minds. Comfort for me has a similar justification: it facilitates mental life. Discomfort handicaps thought; it is difficult when the body is cold and aching to use the mind. Comfort is a means to an end. The modern world seems to regard it as an end in itself, an absolute good. One day, perhaps, the earth will have been turned into one vast featherbed, with man's body dozing on top of it and his mind underneath, like Desdemona, smothered.

APPLIED SCIENCE

HERBERT HOOVER, 1927

Herbert Hoover the engineer became Herbert Hoover the president, and celebrated applied science—one kind of technology—even as he worried that its foundation in basic science was weak.

Business and industry have realized the vivid values of the application of scientific discoveries. To further it, in twelve years our individual industries have increased their research laboratories from less than 100 to over 500. They are bringing such values that they are increasing monthly. Our federal and state governments today support great laboratories, research departments, and experimental stations, all devoted to applications of science to the many problems of industry and agriculture. They are one of the great elements in our gigantic strides in national efficiency. The results are magnificent. The new inventions, laborsaving devices, improvements of all sorts in machines and processes in developing agriculture and promoting health are steadily cheapening cost of production; increasing standards of living, stabilizing industrial output enabling us to hold our own in foreign trade; and lengthening human life and decreasing suffering. But all these laboratories and experiment stations are devoted to the application of science, not to fundamental research. Yet the raw materials for these laboratories come alone from the ranks of our men of pure science whose efforts are supported almost wholly in our universities, colleges, and a few scientific institutions.

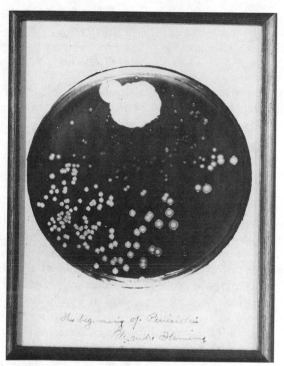

▶ Penicillin, discovered in 1928, inaugurated antibiotic therapy.

THE AUTOMOBILE BOOM

THOMAS C. COCHRAN

A historian speculates on the economic effect of the coming of the automobile.

No one has or perhaps can reliably estimate the vast size of capital invested in reshaping society to fit the automobile. . . . This total capital investment was probably the major factor in the boom of the 1920's and hence in the glorification of American business.

THE DECENTRALIZING POWER
OF THE AUTOMOBILE

ROBERT AND HELEN LYND, 1929

In their pioneering sociological study Middletown, *the Lynds unearthed the beginnings of the technological division between parents and adolescent children that later decades would bemoan.*

RICHARD RHODES

This centralizing tendency of the automobile may be only a passing phase; sets in the other direction are almost equally prominent. "Our daughters [eighteen and fifteen] don't use our car much because they are always with somebody else in their car when we go out motoring," lamented one business-class mother. And another said, "The two older children [eighteen and sixteen] never go out when the family motors. They always have something else on." "In the nineties we were all much more together," said another wife. "People brought chairs and cushions out of the house and sat on the lawn evenings. We rolled out a strip of carpet and put cushions on the porch step to take care of the unlimited overflow of neighbors that dropped by. We'd sit out so all evening. The younger couples perhaps would wander off for half an hour to get a soda but come back to join in the informal singing or listen while somebody strummed a mandolin or guitar." "What on earth do you want me to do? Just sit around home all evening!" retorted a popular high school girl of today when her father discouraged her going out motoring for the evening with a young blade in a rakish car waiting at the curb. The fact that 348 boys and 382 girls in the three upper years of the high school placed "use of the automobile" fifth and fourth respectively in a list of twelve possible sources of disagreement between them and their parents suggests that this may be an increasing decentralizing agent.

FLYING FEELS TOO GODLIKE

CHARLES LINDBERGH, 1927

Charles Lindbergh's historic nonstop solo flight from New York to Paris electrified the world; it was dramatic evidence of how the new technologies were shrinking the globe.

Only eight months ago—it seems as many years—I was sitting in the reality of my mail cockpit, dreaming of a plane that would fly across the sea. Now, I'm in the *Spirit of St. Louis*, on that very flight. The dream has be-

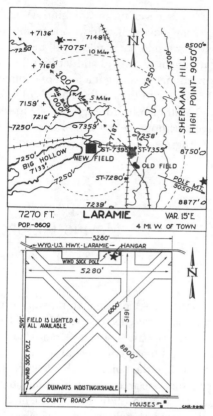

▶ Pilots needed maps: an early Jeppesen approach chart.

come the reality; and the reality, the dream. Such things happen then in fact and not alone in fable. . . . Ideas are like seeds, apparently insignificant when first held in the hand! If a wind or a new current of thought drifts them away, nothing is lost. But once firmly planted, they can grow and flower into almost anything at all, a cornstalk or a giant redwood—or a flight across the ocean. Ideas are even more wonderful than seeds, for they have no natural substance and are less restricted by hereditary form. Whatever a man imagines he can attain, if he doesn't become too arrogant and encroach on the rights of the gods.

Is aviation too arrogant? I don't know. Sometimes, flying feels too god-like to be attained by man. Sometimes, the world from above seems too beautiful, too wonderful, too distant for human eyes to see, like a vision at the end of life forming a bridge to death. Can that be why so many pilots lose their lives? Is man encroaching on a forbidden realm? Is aviation dangerous because the sky was never meant for him? When one obtains too great a vision is there some power that draws one from mortal life forever? Will this power smite down pilot after pilot until man loses his will to fly? Or, still worse, will it deaden his senses and let him fly on without the vision? In developing aviation, in making it a form of commerce, in replacing the wild freedom of danger with the civilized bonds of safety, must we give up this miracle of air? Will men fly through the sky in the future without seeing what I have seen, without feeling what I have felt? Is that true of all things we call human progress—do the gods retire as commerce and science advance?

A DIMINUTIVE
MOVING PICTURE

THE NEW YORK TIMES, 1928

Schenectady, N.Y., Jan. 13.—A diminutive moving picture of a smiling, gesticulating gentleman wavered slowly within a small cabinet in a darkened room of the General Electric Company's radio laboratories this afternoon and heralded another human conquest of space.

Sent through the air like the voice which accompanied the picture, it marked, the demonstrators declared, the first demonstration of television broadcasting and gave the first absolute proof of the possibility of connecting homes through the world by sight as they have already been connected by voice. . . .

The voice of Leslie Wilkins of the General Electric testing department came from the loudspeaker next to one of the cabinets.

"I understand there is an audience in the receiving room now," he said from the alcove where he was broadcasting, "so now we will start."

In the small openings of each of the cabinets appeared the image of his face.

"Now I will take off my spectacles and put them on again," he said. The picture suited his words.

"Here is a cigarette. You can see the smoke," he continued.

The audience saw him breathe out a smoke ring and watched it drift upward across his face.

Louis Dean, the regular announcer of WGY, succeeded Leslie. He was the first to broadcast music with his own picture as he played.

"Ain't She Sweet?" he strummed on his ukelele and his smile flashed through the air to the onlookers.

THE INEVITABILITY OF
THE MACHINE

CHARLES A. BEARD, 1928

*Even before the Great Depression, informed observers like historian Charles A.
Beard felt the need to defend technology against the attacks of critics.*

W hat is called Western or modern civilization by way of contrast with
the civilization of the Orient or medieval times is at bottom a civilization that rests upon machinery and science as distinguished from one
founded on agricultural or handicraft commerce. It is in reality a technological civilization. It is only about two hundred years old, and, far from shrinking in its influence, is steadily extending its area into agriculture as well as
handicrafts. If the records of patent offices, the statistics of production, and
the reports of laboratories furnish evidence worthy of credence, technological civilization, instead of showing signs of contraction, threatens to overcome and transform the whole globe.

Considered with respect to its intrinsic nature, technological civilization presents certain precise characteristics. It rests fundamentally on
power-driven machinery which transcends the physical limits of its human
directors, multiplying indefinitely the capacity for the production of goods.
Science in all its branches—physics, chemistry, biology, and psychology—is
the servant and upholder of this system. The day of crude invention being
almost over, continuous research in the natural sciences is absolutely necessary to the extension of the machine and its market, thus forcing continuously the creation of new goods, new processes, and new modes of life. As
the money for learning comes in increasing proportions from taxes on industry and gifts by captains of capitalism, a steady growth of scientific en-
dowments is to be expected, and the scientific curiosity thus aroused and
stimulated will hardly fail to expand—and to invade all fields of thought
with a technique of ever-refining subtlety. Affording the demand for the
output of industry are the vast populations of the globe; hence mass production and marketing are inevitable concomitants of the machine routine.

For the present, machine civilization is associated with capitalism, under which large-scale production has risen to its present stage, but machine
civilization is by no means synonymous with capitalism—that ever-changing scheme of exploitation. . . . The kind of servile revolt that was so often
ruinous in Greece and Rome is hardly possible in a machine civilization,
even if economic distress were to pass anything yet experienced since the
eighteenth century. The most radical of the modern proletariat want more
of the good things of civilization—not a destruction of technology. . . .

Finally, we must face the assertion that wars among the various nations of machine civilization may destroy the whole order. Probably terrible wars will arise and prove costly in blood and treasure, but it is a strain upon the speculative faculties to conceive of any conflict that could destroy the population and mechanical equipment of the Western world so extensively that human vitality and science could not restore economic prosperity and even improve upon the previous order. . . .

For the reasons thus adduced it may be inferred: that modern civilization founded on science and the machine will not decline after the fashion of older agricultural civilizations; that analogies drawn from ages previous to technology are inapplicable; that according to signs on every hand technology promises to extend its area and intensify its characteristics; that it will afford the substance with which all who expect to lead and teach in the future must reckon. . . .

Such appears to be the promise of the long future, if not the grand destiny of what we call modern civilization—the flexible framework in which the human spirit must operate during the coming centuries. Yet this view by no means precludes the idea that the machine system, as tested by its present results, presents shocking evils and indeed terrible menaces to the noblest faculties of the human race. . . .

To consider for the moment merely the domestic aspects of the question, the machine civilization is particularly open to attack from three sides.

On aesthetic grounds, it has been assailed for nearly a hundred years, England, the classical home of the industrial revolution, being naturally enough the mother of the severest critics—Ruskin, Carlyle, Kingsley, and Matthew Arnold. The chief article in their indictment, perhaps, is the contention that men who work with machinery are not creative, joyous, or free, but are slaves to the monotonous routine of the inexorable wheel. In a sense it is true that, in the pre-machine age, each craftsman had a certain leeway in shaping his materials with his tools and that many a common artisan produced articles of great beauty.

Yet the point can be easily overworked. Doubtless the vast majority of medieval artisans merely followed designs made be master workmen. This is certainly true of artisans in the Orient today. With respect to the mass of mankind, it is safe to assume that the level of monotony on which labor is conducted under the machine regime is by and large not lower but higher than in the handicraft, servile, or slave systems in the past. Let anyone who has doubts on this matter compare the life of laborers on the latifundia of Rome or in the cities of modern China with that of the workers in by far the major portion of machine industries. Those who are prepared to sacrifice the standard of living for the millions to provide conditions presumably favorable to the creative arts must assume a responsibility of the first magnitude. . . .

Frequently affiliated with aesthetic criticism of the machine and science is the religious attack. With endless reiteration, the charge is made that industrial civilization is materialistic. In reply, the scornful might say, "Well, what of it?" But the issue deserves consideration on its merits, in spite of its illusive nature. . . .

If religion is taken in a crude, anthropomorphic sense, filling the universe with gods, spirits, and miraculous feats, then beyond question the machine and science are the foes of religion. If it is materialistic to disclose the influence of technology and environment in general upon humanity, then perhaps the machine and science are materialistic. But it is one of the ironies of history that science has shown the shallowness of the old battle between materialist and spiritist. . . . Doubtless science does make short shrift of a thousand little mysteries once deemed as essential to Christianity as were the thousand minor gods to the religion of old Japan, but for these little mysteries it has substituted a higher and sublimer mystery.

To descend to the concrete, is the prevention of disease by sanitation more materialistic than curing it by touching saints' bones? Is feeding the multitude by mass production more materialistic than feeding it by a miracle? Is the elimination of famines by a better distribution of goods more materialistic than prevention by the placation of the rain gods? At any rate, it is not likely that science and machinery will be abandoned because the theologian (who seldom refuses to partake of their benefits) wrings his hands and cries out against materialism. After all, how can he consistently maintain that Omnipotent God ruled the world wisely and well until the dawn of the modern age and abandoned it to the Evil One because Henry VIII or Martin Luther quarrelled with the Pope and James Watt invented the steam engine?

Arising, perhaps, from the same emotional source as aesthetic and religious criticisms, is the attack on the machine civilization as lacking in humanitarianism. Without commenting on man's inhumanity to man as an essential characteristic of the race, we may fairly ask on what grounds can anyone argue that the masses were more humanely treated in the agricultural civilization of antiquity or the middle ages than in the machine order of modern times. Tested by the mildness of its laws (brutal as many of them are), by its institutions of care and benevolence, by its death rate (that telltale measurement of human welfare), by its standards of life, and by every conceivable measure of human values, machine civilization, even in its present primitive stage, need fear no comparison with any other order on the score of general well-being.

Defining the profession, EB contributor Alfred Douglas Flinn sets high—and ecologically correct—standards.

T he engineer is under obligation to consider the sociological, economical and spiritual effects of engineering and operations and to aid his fellowmen to adjust wisely their modes of living, their industrial, commer-

Telegram to
New York World   Jany. 7 1914

It is such a radical innovation that I cannot at present give an opinion as to its ultimate effect. Some time ago Mr Ford reduced the price of his wonderful touring car to the extent of fifty dollars. The user of the car received the entire benefit. Now he has practically reduced it another fifty dollars, but this time the men who make them get the benefit. Mr Ford's machinery is special and highly efficient. This is what permits these results. This is open to all in nearly every line of business.

Let the public throw bouquets to the inventors and in time we will all be happy

Edison

▶ Thomas Edison's reply when asked about Henry Ford's innovative new assembly line mass-production system.

cial and governmental procedures, and their educational progress so as to enjoy the greatest possible benefit from the progress achieved through our accumulating knowledge of the universe and ourselves as applied by engineering. The engineer's principal work is to discover and conserve natural resources of materials and forces, including the human, and to create means for utilizing these resources with minimal cost and waste and with maximum useful results.

10,000 AUTOMOBILE FRAMES A DAY

SIDNEY G. KOON, 1930

Automation, the cause célèbre of the 1960s, was already gaining ground in the third decade of the century, an observer reports.

Years ago, when I first saw an automatic wire-fence machine, with its processing parts moving with the precision of German soldiers on parade, I was much impressed with the possibilities and the potentialities of automatic machines. Each part handling one of the small sections of wire— vertical in the erected fence—took it from a coil as the cutter snipped it to proper length. The little pieces of wire were then swung around, put into position across the longitudinal wires and the ends quickly coiled around the longitudinals, thus forming the completed fence. All this, of course, was done while the longitudinals were moving slowing and uniformly through the machine.

Such equipment as that, however, presenting merely a few moving parts in a single machine, suffers at once by contrast with a massive group of machines in several production units, such as are employed in the automatic frame plant of the A. O. Smith Corporation, Milwaukee. The impression one gathers after watching that plant at work, turning out its product at the rate of 450 automobile frames an hour, may perhaps be expressed by a statement made by the guide who showed me through the works.

He said that professors in mechanical engineering at universities within striking distance of Milwaukee are handling a psychological problem with their students in the following manner: As the seniors approach the end of the course—as, indeed, they see the finish a short distance ahead and believe that there is nothing more for them to learn—their bumptiousness is taken out of them by a visit to this plant.

In groups of a dozen, a score, or more, they are passed through the plant, where they see the array of automatic, synchronized machines in operation, many of them without a man anywhere near. And, with chastened spirit, they depart, with the idea that perhaps there may be something left to learn, after all.

Naturally, such a plant did not spring into existence over night. It represents the result of some years of development, from the time when the first automobile frame of pressed steel was made, in 1903. It was not, however, until 1920 that this plant was completed. . . .

Fully automatic production is already common practice in bulk manufacturing. What this company has accomplished shows that the principle can be applied successfully and on a large scale to assembly operations.

LEISURE WORTH HAVING?

FLOYD H. ALLPORT, 1931

Enthusiasts of technology heralded a coming era of unlimited leisure. Skeptics such as this Harper's Magazine *essayist asked what people would do with all that free time. Among other issues, Allport examined industrial damage to the environment.*

There . . . arises the problem of how a technological leisure, if it were forthcoming, could be spent. Unless that ruthless destruction, which has so far seemed indispensable to our industrial program, can be checked, by the time we arrive at the goal there will be practically no environment remaining in which free time can be profitably spent. Consider, for example, those primordial human enjoyments, the fellowship with nature and the love of sport. There will be still preserved the mountains and the sea, and probably also the spectacular forests, canyons, and waterfalls of our national reserves. But the local woods and streams, the farms and villages which have been the charm of the countryside for those who cannot travel afar, will be largely despoiled and depopulated through industry. Wild birds and animals are now being destroyed or driven away by growing towns and by the inroads of the automobile. Concrete highways, though affording a fleeting view of the country to many motorists, have injured the beauty of the country itself. Just as walking has given way to less healthful motoring, so rowing and canoeing are succumbing to the noisy, more thrilling speed-boat. Great cities, the homes, we are told, of future civilization, have destroyed unnumbered possibilities for the pursuits of leisure. With the entire seaboard available, millions of city-dwellers must snatch their respite upon a shore as congested as a city street, from which they must return in stifling subways to be buried again where no sight or sound of nature can reach them. In riding along metropolitan elevateds one can see wisps of cherished vegetation hanging from the fire-escapes of hundreds of upper story tenements. Such are the materials for leisure left by a civilization through which leisure itself is to come to pass.

Education in schools and colleges, drifting away from the pure sciences and the liberal arts, is turning in the direction of technical and vocational training. We are training our young people increasingly for that residuum of supervision of machines which is to remain when the goal is reached; but we are neglecting our major responsibility of training for leisure itself. Music and drama are also suffering in character of popular participation, if not in extent. Mechanized production and transmission are eliminating the practicability of music as a common profession. The sensational talking motion picture has discouraged the art of genuine drama, and has replaced troops of actors by the mass production of their sounds and shadows. Thus far the milieu of the technological era has been able to yield scarcely any authentic artistic productions except those of protest directed against the age itself. Great drama is going back to a pre-technological folk culture for its setting.

The same deplorable changes may be noted in social intercourse. Conversations, neighborhood visits, family outings, entertainments at home, sewing circles, literary clubs, and even the well-worn chairs for loungers at the country store were all occasions and places for the enjoyment of yesterday's leisure moments. But in our drive toward technological leisure we have now become too busy to keep up these old contacts. On the highway we used to meet men and women. Whether on foot, on horseback, in buggies, or even on bicycles, it was always *persons* whom we encountered. Now we do not meet individuals, but automobiles (and tomorrow, airplanes)—grim, impersonal machines which we try to crowd past, unmindful of our fellow-being concealed behind the glass and metal. Courtesies and amenities, though natural between men, are lost upon machines. Even such social life as we have preserved has been despoiled of much of its earlier value. Through a strange irony, it was a salesman who told me that he dreaded to go to a certain large city, because while there it would be impossible for him to have any friends. If anyone were to take an interest in him he would at once suspect that that person was trying to sell him something. The decline of social relationships in the broader sense holds true also for family life. Formerly it was unusual to have to seek one's fortune away from the locality in which one was reared. Today it is not the family location which determines the place of residence, but the possibility of obtaining a job or discovering some new field for business enterprise.

The preceding picture, it is true, is highly generalized, and the reader will speedily point to exceptions. Nevertheless, the basic trends of the technological era are those described. If it is necessary, in order to reach our goal, to continue to put technological progress ahead of every other consideration, there will be left scarcely any resources through which our millennium of leisure can be happily spent. We shall have lost not only our work but our play as well. We shall have become strangers to the material of leisure.

PREDICTIONS:
TOWNLESS HIGHWAYS

BENTON MACKAYE AND LEWIS
MUMFORD, 1931

RICHARD RHODES

The introduction of new technology almost always requires a cascade of further adjustments—which force further adjustments in their wake. Here, with a colleague, Lewis Mumford, the preeminent modern American critic of technology, surprisingly elaborates a utopian vision of a limited-access roadway system that he supposes will preserve the character of small towns and enable high-speed travel. The Great Depression and then the Second World War interrupted the development of such a system, which awaited the passage of the Interstate Highway Act in 1956. The system encouraged automobile use far beyond the expectations of highway planners, but bypassed small towns failed rather than flourished and traffic on congested interstate highways slowed to a national average later in the century of less than 25 miles per hour.

It is a commonplace to say that the automobile has revolutionized modern transportation. But the truth of the matter is that this revolution has not got beyond the Kerensky stage.* The motor car has taken the place of the horse-and-buggy, and the motor bus has wiped out the street car in many sections of the country; but motor car and motor bus are still largely crawling along in the ruts laid down by earlier habits and earlier modes of transportation. . . .

[The automobile, like the locomotive] must have a related but independent road system of its own, and this system must be laid down so as to bring into use all the potential advantages of the automobile for both transportation and recreation. This means a kind of road that differs from the original turnpike, from the railroad and, above all, from the greater part of the existing automobile highways. One can perhaps characterize it best by calling it the Townless Highway, to denote its principal feature—the divorce of residence and transport. . . .

*Alexander Feodorovich Kerensky was prime minister of the first, Socialist revolutionary government in what became the Soviet Union. [R.R.]

Not merely must the motor road make up an independent system which by-passes the existing towns; it must be provided with enough land on both sides of the road to insulate it from the surrounding area, whether rural or urban. . . . The through road must be a parkway. . . .

To concentrate the roadside services in definite units, instead of letting them dribble inefficiently along its entire length is an important step; the next is to follow the example of the railroad and keep the road itself absolutely free. . . .

The fewer the intersections the safer and faster will it be for long-distance travel. The only way to dispose once and for all of roadtown is to make it physically impossible to enter or leave the motorway except at properly planned stations. Chairman Edward Bassett, of the National Council on City Planning, has suggested this simple device and given it the name of the freeway. On high-speed arteries, the stations on these freeways would undoubtedly be at considerable distances apart—perhaps as much as ten miles or more—and ordinary traffic would usually cross the express road by the overpass or the underpass. . . .

Let us . . . try to picture the working out of the main elements in the system as they would touch the motorist himself. He awakens after a good sleep: the rumble and wheeze of long-distance traffic is at least a mile from his residence. He glides out with his car onto the relatively narrow local road, which need no longer be wide enough to take care of the heavy cross-country traffic, and he remembers, with a smile, how his local tax bill has gone down since the assessment for the widening of these local roads has been removed and the tax for their upkeep has gone down with the decreased wear and tear. He heads his car for the nearest station on Route No. 1. When he reaches the station he remembers that he is low on gas. As he pauses for a minute to have his tank filled up he watches a group of tourists eating their breakfast on the veranda of the well-equipped restaurant which has supplanted the half a dozen greasy hot-dog incubators that used to be scattered over the roadside. The food at this particular station is good enough to acquire a local reputation, and often people come out from town for a shore dinner; the restaurant itself, turned away from the road, looks out onto a pleasant vista of fields and salt meadows. He now approaches the road, but he must wait for the lights to change before he can turn in from the local road. Now he is off; in a minute the car is doing close to sixty on the flat stretches where the curves have all been smoothed out. With no danger of anyone suddenly cutting across, with no officious advertiser begging him to halt and change his tires or his underwear, or to patronize a hotel in the town he has just left, with unobstructed right of way and unobstructed vision, our motorist has less anxiety and more safety at sixty miles an hour than he used to have in the old roadtown confusion at twenty-five. Even the intersections do not mar his pleasure: they are far enough apart to warrant traffic lights, and unless the red signal is set—in this respect we are at last getting abreast of the railroad!—he speeds past the crossing blithely.

The motorist reaches the country quickly; he sees the country when he is in it. Whether he is traveling for sheer pleasure or to get somewhere, his major purposes are served by the Townless Highway; the motor car has become an honor to our mechanical civilization and not a reproach to it. When our motorist arrives at his destination he is still smiling and fresh; he has been irritated neither by threatened accidents nor by unexpected delays nor by tedious battles with the congestion of Main Street, attempting to rival all the mistakes of Fifth Avenue and Broadway. This is not utopia any more than the efficiency of a limited train on a fine railroad is utopia: it is merely intelligence, effectively applied. A civilization that can achieve the Twentieth Century or the Broadway Limited* will not be content forever to wallow in the confusion and chaos of antiquated motorways and all their ugly accompaniments.

OPERATORS

CARTER GLASS, 1930

Dial telephones would seem to be an obvious improvement over manual telephones—that is, over units that require waiting for an operator to complete the call. In 1930, Virginia Senator Carter Glass and his fellow legislators in the U.S. Senate saw the matter otherwise, and as keen to protect their privileges then as they are today, approved the following resolution (a similar resolution was introduced in the House of Representatives two hours later).

Whereas dial telephones are more difficult to operate than are manual telephones; and

Whereas senators are required since the installation of dial telephones in the Capitol to perform the duties of telephone operators in order to enjoy the benefits of telephone service; and

Whereas dial telephones have failed to expedite telephone service; therefore, be it

Resolved that the sergeant-at-arms of the Senate is authorized and directed to order the Chesapeake & Potomac Telephone Co., to replace with manual telephones, within 30 days after the adoption of this resolution, all dial telephones in the Senate wing of the United States Capitol and in the Senate Office Building.

*Two express interstate passenger trains. [R.R.]

O BRAVE NEW WORLD!

ALDOUS HUXLEY, 1932

*Huxley's enduring dystopia, its title ironically evoking Miranda's innocent first view
of scoundrels washed ashore in Shakespeare's* The Tempest, *centers on this debate
between the Savage (raised by Indians, a reader of Shakespeare) and the Resident
World Controller, Mustapha Mond, concerning the virtues of industrial civilization.*

"But why is [Shakespeare] prohibited?" asked the Savage. . . .
The Controller shrugged his shoulders. "Because it's old; that's the
chief reason. We haven't any use for old things here."

. . ."Why not?"

. . ."Because our world is not the same as Othello's world. You can't
make flivvers without steel—and you can't make tragedies without social instability. The world's stable now. People are happy; they get what they want,
and they never want what they can't get. They're well off; they're safe;
they're never ill; they're not afraid of death; they're blissfully ignorant of
passion and old age; they're plagued with no mothers or fathers; they've got
no wives, or children, or lovers to feel strongly about; they're so conditioned that they practically can't help behaving as they ought to behave. And
if anything should go wrong, there's *soma*. Which you go and chuck out of
the window in the name of liberty, Mr. Savage. *Liberty!*" He laughed. "Expecting Deltas to know what liberty is! And now expecting them to understand *Othello!* My good boy!"

The Savage was silent for a little. "All the same," he insisted obstinately,
"*Othello's* good, *Othello's* better than those feelies."

"Of course it is," the Controller agreed. "But that's the price we have to
pay for stability. You've got to choose between happiness and what people
used to call high art. We've sacrificed the high art. We have the feelies and
the scent organ instead.". . .

The Savage shook his head. "It all seems to me quite horrible."

"Of course it does. Actual happiness always looks pretty squalid in comparison with the over-compensations for misery. And, of course, stability
isn't nearly so spectacular as instability. And being contented has none of the
glamour of a good fight against misfortune, none of the picturesqueness of a
struggle with temptation, or a fatal overthrow by passion or doubt. Happiness is never grand."

"I suppose not," said the Savage after a silence. "But need it be quite so
bad as those twins?" He passed his hand over his eyes as though he were trying to wipe away the remembered image of those long rows of identical
midgets at the assembling tables, those queued-up twin-herds at the entrance to the Brentford monorail station. . . . "Horrible!"

... "The optimum population," said Mustapha Mond, "is modelled on the iceberg—eight-ninths below the water line, one-ninth above."

"And they're happy below the water line?"

"Happier than above it. . . ."

"In spite of that awful work?"

"Awful? *They* don't find it so. On the contrary, they like it. It's light, its childishly simple. No strain on the mind or the muscles. Seven and a half hours of mild, unexhausting labor, and then the *soma* ration and games and unrestricted copulation and the feelies. What more can they ask for? True," he added, "they might ask for shorter hours. Technically, it would be perfectly simple to reduce all lower-caste working hours to three or four a day. But would they be any happier for that? No, they wouldn't. The experiment was tried, more than a century and a half ago. The whole of Ireland was put on to the four-hour day. What was the result? Unrest and a large increase in the consumption of *soma;* that was all. Those three and a half hours of extra leisure were so far from being a source of happiness, that people felt constrained to take a holiday from them. The Inventions Office is stuffed with plans for labor-saving processes. Thousands of them." Mustapha Mond made a lavish gesture. "And why don't we put them into execution? For the sake of the laborers; it would be sheer cruelty to afflict them with excessive leisure. It's the same with agriculture. We could synthesize every morsel of food, if we wanted to. But we don't. We prefer to keep a third of the population on the land. For their own sakes—because it takes *longer* to get food out of the land than out of a factory. Besides, we have our stability to think of. We don't want to change. Every change is a menace to stability. That's another reason why we're so chary of applying new inventions. Every discovery in pure science is potentially subversive; even science must sometimes be treated as a possible enemy. Yes, even science."

... After a little silence, "Sometimes," he added, "I rather regret the science. Happiness is a hard master—particularly other people's happiness. A much harder master, if one isn't conditioned to accept it unquestioningly, than truth." He sighed, fell silent again, then continued in a brisker tone. "Well, duty's duty. One can't consult one's own preferences. I'm interested in truth. I like science. But truth's a menace, science is a public danger. As dangerous as it's been beneficent. It has given us the stablest equilibrium in history. China's was hopelessly insecure by comparison; even the primitive matriarchies weren't steadier than we are. Thanks, I repeat, to science. But we can't allow science to undo its own good work. That's why we so carefully limit the scope of its researches—that's why I almost got sent to an island. We don't allow it to deal with any but the most immediate problems of the moment. All other enquiries are most sedulously discouraged. It's curious," he went on after a little pause, "to read what people in the time of Our

Ford used to write about scientific progress. They seemed to have imagined that it could be allowed to go on indefinitely, regardless of anything else. Knowledge was the highest good, truth the supreme value, all the rest was secondary and subordinate. True, ideas were beginning to change even then. Our Ford himself did a great deal to shift the emphasis from truth and beauty to comfort and happiness. Mass production demanded the shift. Universal happiness keeps the wheels steadily turning; truth and beauty can't. And, of course, whenever the masses seized political power, then it was happiness rather than truth and beauty that mattered. Still, in spite of everything, unrestricted scientific research was still permitted. People still went on talking about truth and beauty as though they were the sovereign goods. Right up to the time of the Nine Years' War. *That* made them change their tune all right. What's the point of truth or beauty or knowledge when the anthrax bombs are popping all around you? That was when science first began to be controlled—after the Nine Years' War. People were ready to have even their appetites controlled then. Anything for a quiet life. We've gone on controlling ever since. It hasn't been very good for truth, of course. But it's been very good for happiness. One can't have something for nothing. Happiness has got to be paid for."

Science Finds—Industry Applies—Man Conforms.

> MOTTO OF THE CHICAGO WORLD'S
> FAIR, 1933

Invention is a delightful and friendly sport and if we did not have competition we would not have inventions, just as you could not have a race unless you have somebody to race with.

> ERNEST ALEXANDERSON, 1930

A SCIENTIFIC SNAKE DANCE

LORING M. BLACK, 1933

New York Congressman Loring M. Black stood up in the House of Representatives one day to skewer with vintage invective the Technocracy fad that originated in the theories of a group of radical young professors at Columbia University.

Mr. Chairman, Technocracy is a word of terrific mule power covering a multitude of miscalculations. It started in a survey of wheels within wheels and wound up as a jumble of wheels within heads. It does a scientific snake dance right under the hard-boiled brow of Nicholas Murray Butler. It is the great Columbia rackety rax. A flock of dithery young scientists wondered what this great world was all about and finished by giving the world the jitters.

Their plan suggests the dreams of an ostrich after a formal dinner in a scrap-iron heap. They poured into a sausage-making machine some Marxian philosophy, Mussolini rantings, single "taxidermy" Volstead gin, and it all came out "baloney." [Laughter.]

The Columbia group viewed with alarm technological unemployment, researched a little bit, became politicians, and pointed with pride at their scheme for economic salvation. Like zealots of the Middle Ages, they promise their faithful an economics of harps and wings and they threaten the heretics with an economics of fire and brimstone. These mechano-messiahs have equaled in their thinking the intellectual accomplishments of Congress in the balmy days of the drys.

Their scheme comprehends a vast American robotage with a great engineering Jehovah sitting in the White House pressing push buttons to set in motion the rest of us as the Jehovah decides whether we shall eat, drink, be merry, or sleep.

Because North America alone has sufficient energy reserves our continent is the only one physically able to obtain the benefits of the plan. The rest of the world is to sink into the sea while we sail merrily along on roller skates. Like all salvationists, wealth and capital are to be immediately destroyed; currency, debts, and lottery tickets are to be burned alive; and everybody is to be given a proportionate share of energy certificates by which they could go places and do things. Of course, all the energy in North America must be computed first. The scientists have struck a snag, because they cannot estimate the horsepower used up in gum chewing, congressional speeches, marathon dances, and research work. It is figured it will take another thousand years to get a fair base for the domestic allotment of energy certificates.

However, in spite of the hysteria of the Technocrats, they have served to give a warning to capitalism that out of itself it must derive compensatory measures for technological unemployment. Congress, under the Constitution, with the cooperation of industrial leaders, can bring about a program of such relief comprising unemployment insurance, old-age pensions, shorter hours of work, and a more equitable distribution of wealth. By law, machinery should be encouraged for its blessings and taxed for its consequential damage. No more important task confronts the new administration than the stirring of science to new inventions and of protecting humanity from the ravages of the machine. The Technocrats preach a desolation under the gilt of promised leisure. Capitalism is merely faltering, but it has a record of achievement which guarantees greater accomplishments under proper initiative. [Applause.]

II. DEPRESSION AND WAR: 1932–1945

IN DEFENSE OF MACHINES

GEORGE BOAS, 1932

In the years of the Great Depression, when machines were commonly blamed for putting men and women out of work, their virtues were in great dispute.

We are first told that though man invented [machines] to be his servants he has become theirs. . . . This argument is a gross exaggeration. Man is no more a slave of his machines now than he has ever been, or than he is to his body, of which they are—as I think Samuel Butler first suggested—an extension. A farmer is certainly as much of a slave to his primitive plow or sickle as a factory hand to his power loom or engine. . . . Steam undoubtedly produces much of the ugliness and dirt of our cities, but we are not for the moment discussing the aesthetic aspects of the question. Why steam is more mechanical than wind or falling water or muscle-driven hammers is somewhat obscure. A sailboat, a rowboat, an inflated goatskin, a log

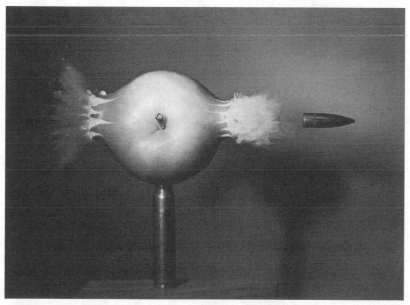

▶ High-speed photography revealed previously unseen realities.

are all equally machines. A linotype, a hand-press, a pen, a reed, a charred stick are all machines. They are all mechanical supplements to man's corporeal inadequacies. . . .

When I have pointed this out in conversation with primitivistic friends I have been invariably charged with sophistry. They have always insisted that my definition of "machine" was too broad. My answer is that the only alternative they offer arbitrarily identifies a machine with a bad machine. . . .

In fact the real clash in opinion is probably ethical. We are—a great many of us—unhappy today and, following a long tradition, we attribute our unhappiness to the economic structure of our civilization. But one can find such outcries of woe as early as eight hundred years before Christ in the works of Hesiod. The crop of cynicism and despair which we uncritically think of as modern is simply human. . . . This unhappiness of ours, which in its literary form expresses itself in tirades against steam, electricity, urban life, manufactures, cannot, therefore, be attributed to machines.

Machines are not the cause either of happiness or unhappiness. They may be present or absent at the time when man is miserable or blissful. . . . It is simply not true that the farmer on his isolated farm in the old days in New England, without radio, telephone, automobile, tractor, reaper, and so on was any happier than the factory hand in Lowell or Lawrence. . . . To be sure factory hands can play a good second to farm hands so far as a dreary life goes. . . . In my boyhood in Providence I used to see men, women, and children trudging to the mills at six-thirty in the morning, tin dinner pails in hand . . . to return at six-thirty at night. They were pale and rickety, God knows, and nothing for the mill owners to look in the eyes. But that does not mean that their contemporaries on the farms were red-cheeked and stocky, effervescing with vitamins, sleeping late in the morning and going to bed early, delighting in robust rural pleasures. . . .

One of the points especially emphasized by the enemies of machines is that they substitute something lifeless for something vital and human. Concretely, this means that a farmer cannot love a tractor or an incubator as he could a horse or a hen. This is very probable, particularly if the farmer started farming with horses and hens. But it is not absolutely certain otherwise. Machines can be as lovable as animals. Who has not known engineers who literally love their locomotives, or boys who care for their radios, speed boats, and automobiles as if they were alive? . . . There is no laying down the law about the inherent lovableness or hatefulness of anything.

. . . It will be said that the old machines, actuated by human muscles rather than by steam or electricity, at least helped a man's creative power. Friends of the machine are constantly being told that hand-weaving is creative whereas machine-weaving is not. The old French artisan, we are told, lived a life of creativity; he stamped things with his own individuality; he projected his personality into his products. The modern American factory hand is passive; he makes nothing; his product is standardized. This, within limits, is true of the factory hand. But it was also true of the artisan. He had certain styles and patterns which he reproduced endlessly, as our great-grandmothers reproduced world without end the same old quilting patterns. That man's products have always been standardized is proved by archeology and the history of taste. . . .

As one digs into this discussion one finds the instinctive hatred that many people have always had for innovation. We do not hate machines, we hate new machines. A woman will object to buying a dress cut by machine, but will not object to buying one sewn by machine. . . . I find myself fuming at automobiles and yearning for the old bicycles. I can remember old folks shaking their heads over telephones as their juniors now curse out the dial phone. I have heard a gardener in France inveighing against chemical fertilizers which *"violent la terre,"* as if horse manure were non-chemical. Sailors in the windjammers railed against the steamboat, and steamboat crews think none too kindly of the johnnies who sail oil-burners. Greek and Roman literature is full of invective against any kind of navigation, for it takes the pine tree off its mountain top and sends men wandering.

Obviously a new machine, like an old one, must be judged on its merits, not on its novelty. But the fact that it is novel should not condemn it. Here are two stalwart platitudes. But think of the fools who objected to anesthesia, to aeronautics, even to cooked foods, because they were not "natural." The question cannot be settled by the wild use of question-begging epithets. We must each establish a system of values for ourselves or absorb that of our social group, and judge machines by it as we do everything else. There is no other way of evaluating anything.

ENERGY FROM
AIR-CONDITIONING

WILLIS R. GREGG, 1933

Wherein the chief of the U.S. Weather Bureau, dedicating Frigidaire's exhibit of an air-conditioned house at the 1933 Chicago World's Fair, meditates on the relationship between climate and capitalistic vigor.

The energetic, hard-hitting tactics of the northerner, who works hard and plays hard because the climate in which he lives inspires and invigorates him to greater activity, has had much to do with the development in our northern States of giant industries and other activities, with a resultant centralization of buying power.

Is it too much to predict that air conditioning of the working and living quarters of other residents of more humid areas may cause more activity in those parts that will open up to use natural resources beyond our imagination?

THE FIRST HUMAN HOPE INDUSTRIALISM HAS OFFERED

ARCHIBALD MACLEISH, 1933

Writing in The Nation *at the bottom of the Depression, the poet and journalist found qualified hope in the increase in productivity that many blamed for widespread unemployment.*

RICHARD RHODES

The real meaning of modern industrialism has no interpretation in terms of fear except for the banking class. For the rest of society it offers hope. Earlier in this century and throughout the latter half of the nineteenth century there was no hope in industrialism but escape. Poets turned aside from it. Philosophers transcended it. Even the Communists who plunged into it to the hair roots, accepting the human categories of economic class it imposed, worshiping the symbolic tractors it produced, and making a religion of the labor it compelled men to perform, were unable to find in industrialism a positive justification and were forced to express their Utopia in negatives—the *end* of capitalism, the *end* of exploitation, the *end* of injustice. But with the third decade of the present century the image changed. Industrialism turned another face. It became apparent that the goal of industrialism was not merely greater and greater production, more and more material, richer and richer bankers, but something quite different from these things. It became apparent that industrialism was moving toward a degree of mechanization in which fewer and fewer men need be, or indeed could be, employed. And that the result of that development must, of physical necessity, be a civilization in which all men would work less and enjoy more. For the alternative to such a choice was a state in which a half or a quarter of the adult population would be unemployed and carried as charity charges by the remainder of the population, and such an alternative was unthinkable. What political or social revolutions must intervene no man could say. But the probable and indicated outcome could be foreseen by any man who was familiar with the figures.

. . . In the last analysis the important question is not whether a man can make 9,000 times or only 550 times as many light bulbs now as he could make in 1914, but whether it is or is not true that the mechanization of industry has reached the point where production can increase without a corresponding increase in the number of wage-earners. For if the answer to that question is in the affirmative, then the direction of growth of industrialism has changed, our civilization has turned a corner, and the ancient conception of human work as the basis of economic exchange and of the right to live is obsolete, since the work of machines and the conversion of non-human energy take the human place.

There is no doubt whatever that the answer is in the affirmative. . . . Comparison of the years 1923 and 1927, by industries, gives overwhelming support to the conclusion.

Industry	Change in Output	Change in Employment
Oil: petroleum refining	84 percent more	5 percent less
Tobacco	53 percent more	13 percent less
Meat: slaughtering, packing	20 percent more	13 percent less
Railroads, 1922–26	30 percent more	1 percent less
Construction, Ohio only	11 percent more	15 percent less
Automobiles, 1922–26	69 percent more	48 percent more
Rubber tires	28 percent more	7 percent more
Bituminous coal	4 percent more	15 percent less
Electricity, 1922–27	70 percent more	52 percent more
Steel	8 percent more	9 percent less
Cotton mills	3 percent more	13 percent less
Electrical equipment	10 percent more	6 percent less
Agriculture, 1920–25	10 percent more	5 percent less
Lumber	6 percent more	21 percent less
Men's clothing	1 percent more	7 percent less
Paper	0	7 percent less
Shoes	7 percent less	12 percent less

. . . It seems impossible to deny the fact of the decline. And if the decline in number of wage-earners in the face of an increase in production is admitted, the case is proved. It would be no answer to argue . . . that the lost population of the factories was taken up in the "services"—the hotels and the bond offices and the gas stations.

For a society which wished to take advantage of the potential benefits of the new industrialism would not force its displaced workers into these selling services to swell a new Coolidge-era expansion, but would decrease workers' hours to spread the leisure resulting from technological advance. It is only the present misfit distribution system which makes it necessary for one man to take to the road selling insurance while nine men left in the factory go on working ten hours a day. The "services" from this point of view are merely a buffer margin to enable the present system to frustrate its own genius in the interest of its creditors. And the benefits from swelling the services are, as the present disaster proves, temporary at best. But even were it otherwise, the opposition would still have failed to prove that the services will take up all the slack. Surveys have been made showing that 1,907,000 new positions were created in the Twenties in medicine and hotels and restaurants and moving-picture theaters and banks and the like as against the 1,485,000 positions estimated to have suffered technological cancellation in industry. But this result neatly omits the population growth in the interval which should have increased the number of wage-earners by 2,000,000. . . .

If the opposition has failed to make its point on the basis of actual employment figures, it has also failed to make its point on the basis of the rate of change. The significant and fundamental fact which no one has denied is that productivity per wage-earner was almost constant for the twenty years from 1899 to 1919, whereas in the eight years following 1919 it increased by approximately *50 percent*. Critics have suggested that a similar or greater increase in productivity occurred at the beginning of the Industrial Revolution. But the eras are not comparable. Changes from handicraft to machine production enormously increased output per worker and displaced thousands of men amid dire forebodings. But mechanization was still only an adjunct to human labor. As production increased with the manufacture of cheaper goods, employment also increased. Only now has man become an adjunct, and an increasingly less important adjunct, of the machine. Only in our time has an increase in production been possible with an actual decrease in number of men employed. [This problem] is the vital problem of our time and the . . . first human hope industrialism has offered. Those who ignore the problem and those who discredit the hope do so at their peril.

STREAMLINING

ROGER BURLINGAME

Streamlining, developed in the 1930s in wind-tunnel work, soon became a fad that defined the look of the era.

An airplane, moving wholly in the air in three dimensions, heads into the air stream. It is propelled by gripping the air with its propeller. The re-

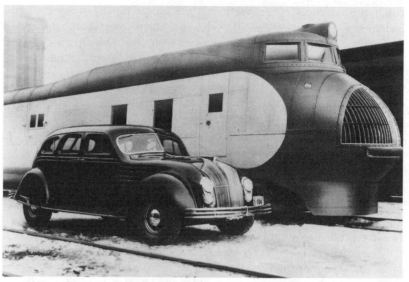

▶ Wind-tunnel studies made streamlining fashionable in the 1930s.

sistance of the air makes its upward motion possible, it climbs upon the air as a train climbs upon the track, a car upon the road. Thus, always the air is resisting it; always an air stream flows round it and past it. The several streams flowing over and under, by the sides, form what are known as the "stream lines"; they enclose a space of dead air which acts as a drag on the craft. Thus design must be adapted to the easiest flow of these streams with as little obstruction as possible, and the space of dead air they enclose in the craft's "wake" must be small.

A branch of mathematics known as aerodynamics grew up round the problems posed by the motion of bodies through the air. Countless thousands of experiments with models in wind tunnels built a mass of data. Finally, a design was worked out on the mathematical basis to overcome superfluous or "parasitic" air resistance, and the effort at the perfect design was called streamlining. To the lay public the streamlines became the lines of the craft.

Students of design studied birds. Then it was assumed that a drop of liquid falling through the air would take a shape offering the minimum of resistance, so the students studied quick photographs of falling drops. The public, learning of these things, became entranced. It saw aesthetic values in copies of efficient nature. It became poetic about the falling drop, calling it the "tear drop." It sighed over its new vision of the bird.

These people failed to note, however, that the horse, the deer, the cockroach, the mouse and other creatures which moved rapidly over the ground had somewhat different designs. What are called streamlines are hard to find in these demonstrations of an efficient evolution. This did not deter the public from believing (when it was suggested by adroit propagandists) that the design of the airplane applied to railroad trains and motor cars would achieve wonderful results of speed and saving of power. . . .

If streamlines had stopped with cars, however, they would have lost much of their symbolism. But, in the 1930s, they spread to buildings, houses, furniture, other things which are rarely required to combat air resistance. Inherent in a building are its static properties. To sacrifice space, comfort, interior efficiency, solidity, in order to give it the outside look of something perpetually in rapid motion, is likely to engender restlessness in the inmate. When this was done the streamline reached its full symbolic significance.

It represents the subjective ideal of speed divorced from utility. Not only can the ideal speed never be attained, but it is self-limiting. The ideal of speed, when it became general, defeated speed itself. . . .

Some of us remembering our old cities know that in the 1890s we reached our destinations more quickly in horse-cars or on foot than we did in the automotive busses and taxis of the 1930s. No spectacle of medieval congestion could equal a New York rush hour in our streamlined era in discomfort and savagery. A Roman, watching the human cattle being herded into a subway train, would marvel at the efficiency with which men were led to the slaughter. Where, he might ask, in his sanguinary excitement, is the Coliseum? But, no, this is speed, to save an hour. But what intense activity will occur in the hour saved?

Men standing at a bar, "killing" time. Women and children fretting over a game, a puzzle, a contest, to kill time. Men and women quarrelling because supper is not ready, but quarrelling later because the evening is so long. They must get a new radio to help kill it. There is more time to kill because the subway, the stove, the can of condensed, cooked food all worked so fast. Because, in a jiffy, the telephone brought the cans from the grocer. Well, the telephone will serve, too, to kill the time saved. We must call up Tom, Jane, Junior. They too will be killing time. What shall we do tonight, we have an hour to kill? Books, papers, magazines are slow. Even the radio is slow. If we are to do a good job killing time, we must do it fast. We might go out in the car.

Millions of people killing time in cars. Give him the horn! This car has power enough to pass every other car on the road! What does that snail think he is doing? Out for the air. In a streamlined, this year's model, with a supercharger. . . . No, it's the car ahead that's the trouble. Now we've missed the light. A cigarette, please, to keep us from doing nothing. Lighter's on

the dash. Broken? Well, time to turn the old bus in, anyway. It's economically sound to turn them in every two years. The new model has a hydra-
matic drive, there'll be nothing to do with your hands. What, then, shall we
do with our hands?

THE MACHINE OUR SERVANT

LEWIS MUMFORD, 1934

Lewis Mumford's Technics and Civilization *commanded wide attention and
debate. Here he predicts the taming of the machine as humankind matures.
Was he right or wrong? Or is it too soon to know?*

In a word, *as social life becomes mature, the social unemployment of machines
will become as marked as the present technological unemployment of men.* Just as
the ingenious and complicated mechanisms for inflicting death used by
armies and navies are marks of international anarchy and painful collective
psychoses, so are many of our present machines the reflexes of poverty, ignorance, disorder. The machine, so far from being a sign in our present civilization of human power and order, is often an indication of ineptitude and
social paralysis. Any appreciable improvement in education and culture will
reduce the amount of machinery devoted to multiplying spurious mechanical substitutes for knowledge and experience now provided through the
channels of the motion picture, the tabloid newspaper, the radio, and the
printed book. So, too, any appreciable improvement in the physical apparatus of life, through better nutrition, more healthful housing, sounder forms
of recreation, greater opportunities for the natural enjoyments of life, will
decrease the part played by mechanical apparatus in salvaging wrecked bodies and broken minds. Any appreciable gain in personal harmony and balance will be recorded in a decreased demand for compensatory goods and
services. The passive dependence upon the machine that has characterized
such large sections of the Western World in the past was in reality an abdi-
cation of life. Once we cultivate the arts of life directly, the proportion occupied by mechanical routine and by mechanical instruments will again
diminish.

Our mechanical civilization, contrary to the assumption of those who
worship its external power the better to conceal their own feeling of impo-
tence, is not an absolute. All its mechanisms are dependent upon human
aims and desires: many of them flourish in direct proportion to our failure
to achieve rational social cooperation and integrated personalities. Hence
we do not have to renounce the machine completely and go back to handi-

craft in order to abolish a good deal of useless machinery and burdensome routine: we merely have to use imagination and intelligence and social discipline in our traffic with the machine itself. In the last century or two of social disruption, we were tempted by an excess of faith in the machine to do everything by means of it. We were like a child left alone with a paint brush who applies it impartially to unpainted wood, to varnished furniture, to the tablecloth, to his toys, and to his own face. When, with increased knowledge and judgment, we discover that some of these uses are inappropriate, that others are redundant, that others are inefficient substitutes for a more vital adjustment, we will contract the machine to those areas in which it serves directly as an instrument of human purpose. The last, it is plain, is a large area: but it is probably smaller than that now occupied by the machine. One of the uses of this period of indiscriminate mechanical experiment was to disclose unsuspected points of weakness in society itself. Like an old-fashioned menial, the arrogance of the machine grew in proportion to its master's feebleness and folly. With a change in ideals from material conquest, wealth, and power to life, culture, and expression, the machine like the menial with a new and more confident master, will fall back into its proper place: our servant, not our tyrant.

CONSISTENT OPTIMISM

ALFRED P. SLOAN, JR., 1934

No depression since man began to use machines effectively has lasted long enough to break down the consistent optimism of generation after generation of inventors and industrialists. This faith has been justified by events. Disregarding temporary dips, and concentrating upon the long swing of industrial civilization, we find that real wages have increased, new wants have been created and supplied, famines and shortages almost erased, and the standard of living so raised that a responsible workingman enjoys a wider range of comfort and culture than did barons in the Dark Ages. Goods and services unknown of old, but now widely used and taken for granted, are the social dividend of individual and group initiative. While the driving motive of these individuals and groups was private and corporate profit, the indirect result has been to spread boons which all civilized men in some degree enjoy.

LIGHTER THAN AIR

R. H. HARRISON

Dirigibles still sailed America in the early 1930s, as this writer reminisced in 1994.

In the early thirties I lived near Canton, Ohio, just twenty miles south of Akron, where [the dirigibles] *Akron* and *Macon* were built. Our father frequently took my brothers and me to Akron on Sunday afternoon to view the progress of construction. At the time, I went to a small country school in the middle of farmland. One spring morning when the clouds were hanging very low, we were out in the schoolyard during morning recess when we heard the sound of engines coming from the north. As we turned toward the sound, a huge dirigible came out of the clouds, heading directly toward us. It was *Macon* on what may have been her first test flight. She was going very slowly, presumably only fast enough to maintain steerageway. And she was so low that as she passed directly over our heads we not only could make out the facial features of the crew members leaning out of the control-car windows but could—and did—actually talk to them. Given the state of instrumentation at the time, I assume now that the crew had to maintain eye contact with the ground to know their altitude. It was an experience that I have never forgotten, and it left me a fan forever of lighter-than-air ships.

THE CAUSES OF THE CAUSES

S. C. GILFILLAN, 1935

Sociologist and sometime museum curator S. C. Gilfillan here proffers a sort of Anthropic Principle of the railroads.

To say that America has railroads today because Stephenson invented them in 1829, (of course he didn't) is the easy thing to teach and think. It is much harder to say that we owe our railroads not only to ten million inventions and discoveries, both before and after that date, but also to such facts as that America now has a population sufficiently numerous, dense, wealthy, educated, sober, diligent, and trustworthy to pay for, staff and use great railroads. In earlier times when Americans were few and poor, part slaves, largely illiterate, and often drunk, much railroading was not to be expected, any more than it is in Venezuela today. In such countries today they possess the invention, but hardly use it. Next, if we are seeking a complete explanation of the railroad, we must go behind those social reasons, to the causes of the causes, such matters as the limitation of births, that has made possible our present prosperity, and the causes of this in turn, in scientific and inventive advances, and also in religious and philosophic change, and all that has been affecting the feminist movement. In all this boundless welter of causation, it may be that the constantly and especially clearly recurring elements of science and invention contribute a majority of the total effect; but who has even started to prove it?

With the inventive and the social aspects of life going forward together, we are at a loss to prove which was more the cause of the other.

PREDICTIONS: ROCKETS THROUGH SPACE

RICHARD VAN RIET WOOLEY, 1936

Reviewing Rockets Through Space *by P. E. Cleator, the first British book on the new science of rocketry, the British Astronomer Royal had this to say.*

The whole procedure . . . presents difficulties of so fundamental a nature that we are forced to dismiss the notion as essentially unpractical, in spite of the author's insistent appeals to put aside prejudice and to recollect the supposed impossibility of heavier-than-air flight before it was actually accomplished. An analogy such as this may be misleading, and we believe it to be so in this case. . . . We are compelled to join the ranks of those whom Mr. Cleator stigmatizes as visionless reactionaries.

OVERALLS

JAMES AGEE, 1936

"During July and August 1936," this novelist and film critic wrote in a preface to his powerful, poetic Let Us Now Praise Famous Men, *"[the photographer] Walker Evans and I were traveling in the middle south of this nation . . . to prepare, for a New York magazine, an article on cotton tenantry in the United States." Out of that experience, as exotic in its way as an anthropological expedition, Agee and Evans fashioned their book, which stands with Walt Whitman's* Leaves of Grass *and William Carlos Williams's* In the American Grain *as enduring testaments to the character of American experience. "Overalls," a prose poem to a common item of modern work clothing that exemplifies the technology and subtle iconography of the human world, appears in a section of the book called simply "Clothing."*

They are pronounced overhauls.
    Try—I cannot write of it here—to imagine and to know, as against

other garments, the difference of their feeling against your body; drawn-on, and bibbed on the white belly and chest, naked from the kidneys up behind, save for broad crossed straps, and slung by these straps from the shoulders; the slanted pockets on each thigh, the deep square pockets on each buttock; the complex and slanted structures, on the chest, of the pockets shaped for pencils, rulers, and watches; the coldness of sweat when they are young, and their stiffness; their sweetness to the skin and pleasure of sweating when they are old; the thin metal buttons of the fly; the lifting aside of the straps and the deep slipping downward in defecation; the belt some men use with them to steady their middles; the swift, simple, and inevitably supine gestures of dressing and of undressing, which as is less true of any other garment, are those of harnessing and of unharnessing the shoulders of a tired and hard-used animal.

They are round as stovepipes in the legs (though some wives, told to, crease them).

In the strapping across the kidneys they again resemble work harness, and in their crossed straps and tin buttons.

And in the functional pocketing of their bib, a harness modified to the convenience of a used animal of such high intelligence that he has use for tools.

And in their whole stature: full covering of the cloven strength of the legs and thighs and of the loins; then nakedness and harnessing behind, naked along the flanks; and in front, the short, squarely tapered, powerful towers of the belly and chest to above the nipples.

And on this façade, the cloven halls for the legs, the strong-seamed, structured opening for the genitals, the broad horizontal at the waist, the slant thigh pockets, the buttons at the point of each hip and on the breast, the geometric structures of the usages of the simpler trades—the complex seams of utilitarian pockets which are so brightly picked out against darkness when the seam-threadings, double and triple stitched, are still white, so that a new suit of overalls has among its beauties those of a blueprint: and they are a map of a working man.

Invention Is a
Great Disturber

National Resources Science
Committee, 1937

In 1937, a commission chaired by Secretary of the Interior Harold L. Ickes reported to President Franklin D. Roosevelt on "the kinds of new inventions which may affect living and working conditions in America in the next 10 to 25 years." The report's remarkably pedestrian findings offer a glimpse of conventional thinking on the effects of technology on national life. They focus almost entirely on economic and domestic issues—not surprising, given the continuing Depression—and ignore, for example, military, public health, cultural and ecological consequences.

1. The large number of inventions made every year shows no tendency to diminish. On the contrary the trend is toward further increases. No cessation of social changes due to invention is to be expected. . . .

2. Although technological unemployment is one of the most tragic effects of the sudden adoption of many new inventions . . . inventions create jobs as well as take them away. . . .

3. No satisfactory measures of the volume of technological unemployment have as yet been developed. . . .

4. The question whether there will be a large amount of unemployment during the next period of business prosperity rests only in part on the introduction of new inventions and more efficient industrial techniques. . . .

5. Aside from jobs, subtracted or added, new inventions affect all the great social institutions; family, church, local community, State, and industry. . . .

6. A large and increasing part of industrial development and of the correlated technological advances arises out of science and research. . . .

7. Advance of many aspects of industry and the correlated technology is dependent upon scientific research and discovery. This fact is made clear by the increasing importance of research laboratories in the great industries. . . .

8. Though the influence of invention may be so great as to be immeasurable, as in the case of gunpowder or the printing press, there is usually opportunity to anticipate its impact upon society *since it never comes instantaneously without signals.* For invention is a process and there are faint beginnings, development, diffusion, and social influences, occur-

ring in sequence, all of which require time. From the early origin of an invention to its social effects the time interval averages about 30 years.

9. While a serious obstacle to considering invention in planning is lack of precise knowledge, this is not irremediable nor the most difficult fact to overcome. Other equally serious obstacles are inertia of peoples, prejudice, lack of unity of purpose, and the difficulties of concerted action.

10. Among the resistances to the adoption of new inventions and hence to the spread of the advantages of technological progress there is specially noted those resistances arising in connection with scrapping equipment in order to install the new. . . .

11. The time lag between the first development and the full use of an invention is often a period of grave social and economic maladjustment, as, for example, the delay in the adoption of workmen's compensation and the institution of "safety first" campaigns after the introduction of rapidly moving steel machines.

THE BURNING OF
THE *Hindenburg*

HERB MORRISON, 1937

In a famous eyewitness radio broadcast on May 6, 1937, the announcer describes the disaster that ended the production of commercial hydrogen-fueled dirigibles.

The back motors of the ship are just holding it, just enough to keep it from—It's burst into flames! Get this started, get this started! It's fla—and it's flashing, it's flashing terrible! Oh my, get out of the way, please! It's burning, bursting into flames a—and it's falling on the mooring mast and all the folks agree that this is terrible, this is one of the worst catastrophes in the world. Ohhhhh! The flames are climbing, ohhh, four or five hundred feet into the sky and it's a terrific crash, ladies and gentlemen, the smoke and the flames, now. And it's crashing to the ground. Not quite to the mooring mast. Oh, the humanity and all the passengers!

PREDICTIONS: FIFTY YEARS FROM NOW IN 1988

ARTHUR TRAIN, JR., 1938

Wherein an essayist for Harper's *attempts to predict what life will be like in the distant future of America.*

Our hero . . . John Doe, born in the year of grace 1938, was in bed and asleep at the time our story begins in 1988. (No synthetic substitute for sleep had then been discovered.) Progress in biology, biochemistry, food technology, and related sciences was responsible for the fact that he was considerably heavier and taller than his forefathers, and also that, although half a century old, he was neither too fat nor too thin, and like Uncle Ned had "plenty of wool on the top of his head, the place where the wool ought to be."

The sounds of the city were filtered at the intake-ducts of the air-conditioning apparatus, and such few persistent discords and jangles as did penetrate into the room were deflected toward the ceiling by the walls which slanted gently upward, like the glass windows of radio broadcasting control rooms, where they were absorbed by special insulation. The entering air passed through a dust filter and was freed from other germs by ultraviolet rays.

Research into the effects of ionization, barometric pressure, condensation nuclei, and the existence of a metastable state of oxygen had made it possible to supply Mr. Doe's room with air as invigorating as that of the seashore or the mountains. Its chemical composition was nicely calculated to give him a maximum of refreshment at night, while during the day its temperature, humidity, and degree of ionization were automatically varied from time to time in order to avoid the soporific effect of monotony. Incidentally, synthetic air, long considered fantastic, was well on the way toward becoming a reality.

Presently, as the radio-controlled clock proclaimed in a soothing voice that it was time to get up (for its direct reading dial showed the hour of seven), the air became sensibly warmer. Heating was provided by the simple process of running the refrigerator mechanism in reverse, although some architects recommended heating coils in the walls or radiant wires in the ceiling.

Although it was dark and rainy outside, the room was gradually flooded with a diffused light. The quantity required was measured out with nice accuracy by the ever-watchful photocell, and on sunny days when clouds passed over the sun, the light in the room would remain constant. This light was provided by a type of gaseous discharge lamp, perhaps employing carbon dioxide, infinitely more efficient than the old-fashioned incandescent filament bulbs, and containing as good a proportion of infrared and ultraviolet rays, as that of the brightest summer sun, which were automatically turned on at intervals.

Meanwhile—the first item in a preselected program from different stations—the television screen faded in on an energetic man in a football sweater who beckoned Mr. Doe to arise and begin his setting-up exercises. In apartment houses these television images were usually "piped" along a coaxial cable (an invention which the public of the '30s had failed to realize was as revolutionary as the telephone itself); but for private homes and for general purposes the old-fashioned system of coaxial cable and linked radio stations had been superseded by the "Yale lock" style of multiple wavelengths using various permutations and combinations to give broader waveband availability of an unlimited number of channels.

The bathroom into which Mr. Doe stepped for his matutinal shower was a prefabricated affair made like an automobile, all the various appliances such as tub, shower, basin, and toilet forming one integrated unit, with special metallic walls for the outer casing. Three identical bathrooms were grouped with it to form a square in the center of the house, so that a minimum of plumbing was required. The old-fashioned system of using thousands of gallons of water to dilute and remove waste, thereby sacrificing its valuable chemical properties, had long ago been superseded by chemical disposal of sewage. The development of new detergents also made it possible to "wash" without water if anyone so desired.

While Doe was slipping into a pair of shorts and a light, three-quarter length rayon fabric smock, which, after all, is all that anyone would need in an air-conditioned home, he haphazardly pushed in various buttons controlling the automatic tuning of his television set so that he might see with his own eyes what was going on in the different parts of the world. He was a man who liked to spend money on gadgets, and the morning paper had been printed out for him by the facsimile recorder while he slept. It was his habit to leave it on just as people in the old days left the radio on, and from the reams of stuff it printed out he would pick what he wanted and throw the rest away. Most of the time, however, he preferred to hear the news rather than read it.

The vegetables and fruit that graced the Doe table out of season had never known the rich soil of a truck garden. Some—possibly the more expensive ones—had been grown in a vegetable factory in the heart of the community center, in a heated tray containing various salts. Others had come in black iron, plastic-coated cans, flash-heated to preserve the natural flavor of the contents, while others, at the other end of the scale, reached his kitchen in a frozen state. Mr. Doe habitually reflected with satisfaction that he never had any trouble getting whatever he wanted whenever he wanted it, and that the real significance of chemically produced crops and other mechanical aids to agriculture was that they permitted an efficient control of the food supply.

His house was situated at a considerable distance from the city, in an "integrated" neighborhood which had been carefully planned by a city planning board. The houses were grouped around a park, and in addition to the school and library there was a central air-conditioning plant and a community center with a television transmission set, an auditorium whose television receiving set boasted color and three-dimensional sound and sight, a trailer camp, all kinds of recreational facilities, the vegetable factory, the poultry factory, and the plant where garbage was converted into fertilizer.

The house itself was somewhat smaller and had smaller rooms than one would have expected of a man with Mr. Doe's means. The large custom-built house had long ago gone the way of the large custom-built automobile. It was a long, low, flat-roofed building made up of a cluster of prefabricated units whose irregular arrangement prevented it from looking monotonous. Unlike the houses of the early part of the century and all preceding eras, whose aim was to give an impression of volume, the whole building was so translucent, neutral, and fragile-looking, so broken into planes by terraces and porches, that it gave the impression of being no more than a part of the out-of-doors which had been etched into the frame with a few strokes of a sharp pointed pencil.

In the construction of the house the use of wood, bricks, and plaster had practically been superseded by panels of beryllium and magnesium alloys; low-grade silicas, or glasslike materials; sheet materials such as asbestos cement, and occasionally plastic which had been developed to a point where its resistance to atmosphere was known. A considerable use was made of moving partitions which made it possible to enclose a small space when privacy was required, and still provide a large space when it was not. The insulation, of "mineral fluff," was of course built into prefabricated panels.

In the various rooms many of the pieces of furniture were made of plastic molded as a unit, while others were made of magnesium alloy. In place of cushions, spongelike synthetic upholstery was used. Some of the most beautiful hangings were of translucent glass fabric.

Outside of a few first editions and beautifully bound volumes with handsome illustrations, Mr. Doe's library contained few books. It consisted chiefly of little drawers filled with thousands of tiny reels of film a few millimeters in width. On his table was a reading machine about the size of a portable typewriter, which projected the tiny photographed pages onto a small screen. Each of those tiny films also carried a sound track, and at his own discretion he could play them on a talking book. Wherever he went Mr. Doe carried a camera hardly bigger than a watch and also a tiny sound-recording device, so that anything he saw or heard during the day he could conveniently remember by mechanical means. The day had not arrived (predicted by Sarnoff back in 1936) when each individual would have his own wavelength and by means of a pocket radio could communicate with anybody anywhere. In Doe's office the principle of mechanical aids to memory was developed to a high state of efficiency. All of his records were "remembered," selected, and analyzed on photoelectric tabulating machines with far greater efficiency than the human brain could achieve and in much less time.

An inventory of the various objects and materials used in Mr. Doe's house would show that the strawboard and fiberboard that lined the walls, the insulating material between them and the outer wall, sometimes the outer wall itself, the synthetic textiles which comprised the clothing of much of the family, and the waterproof materials which protected them if they ever went out in the rain, and all small knick-knacks from ash trays to bottle caps, were made of various types of thermoplastic resin derived from such inexpensive raw materials as soy bean, bagasse, sugar cane, straw, wood pulp, sorghum, linseed, flaxseed, cottonseed hulls, oat hulls, nut shells, Jerusalem artichokes, fruit pits, and skim milk.

We have seen how in Mr. Doe's house the electric eye, or photoelectric tube, coupled with the thyratron tube, which enables it to act on what it sees, automatically measured the amount of illumination necessary to replace the waning light of day. It also performed the functions of a whole corps of servants. It opened the garage door as you drove up, opened the door between the kitchen and the dining room when someone advanced with a tray, opened the door of the refrigerator, and opened and closed windows. But its duties did not end with the fall of day. All night long it was on guard as night watchman, ready to give warning by ringing bells, turning on floodlights, photographing the intruder, paralyzing him with tear gas, and sending for the police.

The roof of the house, as in all houses at that time, was used as a landing field for the family's collection of steep-flight airplanes of assorted sizes, the top story being used as a garage. Doe didn't bother to use his car very often, and in general it was relegated to trips to the community center and to use by the children, playing the role of the station wagon of the late '30s. Its two-cycle motor, smaller, lighter, and more efficient than the old fashioned four-cycle one, could easily drive it along at an average speed of seventy miles an hour on the highly efficient fuels of those days. Such speeds, however, seemed like crawling to Mr. Doe and his friends, who used small steep-flight planes for short hops and giant stratosphere planes for distance flying.

Train qualifies his vision later in his essay, pointing out that his predictions presuppose ideal conditions. In reality, he writes, a man of means isn't likely to want a prefabricated or a modern house, molds for plastic furniture are expensive, television requires so much bandwidth it might never be practical and in any case might be boring (!) and invested capital might not want to allow its factories to be made obsolete by new inventions. He clings to his magnesium furniture and his suburban helicopters and concludes:

If there is any one prediction that can be safely ventured upon, it is that we shall increasingly be obliged to turn to the scientist and to his way of thinking. Our future is in the hands of the technologists. But today we still hold them back and delay the fulfillment of their prophecies.

A banker once defined invention as that which makes his securities insecure.

U.S. NATIONAL RESOURCES SCIENCE
COMMITTEE, 1937

ONE OUNCE OF EUROPEAN GOLD

PAUL B. SEARS, 1939

Embedded in a scientist's popular essay on soil conservation, this almost postmodern critique of technology finds danger in "the effect of change in technology combined with the encrustation of [past] culture-forms."

If we try to examine the world of western Europe and its extensions into other continents we find, first, that the applications of science have been chiefly on the basis of unrestricted individual profit. Through a faulty analysis this is sometimes confused with the profit motive in general, of which it is merely a special case. In the second place, as a result of the motive of unrestricted individual profit, scientific technology has been preoccupied with the elaboration of raw materials into consumers' goods. This involves great technological advances in transportation; but just as it is important not to confuse the various possibilities of the profit motive, so it is essential not to confuse physical transportation with the much more fundamental matter of distribution. The problem of distribution has been essentially ignored in the application of science to modern culture.

It might be objected that the pure scientist is not in any sense concerned with financial profit or even the cultural significance of his work. This may be approximately true so far as the individual is concerned, but fails to carry conviction in view of the immense returns which have accrued from the application of pure science. It is neither accident, philanthropy, nor unworldliness that explains the heavy endowment of pure research in numerous great industrial laboratories. A study of the origin and growth of the concept of pure science among Americans would be a very fruitful and, I guess, very revealing piece of work. One has only to consider the history of various experiment stations and university departments of science to learn that the battle for pure research was essentially a fight to do scientific work that was worth the paper it was written on; in other words, it was a struggle for the right to do scientific work that was good enough to serve some useful end. Often unreasonable pressure from so-called "practical" sources for speedy or showy results had to be met. Where this was done by publication of superficial work, practical needs were ultimately ill-served.

It is necessary to remember too that science arose as commerce was breaking the pattern of feudalism. In quick succession a mercantile society shot through with vestiges of feudalism became a highly industrialized society and then one manipulated in terms of concentrated and fabulous financial power. Feudalism, mercantilism, industrialism, financialism—the landscape of today bears the scars of these four successive and deep-seated phases of human cultural activity.

Feudalism, for example, has persisted in religious, educational, and social relationships. It was transported to the new world in the form of great estates contemptuous of physical labor, it sunk to its nadir in slavery. This slavery, under various names, still persists—north and south, east and west. And the prestige of the feudal lord transformed into reasonably modern idiom still represents the goal of most human striving. Feudal forms still persist in schools; if there is any civil institution much less democratic in certain respects than the average State university I have yet to see it.

In the beginning, feudal estates were subsistence enterprises with mutual obligations running through the whole pattern. But in every great culture that has ever existed, the concentration of wealth has resulted in a degradation of the man who worked the land and the eventual ruin of the land itself. From this fate the mercantile development of the late Middle Ages seemed to offer an escape. Men in towns and cities, through the accumulation of gold, could carry farms and owners about in their pockets. No wonder we developed so much respect for money in spite of the fact that it was and is nothing but a symbol for wealth.

The chain does not end. Goods were the key to wealth. Markets were expanding. Labor was scarce. Rich prizes lay at hand for those who could multiply goods and dispose of them for gold. This became possible through the development of machinery and power. There was no counterbalance offering rewards to those who would have us pause and ask to what end this feverish activity was leading. The final step took place when rapid communication made it possible to centralize financial control of the processes of production, the sources of supply, and the labor of men. Some idea of the power of this change may be secured when we recall that one may now go from Cleveland to New York in less time than it used to take an Ohio farmer to drive to the county seat and visit his banker. Yet the distance a farmer could drive in a half day has determined the number and size of counties in Ohio with all of the attendant organization of institutions, human effort, and values. It would be more logical today to have Ohio divided into not more than a score of counties, but imagine the practical difficulties that would be encountered if an attempt were made to do this.

Cotton growing affords an even more pungent example of the effect of change in technology combined with the encrustation of the successive phases of culture-forms of the past four hundred years. Cotton grown in the Southern States under a feudal social system on land heavily mortgaged is shipped abroad for manufacture in the highly industrialized, low-wage, and slum areas of western Europe. Computing our entire export of the past century to Europe, for each bale of cotton that has been shipped, one hundred and thirty tons of soil have been washed into the Gulf of Mexico. (I am indebted to Robert Montgomery of the University of Texas for this figure.) In exchange, we receive one ounce of European gold as a token for each bale with its investiture of soil. "And then," adds Mr. Montgomery, "we bury the gold in northern Kentucky."

The effect of such confused and irresponsible procedures as this upon the landscape ought to be well known by this time. Certainly it has been presented to the public in a rich diversity of books and magazine articles. Nevertheless, it is characteristic of our confusion that many otherwise intelligent citizens regard this information as calamity howling, special pleading, or a sinister sort of political propaganda for the present [Roosevelt] Administration at Washington. One of the most curious contradictions in the American character is our utter failure to see the connection between the word conservative and the word conservation. Undoubtedly this is due to our persistent habit of reckoning wealth in numerical terms, using the medium of exchange as the reality instead of the vastly more complex thing, the actual wealth of which it is the symbol.

THE WORLD OF TOMORROW

E. B. WHITE, 1939

The New Yorker *essayist visited the New York World's Fair in May 1939 and filed a mixed report, from which these sly excerpts.*

I made a few notes at the Fair, a few hints of what you may expect of To-morrow, its appointments, its characteristics.

In Tomorrow, people and objects are lit not from above but from below. Trees are lit from below. Even the cow on the rotolactor appears to be lit from below—the buried flood lamp illuminates the distended udder.

In Tomorrow one voice does for all. But it is a little unsure of itself; it keeps testing itself; it says, "Hello! One, two, three, four. Hello! One, two, three, four."

Rugs do not slip in Tomorrow, and the bassinets of newborn infants are wired against kidnappers.

There is no talking back in Tomorrow. You are expected to take it or leave it alone. There are sailors there (which makes you feel less lonely) and the sound of music.

The living room of Tomorrow contains the following objects: a broad-loom carpet, artificial carnations, a television radio victrola incessantly pro-ducing an image of someone or something that is somewhere else, a glass bird, a chrome steel lamp, a terracotta zebra, some veneered book cabinets containing no visible books, another cabinet out of which a small newspaper slowly pours in a never-ending ribbon, and a small plush love seat in the shape of a new moon.

In Tomorrow, most sounds are not the sounds themselves but a memory of sounds, or an electrification. In the case of the cow, the moo will come to you not from the cow but from a small aperture above your head.

Tomorrow is a little on the expensive side. I checked this with my cab-driver in Manhattan to make sure. He was full of praise about the Fair but said he hadn't seen it and might, in fact, never see it. "I hack out there, but I got it figured that for me and the wife to go all through and do it right—no

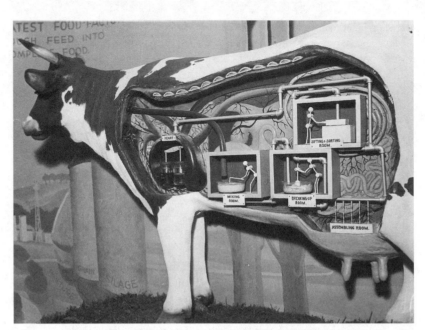

▶ The machine as metaphor: mechanical cow at Century of Progress, Chicago, 1933.

cheapskate stuff—it would break the hell out of a five-dollar bill. In my racket, I can't afford it."

Tomorrow does not smell. The World's Fair of 1939 has taken the body odor out of man, among other things. It is all quite impersonal, this dream. The country fair manages better, where you can hang over the rail at the ox-pulling and smell the ox. It's not only that the sailors can't get at the girls through the glass, but even as wholesome an exhibit as Swift's Premium Bacon produces twenty lovesick maidens in a glass pit hermetically sealed from the ultimate consumer. . . .

So (as the voice says) man dreams on. And the dream is still a contradiction and an enigma—the biologist peeping at bacteria through his microscope, the sailor peeping at the strip queen through binoculars, the eyes so watchful, and the hopes so high. Out in the honky-tonk section, in front of the Amazon show, where the ladies exposed one breast in deference to the fleet, kept one concealed in deference to [Fair director Grover] Whalen, there was an automaton—a giant man in white tie and tails, with enormous rubber hands. At the start of each show, while the barker was drumming up trade, a couple of the girls would come outside and sit in the robot's lap. The effect was peculiarly lascivious—the extra-size man, exploring with his gigantic rubber hands the breasts of the little girls, the girls with their own small hands (by comparison so small, by comparison so terribly real) restrainingly on his, to check the unthinkable impact of his mechanical passion. Here was the Fair, all fairs, in pantomime; and here the strange mixed dream that made the Fair: the heroic man, bloodless and perfect and enormous, created in his own image, and in his hand (rubber, aseptic) the literal desire, the warm and living breast.

TECHNOLOGY IS
RELATIVELY NEUTRAL

TEMPORARY NATIONAL ECONOMIC
COMMITTEE, 1940

Americans struggled in the Depression years to understand the painful mystery of widespread unemployment. Many blamed technology for displacing workers. A congressional committee investigating technology and the concentration of economic power in 1940 assembled this table from U.S. census information. It demonstrates that, to the contrary, technology had created millions of new jobs, and charts the growth industries of an earlier era. Some—block ice, typewriters, asbestos, phonographs—have already all but vanished.

The importance of new industries, which are being constantly created and developed by advancement in science, invention, and technology as sources of new opportunities for employment of labor, is shown in the following table. There it appears that 18 new manufacturing industries alone, which came into existence since 1879, absorbed almost one-seventh of all the labor employed in manufacturing in 1929.

Industry	*Average Number of Wage Earners in 1929*
Electrical machinery, apparatus, and supplies	328,722
Motor vehicles, not including motorcycles	226,116
Motor vehicle bodies and parts	221,332
Rubber tires and inner tubes	83,263
Manufacture of gasoline	39,411
Rayon and allied products	39,106
Manufactured ice	32,184
Aluminum manufactures	21,210
Typewriters and parts	16,945
Refrigerators, mechanical	16,883
Cash registers and adding and computing machines	16,840
Oil, cake and meal, cottonseed	15,825
Aircraft and parts	14,710

Phonographs	14,416
Photographic apparatus and materials	12,967
Motion pict. apparatus except for projection in theaters	10,784
Asbestos products	8,092
Fountain pens	4,508
TOTAL, 18 NEW INDUSTRIES	1,123,314
TOTAL, ALL MFG. INDUSTRIES	8,838,743

In its final report in 1941, the committee explored the influence of technology on the economy, its historic connection to capitalism and its political neutrality.

Technology refers to the use of physical things to attain results which human hands and bodies unaided are incapable of achieving. In this sense, technology reaches back to the beginnings of human culture, has always played a highly significant role in social evolution and will remain a mainstay of civilization.

If the present period is peculiarly technological, it is not solely because of its own technical creations, important as these are, but because it is the recipient of an accumulation of technical resources that have been piling up through the centuries. What was once a thin thread in the evolution of culture has now become a gigantic strand, binding and sustaining a colossal economic system, and transmitting its releases and tensions throughout the entire body of contemporary culture. Technology, an historical development without general plan or purpose, has come to dominate the pattern of modern living. Benefits and disadvantages are consequently intermingled and interspersed within the complexities of the current situation, and await analysis and evaluation from some central viewpoint grounded in considerations of economic health and human well-being. . . .

Capitalism and technology are clearly distinguishable, although the two have been intimately associated during the period of their common history. Prior to this, technology had made gains without capitalist support, and elsewhere capitalism has existed in cultures with a relatively undeveloped technology. In the historical association of the two, each conditioned the other. The capitalist accumulated the necessary funds wherewith technology was exploited, broadened the range of its effects, hastened the productive process for a more rapid return of income, and found new areas for technical applications; the inventor devised new methods and materials of production and thought of new things to be produced. There was a conflict of interest between the two at times, but since the capitalist was the partner with the controlling resources, his will in the long run had the right-of-way.

The fundamental interest of the capitalist in the utilization of technology was private profit. Capitalism has had the effect of speeding up the de-

velopment of machinery and of promoting incessant changes and improvements. This rapid tempo of change, required by the process of competitive
capitalism itself, has undoubtedly done much to make the modern age an
unstable and dynamic one. The rapidity of its technical advances under capitalist stimulus has been so great that the controls necessary to their better
social employment have often lagged behind, so that the machine is today
still employed for medieval and even barbarous purposes. War is not the
only manifestation of this kind, however. Any modern invention of consequence will be found exploited in some measure by those whose social conscience is relatively primitive. Common modes of recreation, prevailing
styles of advertising, the inability of the law to protect the consumer from
flagrant abuses, callous and even inhuman management of the technical resources of the economy itself, all testify to a holdover of crude forms of
competitive individualism which in spirit contradict the more civilized aspirations of science, invention, and engineering.

Modern capitalism is not a planned economy. It grew out of history by
reason of the short-run self-interested purposes of innumerable enterpris-
ers. It derived its peculiar characteristics from the conditions and circumstances of its origin and development. But it did in time evolve an ideology
with the help of great economic thinkers who gave it a social meaning and
justification, and it had a certain congeniality with the principles of historical democracy. . . .

Technology is relatively neutral; the more dynamic forces lie within the
economic system that controls it. If this system is socially wholesome, its
employment of technology will be socially advantageous; if it is less than
this, its influence will be uneven—rendering benefits here, disadvantages
there, as the prevailing cluster of conflicting economic forces may decide.

AIR POWER WILL
REPLACE SEA POWER

JOHN PHILIPS CRANWELL, 1940

The factor which seems to have been disregarded in the discussions of
the effect of air attack on Sea Power is that Air Power, because the
plane is a means of carrying on commerce, is so closely akin to Sea Power as
to be nearly identical with it. Aircraft and seacraft are basically vehicles of
transportation. So long as the commerce of the world is carried exclusively
in ships which sail on the surface of the waters, Sea Power will remain the
most potent single factor in the conflicts of maritime states. But as aerial

commerce increases in quantity and by that much decreases seaborne trade, the influence of Sea Power on history will be replaced by the influence of Air Power on history. The logical conclusion is that in the end Air Power will replace Sea Power, not because military aircraft will drive navies from the sea, but because commerce will leave the water and take to the air. . . .

It will be urged of course that years will pass before the commerce of the sea will give place to that of the air, and that until that time comes Sea Power remains the controlling factor in the lives of states which depend on foreign trade for supplies and raw materials. True; but how many years? It must be remembered that in little more than a quarter of a century of flying both great oceans have been spanned not only by isolated planes with crews of adventurers, but also by regular services which carry passengers, mail, and express on definite schedules and over routes which, if not so well traveled as the shipping lanes below, are at least as clearly defined. It must be remembered too that planes built in this country for England are being flown there and not carried by ship. Is it too much to expect that, with the rapid growth in the size, speed, and carrying capacity of aircraft, within another twenty-five years nearly all the passengers and much freight will go by air? Twenty-five years is not a long time in the lives of nations. Even though the liners of the air never attain the size of the *Queen Mary* or the *Normandie*, even though they never reach that of the humble freighter, they already make much faster and, therefore, more frequent trips. Numbers can offset size to a large extent, and when to numbers is added speed, the advantage now held by the large ship over the relatively smaller plane begins to disappear.

HOW THEY COMB THEIR HAIR

ORRIN DUNLAP, 1940

NBC televised the 1940 national convention of the Republican party. A New York Times correspondent saw immediately that the new medium would change politics.

Sincerity of the tongue and facial expression gain in importance. . . . Naturalness is the keystone of success. . . . The sly, flamboyant or leather-lunged spellbinder has no place on the air. Sincerity, dignity, friendliness and clear speech . . . are the secrets of a winning telecast. More than ever, the politician must picture himself in the living room, chatting heart-to-heart with a neighbor. . . . How they comb their hair, how they smile and how they loop their necktie become new factors in politics.

A RADIOACTIVE SUPERBOMB

OTTO FRISCH AND RUDOLF PEIERLS,

1940

*Two Jewish physicists escaped from Nazi Germany, working at the University of
Birmingham in Great Britain, realized in the winter of 1939–1940 that the recent
discovery of nuclear fission made it possible to create an atomic bomb. Frightened
that Germany might already be at work on such a fearsome weapon, Frisch and
Peierls prepared a two-part report for the British government which began the
bureaucratic process that led to the invention and manufacture of the first atomic
bombs jointly by the United States and Great Britain in time to end the Second
World War in August 1945. The second part, "On the Construction of a 'Super-
Bomb'; Based on a Nuclear Chain Reaction in Uranium," is highly technical; the
first, "Memorandum on the Properties of a Radioactive 'Super-bomb,'" transcribed
here, summarizes the probable advantages and disadvantages of building such a
momentous weapon. The memorandum is especially noteworthy for its imaginative
realism: it anticipates fallout, civilian casualties and the defense that came to be
known as deterrence. In the long run, the technology first sketched in the Frisch-
Peierls report almost certainly catalyzed an end to world-scale war by making
such convulsive conflict suicidal—surely the most important consequence of
technological progress in the twentieth century.*

The attached detailed report concerns the possibility of constructing a 'super-bomb' which utilizes the energy stored in atomic nuclei as a source
of energy. The energy liberated in the explosion of such a super-bomb is about
the same as that produced by the explosion of 1,000 tons of dynamite. This energy is liberated in a small volume, in which it will, for an instant, produce a
temperature comparable to that in the interior of the sun. The blast from such
an explosion would destroy life in a wide area. The size of this area is difficult
to estimate, but it will probably cover the center of a big city.

In addition, some part of the energy set free by the bomb goes to produce radioactive substances, and these will emit very powerful and dangerous radiations. The effects of these radiations is greatest immediately after
the explosion, but it decays only gradually and even for days after the explosion any person entering the affected area will be killed.

Some of this radioactivity will be carried along with the wind and will
spread the contamination; several miles downwind this may kill people. . . .

We do not feel competent to discuss the strategic value of such a bomb,
but the following conclusions seem certain:

1. As a weapon, the super-bomb would be practically irresistible. There
is no material or structure that could be expected to resist the force of the

explosion. If one thinks of using the bomb for breaking through a line of fortifications, it should be kept in mind that the radioactive radiations will prevent anyone from approaching the affected territory for several days; they will equally prevent defenders from reoccupying the affected positions. The advantage would lie with the side which can determine most accurately just when it is safe to re-enter the area; this is likely to be the aggressor, who knows the location of the bomb in advance.

2. Owing to the spread of radioactive substances with the wind, the bomb could probably not be used without killing large numbers of civilians, and this may make it unsuitable as a weapon for use by this country. (Use as a depth charge near a naval base suggests itself, but even there it is likely that it would cause great loss of civilian life by flooding and by the radioactive radiations.)

3. We have no information that the same idea has also occurred to other scientists but since all the theoretical data bearing on this problem are published, it is quite conceivable that Germany is, in fact, developing this weapon. . . .

4. If one works on the assumption that Germany is, or will be, in the possession of this weapon, it must be realized that no shelters are available that would be effective and could be used on a large scale. The most effective reply would be a counter-threat with a similar bomb. Therefore it seems to us important to start productions as soon and as rapidly as possible, even if it is not intended to use the bomb as a means of attack. Since the separation of the necessary amount of uranium is, in the most favorable circumstances, a matter of several months, it would obviously be too late to start production when such a bomb is known to be in the hands of Germany, and the matter seems, therefore, very urgent.

MACHINERY HAS DESTROYED THE PEACE

Roy Helton, 1941

Heralding the approach of an "anti-industrial revolution," this essayist bitterly excoriates "machinery"—technology—for driving the world to war but inconsistently distinguishes technology he considers benevolent, such as medicine and public health, from technology he considers malign. This essay was published in December 1941, the same month the United States entered the Second World War following the Japanese attack on Pearl Harbor.

Productive machinery is, in a sense, the crowning achievement of half a million years of desperate struggle and of infinite craft and contrivance.

▶ Technology applied to mass killing: crematoria at Buchenwald death camp.

Yet it is a jewel with but a single facet. As distinguished from a work of art, with which it is comparable as representing a fruition of ages of effort, productive machinery expresses only one trait in man, and that one trait is his craving for power. Power is its achievement. Power is what it is for—power over nature, power over people, power over time, and power over space.

Now power is good, but so is love good, and so is beauty good, and so is quiet good, and so is truth good, but none of them is adequate alone as a basis for civilization. Yet for the past hundred and fifty years we have been attempting to build a way of life on but one of man's many hungers, and today we are compelled to face the fact that the attempt has failed. For productive machinery has not drained off and sublimated man's hunger for power. It has fed that hunger, and lent it teeth and claws, and also has sanctified it. . . .

A man with a sword who kills an enemy on the field of battle can assign to the act only one of two responsibilities—his own skill or his cause. But a man with a machine gun can visibly wreak such a destruction of human life and create such widespread havoc and suffering that it is no longer adequate for him, being human himself, to embody the responsibility for what he is doing. So it is not the machine gunner but the machine gun, as symbolizing civilization, and the whole intellectual and technical effort of humanity, which kills fifty men a minute on a crowded front, with only one man's finger on the trigger. The gunner has, in fact, less of a moral problem to haunt him than the wielder of any single bayonet aimed at any single breast.

The bombing of civilians is a comparatively new phase of war. Does it represent an increase of savagery, a reversion to ancestral cruelty? Not at all. No one feels that. It represents merely an increase in man's ability to transfer the effect of machinery back to the ideals of his civilization.

Every day in our papers we read dreadful things about war. What Ger many is doing to the stored culture and beauty and to the life and happiness of a land once known as Merrie England. And what England is doing in return to the coast cities of Germany and France. The means by which these wonders of destruction are achieved are the supreme triumphs of our will to power. Yet every bomb which falls is a blow at the civilization of power, for it is not merely churches and houses and factories that are being destroyed, but also our faith in the way of life which has finally led to such atrocity. . . .

The great and enduring triumphs of modern science are not products of the hunger for power. They are products of humanity, of a struggle not against space, but against suffering, not for speed or sensation, but for health and sensibility. The Anti-Industrial Revolution must accept all the industrialism needed for sanitation, for medical progress, and for personal hygiene, and go on from there to create a standard of living based on motions not solely rotatory. I believe in the desire for freedom as inherent in the nature of man, and I consequently believe that this revolution will succeed, and that with its success the motives which gave rise to communism and Nazism in the first half of the twentieth century and to the orgy of wars and hatreds which have occupied our lives will die of malnutrition. So soon as modern man perceives that if machine ideals continue to dominate his world the large advantages he has planned for himself must be waived for collective power and collective competition, this change may proceed at accelerated speed. . . .

Machinery has destroyed the peace of the earth. Machinery has led civilization into a feverish decadence. Machinery cannot cure the diseases created by the nature of the impulses which have given it this awful power to disorganize the whole life of man. The only cure lies in the discipline of machinery and its relegation to a minor function.

Some industrialists and many inventors and technicians, who have a sacerdotal interest in industrialism *per se*, will deny every premise on which these conclusions are based, just as they are compelled to repeal the law of diminishing returns in planning their technological future. Industry, they tell us, will yet save man and make for us a wonderful world—a kind of Santa Claus's toy shop—for that is the complete nature of the dream. Mechanical industry will save man. But from what? From the elements? That has been done. From hunger? That also—and both by simpler devices—and by them nearly as dependably for the population they were adapted to. It is not from these dangers man is to be saved by machinery. But obviously from machinery itself. Machinery will outwit machinery and do better sometime than it has done in the world it has made and possessed.

By what? By new and more devilish devices? By new and more curious patents to tie our shoes in a new way, or to dress us in seamless garments?

Or by what? Faster motor cars? Airplanes on every roof, bearing bonbons and not bombs? But why not bombs? What has industrialism yet done to transform and mollify the human spirit? It has had fifty clear triumphant years to show its worth to the human spirit. I am casting the unpopular vote of No Confidence in its result.

BENEFICIAL INVENTIONS AND DIABOLICAL PURPOSES

ORVILLE WRIGHT, 1942

Writing to Henry Ford, one of the inventors of the airplane attributes his invention's destructive potential to human villainy rather than inherent versatility—an early and extended version of the popular slogan "Guns don't kill people: people kill people."

I quite agree with you that the aeroplane will be our main reliance in restoring peace to the World. The use of a beneficial invention for diabolical purposes, as in the present war, calls to mind a story I heard related fifty years ago by a missionary back from China. The Chinese gathered their grain by seizing the grain with one hand and cutting off the stalks with shears held in the other. The missionary thought he would be making a great contribution to their welfare by introducing the use of the scythe. More grain could be cut in an hour with a scythe than could be cut in a day with the shears. So he had a scythe shipped to him from America, and invited the natives to come to see it demonstrated. The Chinese of the neighborhood turned out in a crowd. The demonstration was a great success, and the spectators were impressed and very enthusiastic. The next morning, however, a delegation came to see the missionary. The scythe must be destroyed at once. What, they said, if it should fall into the hands of thieves; a whole field could be cut and carried away in a single night. So the use of the scythe could not be adopted! Apparently it didn't occur to them that the way to avoid such a situation would be to stop the thieves instead of to stop the use of the scythe.

> Give them the third best to go on. The best never comes, and the second best comes too late.
>
> ROBERT WATSON-WATT, BRITISH RADAR
> PIONEER, EXPLAINING HIS RADAR-
> DEVELOPMENT POLICY

SCIENCE AT WAR: I

V. B. WIGGLESWORTH

During the war, pure scientists found themselves working on practical problems—that is, on technology. The experience was enlightening, a biologist remembers, if not always comfortable.

In the pure science to which they were accustomed, if they were unable to solve problem A they could turn to problem B, and while studying this with perhaps small prospect of success they might suddenly come across a clue to the solution of problem C. But now they must find a solution to problem A, and problem A alone, and there was no escape. Furthermore, there proved to be tiresome and unexpected rules which made the game unnecessarily difficult: some solutions were barred because there was not enough of the raw material available: others were barred because the materials required were too costly; and yet others were excluded because they might constitute a danger to human life or health. In short, they made the discovery that applied biology is not "biology for the less intelligent," it is a totally different subject requiring a totally different attitude of mind.

SCIENCE AT WAR: II

J. ROBERT OPPENHEIMER

The director of the laboratory at Los Alamos where the first atomic bombs were designed observed a sharper conflict between the scientist and the developer—one that his own executive skills enabled him to avoid.

The scientist is irritated by the practical preoccupations of the man concerned with development, and the man concerned with development thinks that the scientist is lazy and of no account and is not doing a real job anyway. Therefore the laboratory very soon gets to be all one thing or all the other.

▶ Prewar home electronics: a vacuum-tube "portable" radio with molded plastic case removed.

PLASTICS GO TO WAR

JOSEPH L. NICHOLSON AND
GEORGE R. LEIGHTON, 1942

Plastics had been touted as the wave of the future and trivialized as the stuff of gadgets and junk, these observers note, but their greater significance lay in their syntheticism: to some degree they could be constructed to order. The war found use for them in quantity.

It is conceivable that plastics may one day become a dominant material, just as steel did in the immediate past. Or, to put it more strictly (since there are so many kinds of plastics for so many purposes), they may become dominant as metals in general have been dominant from times remote.

But aside from any potential revolutionary *function* or performance, synthetics are revolutionary in their origin. They are substance whose molecules are not constructed by nature but constructed to order by man. In the older substances nature arranged the atoms into molecules. By combining the atoms of disparate substances, as of the gases hydrogen and oxygen, nature formed a new substance, water. Man, in synthetics, does as nature has always done: he takes all the atoms found in hydrogen, oxygen, and carbon and makes combinations of them different from those in nature, to form new molecules not found in nature. In this way new substances are created; by combining them it is possible to form one material that is lighter, another that is stronger, another that will stand heat better and so on, than any substance yet found in nature. Man can make a list of the properties which he would like to find embodied in a new material, and—within limits—he can custom-build that material as he never could before in all history. Without some recognition of this radical difference between synthetics and the older materials, without some understanding of what it means for man to be free from dependence on the materials that nature gives him—whether ores, wood, plants, or what not—it is not possible to grasp the significance that plastics may have for our future. . . .

It was the war that, in a sort of public way, proved the plastics' case. When it first dawned on the American manufacturer that a large proportion of the nation's raw materials were going to be diverted from civilian use to the needs of the war machine he looked to the plastics industry for his salvation. He had been told that plastics were made from such simple ingredients as "air, coal, and water," that they were wonder materials—look at what a giant industry the manufacture of synthetic fibers had become!—and that the industry was waiting with open arms to receive him.

For a while the industry itself may have believed this too, but not for long. The shortage of the plastics raw materials was as acute as the metals shortage; the materials were necessary for smokeless powder and other essentials. Priority control was placed over formaldehyde, phenol, acetic acid, numerous plasticizers, and eventually over some of the plastics themselves. And even when plastics were comparatively plentiful it became almost impossible to obtain steel for new molds without a high priority number. . . .

Because of the war plastics have been turned to new uses and the adaptability of plastics demonstrated all over again. All the glamour-magic talk may have been very fine from the advertising point of view; now the facts of the adaptability of plastics are really being driven home.

. . . The Quartermaster Corps is using plastics for canteen closures, the Ordnance Department is using plastics for M52 trench mortar fuses, for pistol grips and bomb detonators, the Navy is using enormous amounts of plastics for insulating electrical parts on ships and in aircraft. Both the Army and the Navy use plastics in almost every form in their aircraft—as fabricated sheets in bomber noses and gun turrets, as molded parts for control boards and instrument housings, as extruded strips for tubing, as foil for electrical insulation, as resin-impregnated plywood or canvas for structural parts, and as resin lacquers for finishing. The fact is that the number of applications of plastics in aircraft construction is increasing so rapidly that the all-plastics plane may precede the all-plastics automobile. All these developments are important now; the fact that they open new ways for the peacetime use of plastics is important also.

There is no use in saying the sky's the limit for plastics; they cannot be used for everything. But that the advance of plastics will be extraordinary is certain. The industry is out of its infancy and the effort expended in research and investigation is steadily mounting. The annual summation of this research, both in chemistry and in the technics of shaping, as shown, for example, in *Plastics Catalog*, the industry's annual, is evidence of the vigorous development in the whole field. There is scarcely an expanding force in American industry in which plastics are not involved: communication—radio, telephone, and television; in the automotive field, and in aviation.

The use of plastic coatings to make other materials corrosion-proof—as in metal cans and water tanks—is already accepted. The wider use of extruded plastic pipe is a certainty. Still other possibilities are opened with the molding of ever larger shapes. Coffins, for example, have been made of plastic in England. What this trend may mean in furniture making, provided cost is reduced, is obvious. Plastic canes for weaving seat coverings that may be washed off with soap and water have proven more durable than the natural cane coverings used in New York subway cars.

It is extremely difficult in enumerating examples to avoid giving the impression that after the war we may look for an all-plastics world. There is no such possibility. In many cases where plastics could be used, trial will show that for one reason or another, the application is impractical. This fact cannot obscure still another fact: a way has been found to devise materials to meet precise specifications. That is a great step toward the practical mastery of the physical world we live in.

RATIONS

HARLAND MANCHESTER, 1943

Millions of U.S. soldiers and marines experienced the application of new technology to food packaging during the Second World War. This Atlantic Monthly writer's chronicle of the adaptation of military gear to war conditions for "Private Jones, U. S. Army," is typically enthusiastic and uncritical. The consumers of military rations—especially the infamous foxhole K rations—would report more ambivalent responses to foods tortured with dehydration and compression into insoluble Lilliputian bricks.

Perhaps the most spectacular job of the Quartermaster Corps is the provision of nutritious, compact, appetizing rations for troops all the way from base camps to lonely outposts. Complete meals wrapped in immer-sion-proof, indestructible packages have been designed for various climates, and may be dropped by parachute to an arctic outpost or floated ashore on the tide to a tropical island. . . .

A varied diet helps to keep men from flying off the handle. All the ration units packaged by the Quartermaster Corps are thoroughly tested by groups of volunteers, and one of the things the food experts watch for is the kind of grouch that develops when a man gets tired of his meals.

Cigarettes and candy rank high among morale factors, yet during World War I men had to buy them at canteens, and went without when their money was gone. It was Quartermaster General E. B. Gregory who conceived the idea of *issuing* these things to men at the front. Cigarettes and candy are included in each food ration package, and there is also an accessory packet containing an assortment large enough to supply two hundred men for a day.

Ration packages for all purposes, developed and tested by the research group, are now turned out in million lots by the food industries. There are special rations for the jungle and for high-altitude flying, and low-protein rations for lifeboats, since proteins absorb water and dehydrate the system. The farther a man advances toward the firing line, the lighter and more concentrated his rations become. Fresh foods make up much of the "A" rations served in home training camps; canned foods, supplemented by local produce, comprise the "B" ration served in permanent mess halls abroad, and other rations are designed for the individual soldier in action who carries his food on his back.

The "K" ration—the most compact of the lot—is composed of three cellophane-wrapped packages of pocket size, labeled "Breakfast," "Dinner," and "Supper," and weighing altogether about 2 1/2 pounds. Each of the packages contains meat or a protein substitute, two kinds of biscuits, a drink made from soluble powder, a concentrated sweet, and a few cigarettes. A man may have ham and eggs and coffee for breakfast, cheese sandwiches and lemonade at noon, and at night a cup of bouillon and meat of one of eight different varieties.

These pocket meals are used as emergency rations for troops on their own, but tests indicate that, in a pinch, men could live on them indefinitely without getting malnutrition ailments. Military men say that this light ration more than doubles the effective fighting range of an independent detachment.

Dehydration is one of the most effective weapons in the Quartermas-ter Corps's battle of rations. Only two years ago the process was experimental, but this year the Army will use an estimated 200,000 tons of the quick-dried foods. The water extracted from this food weighs a million tons—cargo for a hundred 10,000-ton freighters which now can be used for munitions and men. Dehydrated foods used by the Army to date include milk, eggs, cheese, vegetables, soups, and beverage powders; meats and other items are on the way.

Dehydration saves weight bulk, but a further step, compression, squeezes out the air and adds an important new saving in shipping space. A block of dehydrated, compressed potatoes the size of a shoebox will make mashed potatoes for fifty men. Cranberries squeezed into the size of a small building brick will make sauce for a hundred, and experiments are being made with strawberries and other fruits. In some cases, idle machines which once made civilian goods have been converted to food compression. For instance, a Massachusetts tile-manufacturing firm uses its equipment to cram five dozen eggs into a space the size of a bar of laundry soap. After dehydration and compression, the food bricks are machine-wrapped in cellophane and sealed by heat to keep out air and moisture.

Even more compact and nutritious "pocket dinners" are on the way, and their future is not limited to warfare. They may feed starved millions after the victory, and they will offer a means of quick relief to areas devastated by flood or famine.

THE BUSINESS OF
THE FUTURE

EDWIN H. LAND, 1944

The founder of Polaroid presented this paper at a forum on the future of industrial research.

I believe quite simply that the small company of the future will be as much a research organization as it is a manufacturing company, and that this new kind of company is the frontier for the next generation.

The business of the future will be a scientific, social and economic unit. It will be vigorously creative in pure science, where its contributions will compare with those of universities. Indeed, it will be expected that the career of the pure scientist will be as much in the corporation laboratory as in the university.

Internally this business will be a new type of social unit. There will be a different kind of boundary between management and labor. All will regard themselves as *labor* in the sense of having as their common purpose learning new things and applying that knowledge for public welfare. The machinist will be proud of and informed about the company's scientific advances; the scientist will enjoy the reduction to practice of his basic perceptions.

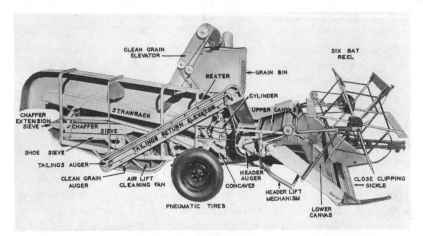

CLEAN GRAIN ELEVATOR — SIX BAT REEL — GRAIN BIN — BEATER — CYLINDER — CHAFFER EXTENSION SIEVE — STRAWRACK — CHAFFER SIEVE — TAILINGS RETURN ELEVATOR — UPPER CANVAS — SHOE SIEVE — TAILINGS AUGER — CLEAN GRAIN AUGER — AIR LIFT CLEANING FAN — CONCAVES — HEADER AUGER — CLOSE CLIPPING SICKLE — HEADER LIFT MECHANISM — PNEUMATIC TIRES — LOWER CANVAS

▸ The combine, combining reaping and threshing in one mobile system, automated harvesting after World War II and emptied the countryside of farmers.

Economically such small scientific manufacturing companies can, I believe, carry us quickly into the next and best phase of the Industrial Revolution. . . .

And year by year our national scene would change in the way, I think, all Americans dream of. Each individual will be a member of a group small enough so that he feels a full participant in the purpose and activity of the group. His voice will be heard and his individuality recognized. . . . These small groups will be located on the periphery of large cities and distributed throughout the countryside. Thus, the worst phase of the Industrial Revolution—the slums of Charles Dickens which still disgrace nearly every one of our large cities—will be gone.

. . . This new company will start by contemplating all of the recent advances in pure science and in engineering. Its staff will be alive to the significance of newly available polyamide molecules, the cyclotron, radar technics, the details of new processes for color photography, and recent advances in enzymology. A group of fifty good scientists contemplating one of these fields and inspired by curiosity about them and a determination to make something new and useful, can invent and develop an important new field in about two years. This new field will be a monopoly for the group—a monopoly in the best sense of the word—because it will derive from justifiable patents on important inventions, and from know-how deliberately acquired by the group. . . .

President [James Bryant] Conant of Harvard urged a few weeks ago . . . that industry not destroy the sources from which they derive their good men. President Conant fears that industry may offer such high salaries to competent scientists that they will leave the universities. It seems to me that the sensible response for industry to make is this: "We will not take the good men away from the universities, but we should like to share them. We feel that the industrial environment can be as stimulating to the development of pure science as the university has been. We should like to bring into industry the kind of professional ethics that characterize the relationship between pure scientists. We should like to have knowledge of the scientific method permeate our organizations. Let us share your scientists and we will teach them many new aptitudes."

Any honest scientist will recognize that there are fads and trends in the pursuit of pure science and that many competent young men waste many years in activities in which they learn little and contribute little. Industry can provide a much larger field of inquiry for pure science and much greater human stimulus to many of the young scientists than are now provided by the university. In short, a continuum between pure science in the university and pure science in industry should stimulate and enrich our social system.

Finally, the small business that incorporates its own research department is adaptable, mobile, socially integrated and profitable. It finds new markets for big business and molds the product of big business to the transient demands of an evolving technical society. The small scientific business is individualized yet organized, free from political domination but a successful contributor to the solution of our great social problems.

PROGRESS AND
THE SERVANT PROBLEM

ALLAN G. B. FISHER, 1944

A New Zealand economist attached to his country's legation in Washington during the Second World War considers an intricate impediment to material progress in the world.

M onopolistic resistance to changes in economic structure is a commonplace. All of us have heard of the purchase and suppression of inventions by corporations whose product is threatened; all of us know how unions have spread-the-work and feather-bedded useless jobs. But the resistances go a great deal deeper. For the higher income groups, material progress of a general character often demands significant changes in some of the elements of real income to which traditionally they have come to attach great importance. These changes are annoying to them, and if they feel that the loss of certain customary personal conveniences is inadequately compensated by the increased opportunities for the enjoyment of new things, they will probably add the weight of their influence to the other resistances to structural adaptation.

Personal services, for example, traditionally have been paid at low rates. Cheap personal services are always most abundantly available in countries with a low general level of income. Unskilled labor, male or female, is relatively plentiful in such communities, whereas in wealthier countries there are numerous alternative outlets for low-paid labor, whose price thus tends to rise. In normal times the proportion of working women employed in domestic service in the United States is only half the corresponding proportion in Great Britain. This is not a matter of chance, but an inexorable consequence of the more rapid material progress of this country. If the general income level rises, wealthy people accustomed to a great deal of personal service soon begin to complain that they cannot get what they want, or indeed, as they are apt to put it, what they ought to have. During the war we have heard of people regretting the "good old days" of the Great Depression, when personal services were abundant and cheap! Such complaints are really identical with complaints that production is becoming more efficient, for that is the most significant explanation of rising incomes. The amount of time spent by members of the middle classes in all highly developed countries in discussions of "the servant problem" is merely a reflection of a widespread social outlook which finds it difficult to stomach the inevitable consequences of material progress. From many people with great economic and political influence, material progress demands significant changes in their mode of life. Some would find the changes, once made, quite agreeable. But timid people often dislike the prospect of being obliged to make them. If they successfully resist the structural changes which the economic system needs, they are at the same time checking the general improvement in standards of living which ought to be the normal consequence of scientific and technical progress.

THE RADIO AGE: II

BERNARD B. SMITH, 1945

By the middle of the Second World War, the "insidious dangers of radio advertising" that Printer's Ink denounced in 1923 had become golden opportunities for investors, who competed with each other for the privilege of controlling a unique frequency on the limited radio band (not everyone, it seems, was preoccupied with fighting the war). Here a public-interest lawyer describes consequences for the commons that politicians and reformers continue to debate today.

During 1944 thirty-two radio stations sold for a total of more than ten million dollars. Most of them were small stations. Most of them sold at prices which represented fantastically high profits—so high, indeed, that the Federal Communications Commission got worried and asked Congress what to do about it. . . .

The reason for these high prices is clear enough. The sale of radio time to advertisers and others is big business, and has been getting bigger and bigger. In 1927 the gross time sales of all broadcasters were less than $5 million; in 1932 (in spite of the depression) they were nearly $62 million; in 1937 they had jumped to more than $144 million; and in 1942 they totaled almost $200 million. From 1942 to 1944 they almost doubled, jumping the figure to nearly $400 million last year. . . .

But radio frequencies cannot be sold. The Supreme Court has held that *a license to broadcast on a radio frequency does not constitute property.* Congress, by the Federal Communications Act of 1934, set up the Federal Communications Commission specifically "to maintain the control of the United States over all the channels of interstate and foreign radio transmission; and to provide for the use of such channels, *but not the ownership thereof,* by persons for limited periods of time, under licenses granted by Federal authority." (Italics ours.)

Congress provided that licenses to broadcast were to be granted free by the FCC "if public convenience, interest, or necessity will be served thereby," but not for a longer term than three years. It further provided that such licenses could not be transferred without FCC approval and could not be renewed except on the same terms which apply to the granting of original applications.

But the purchasers of radio stations know what they are doing when they pay huge prices for a frequency which cost the original licensee not a penny. They know that, whatever the law may be, the FCC almost invariably renews radio station licenses; that out of 9,000 renewal applications, 98 percent have been granted without so much as a hearing and only a handful of the others have been denied. They know, in other words, that for all practical purposes the purchase of a radio station gives them a perpetual right to a channel of radio transmission which in fact and in law belongs to the people of the United States. . . .

Few, if any, stations provide the percentage of free public service sustaining programs which they promised when they were applying for a frequency. If they did, it is unlikely that stations would sell for the inordinately high prices which have been troubling the FCC and Congress. For the plain truth is that the enormous increase in revenue from radio time sales—from $200 million to $400 million in two years—was brought about mainly not by charging higher prices for radio time but by dropping public service programs in favor of commercially sponsored ones. One by one the sustaining public service programs disappear or are shifted to undesirable hours when few listeners are on hand. In the typical week ending November 20, 1944, during the peak listening hours from 8:00 to 10:30 P.M., not a single program of this type was broadcast over either of the two principal networks [CBS and NBC]. . . .

As [FCC] Commissioner [Clifford] Durr said recently, in an article in the *Public Opinion Quarterly:*

> The only barriers to the complete occupation of the air by advertisers, and the consequent total elimination of public-service programs, are self-restraint on the part of the broadcasters and networks themselves—somewhat fortified, perhaps, by the complaints of their listeners—and the public-interest provisions of the Communications Act. . . .

At no time in history has an informed public opinion been more necessary than it is today, and the radio could be a powerful vehicle for public education and information. Surely the nation has a right to insist that those who are licensed to use its radio frequencies shall contribute to this end.

TIMBER!

ROY A. H. THOMPSON, 1945

If that postwar building boom fails to get under way quickly and on the sweeping scale which millions of job-hunters and overcrowded families are hoping for, one of the reasons will be a shortage of lumber. It is still the most important of all homebuilding materials, and America is running out of it fast.

During the war—and long before it, for that matter—the "inexhaustible virgin forests" which we learned about in our grade-school geography books have been exhausted at an alarming rate. In the thirty years before the war, almost 40 percent of all the nation's standing saw timber disappeared. After Pearl Harbor the rate of cutting rose sharply, because wood is a prime military resource. Some twelve hundred different items of equipment—ranging from barracks to photographic film—are made from wood, and altogether the armed forces use a greater tonnage of forest products than of steel. Consequently, the drain on our forests increased to about seventeen billion cubic feet a year, exceeding by 50 percent the annual growth.

We will come out of the war with considerably less than 100 million acres of virgin timber still standing, out of 462 million acres of potential commercial forest growing land. About a third of what is left lies in mountains so rugged and remote that it cannot be harvested economically—and still more is poor quality in comparison with the trees already gone. Most of the remaining high-quality virgin timber is located in a relatively small area of the Pacific Coast states, and can be hauled to the big eastern markets only at high cost.

Another dark touch is added to this already gloomy picture by the fact that wood is one material we simply can't do without. It goes into thousands of essential products—paper, clothing, automobile tires, railroads, many chemicals—and all the ingenuity of the plastics-inventors has not yet produced adequate substitutes. On the contrary, many of the new plastics are derived from wood. . . .

In my opinion, based on almost forty years experience in financing the lumber, paper, and wood-pulp industries, the blame for the destruction of America's forests does not rest on any particular shoulders. It is neither fair nor useful to berate the lumber companies or the wood-using industries. Until fairly recently, neither the government nor private businessmen knew either the extent or the utility of our forests. Nor did they realize back in 1880 that the population of the United States would almost treble within sixty years while the demand for wood products would jump even more rapidly. (How many industries—or government agencies—are now planning sixty years ahead?) Consequently, farmers cut trees lavishly for houses, barns, fuel, rail fences, or just to clear the land. Lumbermen cut equally lavishly, often clearing away every tree so that today some 77 million acres of once-rich woodland are ruined and produce virtually no second growth.

Tree farming, which is generally practiced in Europe in order to produce a steady, perpetual yield from nearly every forest, was almost unknown here until recently; and in the experience of the American lumberman it did not seem necessary. Always there was more and often better timber to be had a little further on. Many people believed—and still believe—that our forests were so vast that there was no need for worry or to mend our destructive methods of cutting. Even today some lumbermen like to point out that we deplete only about two percent of our standing raw timber for lumber every year. They forget that the drain is almost as much more in the cut for pulp, fuel, and other uses; that there are further heavy losses from fire, insects, and storms; and that much of our remaining timber cannot be reached at reasonable cost. And our fine virgin timber is being cut *five* times as fast as it grows. . . .

The future of the lumber industry looks far from promising. There is no question that the major timber stands of the Northwest which can be harvested at reasonable cost are now almost gone. I believe I am conservative in estimating that two-thirds of the sawmill capacity of Puget Sound will be dismantled in a relatively few years because of lack of logs. In the Columbia River Basin, the remaining forests will not maintain half the present

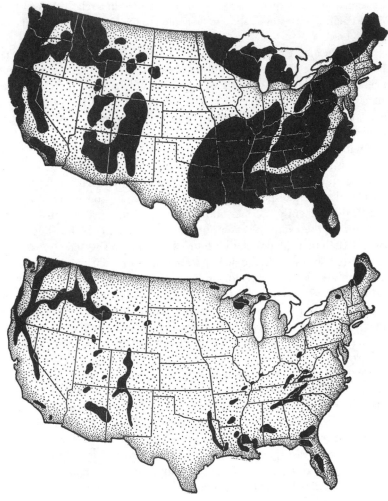

▶ Spending the inheritance: remaining virgin forests in 1850 and 1945.

mill capacity, and in the future Oregon's major cut will come from the southern part of the state where the timber is of poorer quality.

As the cutting of fine timber comes to an end, the reduced output of lower-grade lumber will be handled principally by the independent operators, using improved trucks and new logging techniques to assemble logs from small, isolated stands. Even they will be dependent to a considerable extent on government aid, in the form of road construction into the areas where private owners cannot afford to build access routes. . . .

We can salvage what is left of our forests, and restore much that has been destroyed; and we probably shall, simply because no country can afford to throw away a resource so vital to its industries and its national defense.

It is not a theoretical problem. Commercial forests can be grown for profit. They have been grown commercially in Europe for centuries. We know how long it takes a tree to grow from seed to harvest-size, and we know what trees grow best in different locations. We have skilled foresters, and we can train as many more as we need—as the [Depression-era] Civilian Conservation Corps program demonstrated. We know what it costs to plant trees and care for them; and although the harvest is long in coming, we know that the selling price shows a sturdy profit.

The main problem is financing. A study of the financial resources of the lumber, paper, and pulp industries as a whole proves conclusively that they do not have the capital to engage in scientific tree farming. If they had the large sums required, it would pay them to plant forests and manage them carefully for fifty to seventy years in the South and from eighty to a hundred years in the Northwest, before beginning to harvest the crop. . . . Only a handful of the richest operators have the money to embark on such long-term investments; most of them also have timber reserves to keep them going during the decades before the new plantings mature. These firms should be lauded and encouraged—but at best they can handle only a tiny fraction of the job. . . .

Even under the most favorable conditions, however, private enterprise could never shoulder the entire responsibility for conserving and rebuilding America's forests. A large part of the job must be handled by the government, because in many areas it alone can supply the funds and management personnel and stand the long wait for the harvest and the profits. As the states build up the necessary capital and organization, they properly can take over large acreages. But for the beginning—which is now—it is a federal job. . . .

If such a program were put into operation promptly, it would mean a stable population and a permanent foundation for local forest industries. In the long run, it would probably turn out to be a paying proposition in the saving of relief costs alone.

PREDICTIONS:
GEOSYNCHRONY

ARTHUR C. CLARKE, 1945

The physicist and science-fiction writer's October 1945 proposal of communications satellites in geosynchronous orbit, published in Wireless World, *is well known. This selection is taken from the first incarnation of the idea, a memo Clarke wrote on May 25, 1945. A historian who interviewed him in 1981 recalls: "His article was greeted with balanced measures of derision, amusement and skepticism. Indeed, Clarke . . . explained to me that he did not bother to attempt to patent the idea at that time, because he did not anticipate the invention of the transistor, which allowed the size of communication satellites to shrink to reasonable sizes and to be unmanned." Geosynchronous orbits are now called Clarke orbits.*

The space-station was originally conceived as a refueling depot for ships leaving the Earth. As such it may fill an important though transient role in the conquest of space, during the period when chemical fuels are employed. . . . However, there is at least one purpose for which the station is ideally suited and indeed has no practical alternative. This is the provision of world-wide ultra-high-frequency radio services, including television.

In the following discussion the world "television" will be used exclusively but it must be understood to cover all services using the u.h.f. spectrum and higher. It is probable that television may be among the least important of these as technical developments occur. Other examples are frequency modulation, facsimile (capable of transmitting 100,000 pages an hour), specialized scientific and business services and navigational aids.

Owing to bandwidth considerations television is restricted to the frequency range above 50 Mc/sec. and there is no doubt that very much higher frequencies will be used in the immediate future. . . . Waves of such frequencies are transmitted along quasi-optical paths and accordingly receiver and transmitter must lie not far from the line of sight. Although refraction increases the range, it is fair to say that the service radius for a television station is under 50 miles. . . . *As long as radio continues to be used for communication, this limitation will remain, as it is a fundamental and not a technical consideration.*

. . . [This problem] can be solved by the use of a chain of space-stations with an orbital period of 24 hours, which would require them to be at a distance of 42,000 Km from the center of the earth. There are a number of possible arrangements for such a chain but that shown is the simplest. The stations would lie in the earth's equatorial plane and would thus always remain fixed in the same spots in the sky, from the point of view of terrestrial observers. Unlike all other heavenly bodies they would never rise nor set. This would greatly simplify the use of directive receivers installed on the earth.

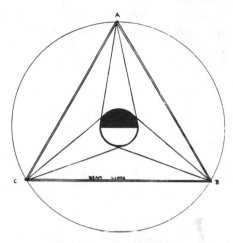

► Arthur C. Clarke's original drawing of a geosynchronous satellite system, May 25, 1945.

The following longitudes are provisionally suggested for the stations to provide the best service to the inhabited portions of the globe, though all parts of the planet will be covered.

30 E—Africa and Europe.
150 E—China and Oceana.
90 W—The Americas.

Each station would broadcast programs over about a third of the planet. Assuming the use of a frequency of 3,000 megacycles, a reflector only a few feet across would give a beam so directive that almost all the power would be concentrated on the earth. . . .

The stations would be connected with each other by very-narrow-beam, low-power links, probably working in the optical spectrum or near it, so that beams less than a degree wide could be produced.

The system would provide the following services which cannot be realized in any other manner:—

a) Simultaneous television broadcasts to the entire globe, including services to aircraft.

b) Relaying of programs between distant parts of the planet. . . .

The receiving equipment at the earth end would consist of small parabolas perhaps a foot in diameter with dipole pickup. . . . They would be aimed toward the station with the least zenithal distance and once adjusted need never be touched again. . . .

No communication development which can be imagined will render the chain of stations obsolete and since it fills what will eventually be an urgent need, its economic value will be enormous.

For completeness, other major uses of the station are listed below:—

a) *Research*—Astrophysical, Physical, Electronic.

These applications are obvious. The space-station would be justified on these grounds alone, as there are many experiments which can only be conducted above the atmosphere.

b) *Meteorological.*

The station would be absolutely invaluable for weather forecasting as the movement of fronts, etc. would be visible from space.

c) *Traffic.*

This is looking a good deal further ahead, but ultimately the chain will be used extensively for controlling and checking, possibly by radar, the movement of ships approaching or leaving the earth. It will also play an extremely important role as the first link in the solar communication system.

MANHATTAN PROJECT SUCCESS

LIEUTENANT GENERAL
LESLIE R. GROVES

From 1942 to 1945, General Groves directed the Manhattan Project to build the first atomic bombs, a technological tour de force that continues to be evoked whenever the nation is considering new technological initiatives. Here he distills the essence of the Project's managerial principles.

Looking back, I think I can see five main factors that made the Manhattan Project a successful operation.

First, we had a clearly defined, unmistakable, specific objective. Although at first there was considerable doubt about what it was. Consequently the people in responsible positions were able to tailor their every action to its accomplishment.

Second, each part of the project had a specific task. These tasks were carefully allocated and supervised so that the sum of their parts would result in the accomplishment of the overall mission. This system of compartmentalization had two principal advantages. The most obvious of these was that it simplified the maintenance of security. But over and above that, it required every member of the project to attend strictly to his own business. The result was an operation whose efficiency was without precedent.

Third, there was positive, clear-cut, unquestioned direction of the project at all levels. Authority was invariably delegated with responsibility, and this delegation was absolute and without reservation. Only in this way could the many apparently autonomous organizations working on the many apparently independent tasks be pulled together to achieve our final objective.

Fourth, the project made a maximum use of already existing agencies, facilities and services—governmental, industrial and academic. Since our objective was finite, we did not design our organization to operate in perpetuity. Consequently, our people were able to devote themselves exclusively to the task at hand, and had no reason to engage in independent empire-building.

Fifth, and finally, we had the full backing of our government, combined with the nearly infinite potential of American science, engineering and industry, and an almost unlimited supply of people endowed with ingenuity and determination.

A COMMON PROBLEM

J. ROBERT OPPENHEIMER, 1945

At Los Alamos, where the first atomic bombs were designed and assembled, on a stormy November night three months after the war was won, the theoretical physicist who had directed the work assessed its meaning. Fifty years later, I spoke in the same hall to many who had been in that audience and their descendants, and found in that further fifty years no more profound text on which to base my commemoration than this verbatim transcript.

The real impact of the creation of the atomic bomb and atomic weapons—to understand that, one has to look further back—look, I think, to the times when physical science was growing in the days of the renaissance, and when the threat that science offered was felt so deeply throughout the Christian world. The analogy is, of course, not perfect. You may even wish to think of the days in the last century when the theories of evolution seemed a threat to the values by which men lived. The analogy is not perfect because there is nothing in atomic weapons—there is certainly nothing that we have done here or in the physics or chemistry that immedi-

▶ Technology applied to mass killing: the first atomic bomb, tested July 16, 1945.

ately preceded our work here—in which any revolutionary ideas were in-volved. . . . It is in a quite different way. It is not an idea—it is a development and a reality—but it has in common with the early days of physical science the fact that the very existence of science is threatened, and its value is threatened. This is the point that I would like to speak a little about.

I think that it hardly needs to be said why the impact is so strong. There are three reasons: one is the extraordinary speed with which things which were right on the frontier of science were translated into terms where they affected many living people, and potentially all people. Another is the fact, quite accidental in many ways, and connected with the speed, that scientists themselves played such a large part, not merely in providing the foundation for atomic weapons, but in actually making them. In this we are certainly closer to it than any other group. The third is that the thing we made—partly because of the technical nature of the problem, partly because we worked hard, partly because we had good breaks—really arrived in the world with such a shattering reality and suddenness that there was no op-portunity for the edges to be worn off.

In considering what the situation of science is, it may be helpful to think a little of what people said and felt of their motives in coming into this job. One always has to worry that what people say of their motives is not ade-quate. Many people said different things, and most of them, I think, had some validity. There was in the first place the great concern that our enemy might develop these weapons before we did, and the feeling—at least in the early days, the very strong feeling—that without atomic weapons it might be very difficult, it might be an impossible, it might be an incredibly long thing to win the war. These things wore off a little as it became clear that the war would be won in any case. Some people, I think, were motivated by curiosity, and rightly so; and some by a sense of adventure, and rightly so. Others had more political arguments and said, "well, we know that atomic weapons are in principle possible, and it is not right that the threat of their unrealized possibility should hang over the world. It is right that the world should know what can be done in their field and deal with it." And the people added to that that it was a time when all over the world men would be particularly ripe and open for dealing with this problem because of the immediacy of the evils of war, because of the universal cry from everyone that one could not go through this thing again, even a war without atomic bombs. And there was finally, and I think rightly, the feeling that there was probably no place in the world where the development of atomic weapons would have a better chance of leading to a reasonable solution, and a smaller chance of leading to disaster, than within the United States. I believe all these things that people said are true, and I think I said them all myself at one time or another.

But when you come right down to it the reason that we did this job is because it was an organic necessity. If you are a scientist you cannot stop such a thing. If you are a scientist you believe that it is good to find out how the world works; that it is good to find out what the realities are; that it is good to turn over to mankind at large the greatest possible power to control the world and to deal with it according to its lights and its values. . . .

It is not possible to be a scientist unless you believe that it is good to learn. It is not . . . possible to be a scientist unless you think that it is of the highest value to share your knowledge, to share it with anyone who is interested. It is not possible to be a scientist unless you believe that the knowledge of the world, and the power which this gives, is a thing which is of intrinsic value to humanity, and that you are using it to help in the spread of knowledge, and are willing to take the consequences. . . .

There are many people who try to wiggle out of this. They say the real importance of atomic energy does not lie in the weapons that have been made; the real importance lies in all the great benefits which atomic energy, which the various radiations, will bring to mankind. There may be some truth in this. I am sure that there is truth in it, because there has never in the past been a new field opened up where the real fruits of it have not been invisible at the beginning. I have a very high confidence that the fruits—the so-called peacetime applications—of atomic energy will have in them all that we think, and more.

There are others who try to escape the immediacy of this situation by saying that, after all, war has always been very terrible; after all, weapons have always gotten worse and worse; that this is just another weapon and it doesn't create a great change; that they are not so bad; bombings have been bad in this war and this is not a change in that—it just adds a little to the effectiveness of bombing; that some sort of protection will be found. I think that these efforts to diffuse and weaken the nature of the crisis make it only more dangerous. I think it is for us to accept it as a very grave crisis, to realize that these atomic weapons which we have started to make are very terrible, that they involve a change, that they are not just a slight modification: to accept this, and to accept with it the necessity for those transformations in the world which will make it possible to integrate these developments into human life. . . .

It is clear to me that wars have changed. It is clear to me that if these first bombs—the bomb that was dropped on Nagasaki—that if these can destroy ten square miles, then that is really quite something. It is clear to me that they are going to be very cheap if anyone wants to make them; it is clear to me that this is a situation where a quantitative change has all the character of a change in quality, of a change in the nature of the world. . . . I think the advent of the atomic bomb and the facts which will get around that they are not too hard to make—that they will be universal if people wish to make them universal, that they will not constitute a real drain on the economy of any strong nation, and that their power of destruction will grow and is already incomparably greater than that of any other weapon—I think these things create a new situation, so new that there is some danger, even some danger in believing, that what we have is a new argument for arrangements, for hopes, that existed before this development took place. By that I mean that much as I like to hear advocates of a world federation, or advocates of a United Nations organization, who have been talking of these things for years—much as I like to hear them say that here is a new argument, I think that they are in part missing the point, because the point is not that atomic weapons constitute a new argument. There have always been good arguments. The point is that atomic weapons constitute also a field, a new field, and a new opportunity for realizing preconditions. I think when people talk of the fact that this is not only a great peril, but a great hope, this is what they should mean. I do not think they should mean the unknown, though sure, value of industrial and scientific virtues of atomic energy, but rather the simple fact that in this field, because it is a threat, because it is a peril, and because it has certain special characteristics . . . there exists a possibility of realizing, of beginning to realize, those changes which are needed if there is to be any peace.

Those are very far-reaching changes. They are changes in the relations between nations, not only in spirit, not only in law, but also in conception and feeling. I don't know which of these is prior; they must all work together, and only the gradual interaction of one on the other can make a reality. I don't agree with those who say the first step is to have a structure of international law. I don't agree with those who say the only thing is to have friendly feelings. All of these things will be involved. I think it is true to say that atomic weapons are a peril which affect everyone in the world, and in that sense a completely common problem, as common a problem as it was for the Allies to defeat the Nazis. . . .

The point I want to make, the one point I want to hammer home, is what an enormous change in spirit is involved. There are things which we hold very dear, and I think rightly hold very dear; I would say that the word democracy perhaps stood for some of them as well as any other word. There are many parts of the world in which there is no democracy. There are other things which we hold dear, and which we rightly should. And when I speak of a new spirit in international affairs I mean that even to these deepest of things which we cherish, and for which Americans have been willing to die—and certainly most of us would be willing to die—even in these deepest things, we realize that there is something more profound than that; namely, the common bond with other men everywhere. . . .

I can think of an analogy, and I hope it is not a completely good analogy: in the days in the first half of the nineteenth century there were many people, mostly in the North, but some in the South, who thought that there was no evil on earth more degrading than human slavery, and nothing that they would more willingly devote their lives to than its eradication. Always when I was young I wondered why it was that when Lincoln was President he did not declare that the war against the South, when it broke out, was a war that slavery should be abolished, that this was the central point, the rallying point, of that war. Lincoln was severely criticized by many of the Abolitionists as you know, by many then called radicals, because he seemed to be waging a war which did not hit the thing that was most important. But Lincoln realized, and I have only in the last months come to appreciate the depth and wisdom of it, that beyond the issue of slavery was the issue of the community of the people of the country, and the issue of the Union. . . .

I don't have very much more to say. There are a few things which scientists perhaps should remember. . . . I think that we have no hope at all if we yield in our belief in the value of science, in the good that it can be to the world to know about reality, about nature, to attain a gradually greater and greater control of nature, to learn, to teach, to understand. I think that if we lose our faith in this we stop being scientists, we sell out our heritage, we lose what we have most of value for this time of crisis.

But there is another thing: we are not only scientists; we are men, too. We cannot forget our dependence on our fellow men. I mean not only our material dependence, without which no science would be possible, and without which we could not work; I mean also our deep moral dependence, in that the value of science must lie in the world of men, that all our roots lie there. These are the strongest bonds in the world, stronger than those even that bind us to one another, these are the deepest bonds—that bind us to our fellow men.

PREDICTIONS: ICBMs

VANNEVAR BUSH, 1945

Testifying before the Special Senate Committee on Atomic Energy in November, the wartime science czar allowed his hope that physics might limit long-distance warfare to cloud his technical judgment.

L et me say this: There has been a great deal said about a 3,000-mile high-angle rocket. In my opinion such a thing is impossible and will be impossible for many years. The people who have been writing these things that annoy me have been talking about a 3,000-mile high-angle rocket shot from one continent to another carrying an atomic bomb, and so directed as to be a precise weapon which would land on a certain target such as this city.

I say technically I don't think anybody in the world knows how to do such a thing and I feel confident it will not be done for a very long period of time to come. I think we can leave that out of our thinking. I wish the American public would leave that out of their thinking.

III. Postwar Boom: 1945-1970

A VISION OF HYPERTEXT

VANNEVAR BUSH, 1945

The wartime director of the Office of Scientific Research and Development took time out in spring 1945, when Germany was finally defeated and the first atomic bomb was being assembled in New Mexico, to dream of a path through the thicket of documents the ages had accumulated. The instrument of this vision of hypertext, he imagined, would be a "memex": a Babbage-like electromechanical predecessor to the computer and Internet to come.

When the user is building a trail, he names it, inserts the name in his code book, and taps it out on his keyboard. Before him are two items to be joined, projected onto adjacent viewing positions. At the bottom of each there are a number of blank code spaces, and a pointer is set to indicate one of these on each item. The user taps a single key, and the items are permanently joined. In each code space appears the code word. Out of view, but also in the code space, is inserted a set of dots for photocell viewing; and on each item these dots by their positions designate the index number of the other item.

Thereafter, at any time, when one of these items is in view, the other can be instantly recalled merely by tapping a button below the corresponding code space. Moreover, when numerous items have been thus joined together to form a trail, they can be reviewed in turn, rapidly or slowly, by deflecting a lever like that used for turning the pages of a book. It is exactly as though the physical items had been gathered together from widely separated sources and bound together to form a new book. It is more than this, for any item can be joined into numerous trails.

The owner of the memex, let us say, is interested in the origin and properties of the bow and arrow. Specifically he is studying why the short Turkish bow was apparently superior to the English long bow in the skirmishes of the Crusades. He has dozens of possibly pertinent books and articles in his memex. First he runs through an encyclopedia, finds an interesting but sketchy article, leaves it projected. Next, in a history, he finds another pertinent item, and ties the two together. Thus he goes, building a trail of many items. Occasionally he inserts a comment of his own, either linking it into the main trail or joining it by a side trail to a particular item. When it be-

171

comes evident that the elastic properties of available materials had a great deal to do with the bow, he branches off on a side trail which takes him through textbooks on elasticity and tables of physical constants. He inserts a page of longhand analysis of his own. Thus he builds a trail of his interest through the maze of materials available to him.

And his trails do not fade. Several years later, his talk with a friend turns to the queer ways in which a people resist innovations, even of vital interest. He has an example, in the fact that the outranged Europeans still failed to adopt the Turkish bow. In fact he has a trail on it. A touch brings up the code book. Tapping a few keys projects the head of the trail. A lever runs through it at will, stopping at interesting items, going off on side excursions. It is an interesting trail, pertinent to the discussion. So he sets a reproducer in action, photographs the whole trail out, and passes it to his friend for insertion in his own memex, there to be linked into the more general trail.

Wholly new forms of encyclopedia will appear, ready-made with a mesh of associative trails running through them, ready to be dropped into the memex and there amplified. The lawyer has at his touch the associated opinions and decisions of his whole experience, and of the experience of friends and authorities. The patent attorney has on call the millions of issued patents, with familiar trails to every point of his client's interest. The physician, puzzled by a patient's reactions, strikes the trail established in studying an earlier similar case, and runs rapidly through analogous case histories, with side references to the classics for the pertinent anatomy and histology. The chemist, struggling with the synthesis of an organic compound, has all the chemical literature before him in his laboratory, with trails following the analogies of compounds, and side trails to their physical and chemical behavior.

The historian, with a vast chronological account of a people, parallels it with a skip trail which stops only on the salient items, and can follow at any time contemporary trails which lead him all over civilization at a particular epoch. There is a new profession of trail blazers, those who find delight in the task of establishing useful trails through the enormous mass of the common record. The inheritance from the master becomes, not only his additions to the world's record, but for his disciples the entire scaffolding by which they were erected.

Thus science may implement the ways in which man produces, stores, and consults the record of the race.

If warfare is to consist of a few teams of professors pushing buttons, why have an Army and Navy at all?

EDITORIAL, *Life* MAGAZINE, 1945

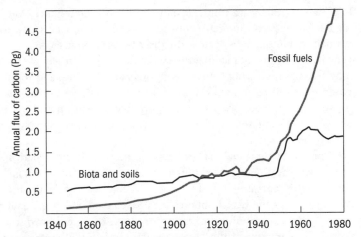

▶ Annual net release of carbon into atmosphere from changes in land cover vs. release from burning of fossil fuels, 1850–1980 (1 Pg = 10^{15} grams).

WEIGHING UP DDT

V. B. WIGGLESWORTH, 1945

Long before Rachel Carson published Silent Spring, *the complex effects of DDT on the environment had begun to reveal themselves to scientific observers. Many today have heard of DDT's dangers; few are aware that it served widely in the Second World War to prevent epidemic malaria and typhus. Here a British entomologist weighs up the benefits and emerging risks. Notice that he has to define the unfamiliar word "ecology."*

In 1940 the Swiss firm of J. R. Geigy, A. G., of Basle discovered the insecticidal properties of a chemical that is known as 2, 2-bs (parachlorophenyl) 1, 1, 1-trichloroethane or more familiarly as dichloro-diphenyltrichloroethane, later abbreviated to DDT. The firm took out patents to cover the manufacture of this chemical and its use as an insecticide, and early in 1942 the British and American branches of the firm brought these patents to the notice of the entomologists in the two countries, who were seeking a substitute for [the vegetable insecticide] pyrethrum.

In England we read those patents and were frankly skeptical. It seemed to us that too much was claimed. The new insecticide appeared to be so exactly what we wanted that it looked too good to be true. But clearly the stuff should be tested. . . .

The Swiss claims were fully substantiated. And immediately, on both sides of the Atlantic, entomologists got to work to discover how best to use the new material in the fight against the disease-carrying insects. . . .

Improved methods of manufacture were devised, chemical plants were set up by the manufacturing firms in America and England, and soon after the requirements were defined the supplies were available. As the months went by, more uses were discovered; requirements went up and up; until, with an output of some scores of tons a month, the military demands could still scarcely be fully met. . . .

The control of rural malaria, so long the despair of the malariologist, may at last become a reality. Preliminary trials in West Africa and in India have been highly encouraging. This method figures largely in the "Extended Program" of the U.S. Public Health Service, in which the energetic campaign for the control of malaria in war areas is being enlarged to cover all the malarious regions in the United States; and in the control of rural malaria by the Health and Safety Division of the TVA. . . .

Almost all the vast output of DDT during the war years was earmarked for service users. But enough has been made available for trials to prepare the way for what will soon be its main applications—against insects of veterinary importance, against the pests of growing crops, and perhaps against forest insects. . . .

[But] perhaps DDT is like a blunderbuss discharging shot in a manner so haphazard that friend and foe alike are killed. . . .

The very small quantities of insecticide which need to be applied on water to kill mosquito larvae do not normally kill fish; but they kill a variable proportion of the other aquatic insects. This result may not matter very much. But it may have far-reaching effects which it is impossible at present to predict. The broadcasting of sprays from airplanes in order to kill the flying mosquito is still more prone to upset the balance of nature. These sprays settle on the vegetation and kill vast numbers of insects of all sorts. Without careful study, it is impossible to guess what the ultimate results of this process may be.

Still more drastic are the effects of the highly concentrated sprays of DDT that are being used against the forest pests. These sprays kill other kinds of insects, in addition to the pests against which they are directed. They settle into the streams and kill the aquatic insects. Some fish, such as the speckled trout, are reported to have been killed when they fed on poisoned insects. Crayfish and tadpoles are killed either by contact with the poison or by eating contaminated food. These again are important sources of food for fish, which must suffer when they are destroyed.

There seems little doubt that the general insect population in the forests is greatly reduced and that the natural waters may be nearly sterilized and lifeless. It has even been suggested that nesting birds may suffer from the shortage of insects and be unable properly to feed their young. Only careful experiments can prove whether birds will suffer or not. . . .

It is obvious enough that DDT is a two-edged sword. We can see how seriously it may upset the balance locally between insect enemies and friends. . . .

Chemicals which upset the balance of nature have been known before. DDT is merely the latest and one of the most violent. It can bring about within a single year a disturbance that it would take other chemicals a good many years to produce. It has thus focused attention on the problem and has provided a great stimulus to entomologists to reflect upon what is happening.

One conclusion they have come to is that we know far too little about the interaction of pests with their physical environment and with the other insects around them. We need to know far more about their ecology—that is, about their natural history studied scientifically. . . .

We want to learn how to apply the insecticide at such a time or in such a way as to touch the pest and not its enemies. We want to choose insecticides which discriminate between friend and foe.

This is probably where the future of insecticides lies—in the development of materials with a selective action. It may well be that in the long run an insecticide which kills 50 percent of the pest insect and none of its predators or parasites may be far more valuable than one which kills 95 percent but at the same time eliminates its natural enemies.

ATOMIC MORALITY

VANNEVAR BUSH

The morality of atomic-bombing Hiroshima and Nagasaki continues to be debated. The science administrator on whose watch the deeds were done understood immediately, and said publicly in 1949, that something else than destructiveness was at issue.

It was indeed the bizarre nature of the bomb, and the uncanny sort of future it suggested, rather than its actual results in the war, that impressed people. The fire raids upon Japan were much more terrible, they reduced a far greater area of the frail Japanese cities to ashes, they caused far greater casualties among civilians—panic, the crush of mobs, and horrible death; yet they occurred almost unnoticed and created few later arguments. The moral question was hardly raised regarding the fire raids, yet that question is

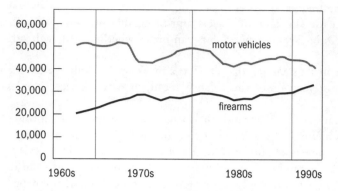

▶ Yearly U.S. deaths from motor-vehicle vs. firearm injuries, 1950–1994.

substantially identical in the two cases. It was fear of the future that concentrated attention on the atomic bomb.

FLYING BLIND

GEORGE A. LUNDBERG, 1945

A professor of sociology at the University of Washington and former president of the American Sociological Society defends science and its handmaiden technology from the fearful reaction triggered by the invention of the atomic bomb.

The idea of a moratorium on scientific development was advanced long before the atomic bomb. Even scientists, appalled at some of the social results of their handiwork, have in times past come out for a moratorium. More frequently the thought is advanced by writers living in New York apartments who would not know how to get downstairs if the elevators stopped running. Actually, very few people seriously question the importance of the continued advancement of science. They readily agree with Dr. Vannevar Bush that, when put to practical use, scientific advances may mean more jobs, higher wages, shorter hours, more abundant crops, and more leisure, as well as higher standards of living, prevention or cure of disease, conservation of natural resources, and defense against aggression. . . .

But granting all the arguments for more industrial, biological, and medical research . . . one crucial question remains: Will it solve or even facilitate solutions of the problems of human relations?

. . . Can science save us? Yes. But we must not expect physical science to solve social problems. We cannot expect penicillin to solve the employer-employee struggle, nor can we expect better electric lamps to illumine darkened intellects and emotions. We cannot expect atomic fission to reveal the nature of the social atom and the manner of its control. If we want results in improved human relations we must direct our research at these problems.

To those who are still skeptical and unimpressed with the promise of social science, we may address this question: What alternatives do you propose that hold greater promise? If we do not place our faith in social science, to what shall we look for social salvation? . . . The dominant current faith is a moralistic-legalistic thoughtway, sharply at variance with our analytic attitude toward the rest of nature. The pathetic faith and hopeful trumpetings about so frail an instrument as the San Francisco Charter [of the United Nations] provides a perfect example. Millions have been led to attach an entirely absurd significance to it as a preventive of war.

Most thinking people admit that the remedy for world discord lies quite elsewhere, namely, in discovering and altering the conditions that produce wars, and, by inference, in the scientific analysis which is indispensable to such discovery and remedy. It is true, also, that a social and economic council is provided [in the U.N. charter] presumably to look into such matters. The crucial consideration is the degree to which that council will be in a position to undertake research comparable to that which produced the atomic bomb. We shall see whether two billion dollars and the ablest scientific talent, unhampered by political controls, are allocated to such purposes.

To put it bluntly, in the present state of development of the social sciences, centralized administration of large national or international societies can be carried out only by precisely the methods that have thus far characterized such regimes, namely, ruthless suppression of all opposition and wholesale starvation or other deprivation when the success of a "plan" takes priority over human life. A leader, however admirable in ability and intentions, attempting to administer centrally a large society today is somewhat in the position of a pilot trying to fly a B-29 [bomber] without an instrument board or charts. That is to say, it cannot be a very smooth flight. If he succeeds at all, it will be at the expense of much wreckage of men and materials.

Successful piloting depends directly upon the adequacy and accuracy of the instruments in the machine, the charts by which a course can be pursued or modified, and the training of the pilot to read both aright. Only as a result of the development of the basic physical sciences can a B-29 either be built or flown. Only through a comparable development of the social sciences can a workable world society be either constructed or administered.

APPROPRIATIONS

GEORGE B. DYSON

The ENIAC, traditionally characterized as the first electronic computer, began calculating artillery firing tables in late 1945 but soon was applied to a very different purpose: calculating the hydrodynamics of the Super, Edward Teller's first (and failed) design for a hydrogen bomb. The son of theoretical physicist Freeman Dyson explores the origins of the ENIAC and finds that an evolutionary appropriation made it adaptive.

The production of firing tables still demanded equal measures of art and science when [mathematician John] von Neumann arrived on the scene at the beginning of World War II. Shells were test-fired down a range equipped with magnetic pickup coils to provide baseline data. Then the influence of as many variables as could be assigned predictable functions was combined to produce a firing table composed of between two thousand and four thousand individual trajectories, each trajectory requiring about 750 multiplications to determine the path of that particular shell for a representative fifty points in time.

A human computer working with a desk calculator took about twelve

▸ Computer programming circa 1946: the ENIAC.

hours to calculate a single trajectory; the electromechanical differential analyzer at the [army] Ballistic Research Laboratory (a ten-integrator version of the machine that Vannevar Bush had developed at MIT) took ten or twenty minutes. This still amounted to about 750 hours, or a month of uninterrupted operation, to complete one firing table. Even with double shifts and the assistance of a second, fourteen-integrator differential analyzer (constructed at the Moore School of Electrical Engineering at the University of Pennsylvania in Philadelphia), each firing table required about three months of work. The nearly two hundred human computers working at the Ballistic Research Laboratory were falling hopelessly behind. "The number of tables for which work has not been started because of lack of computational facilities far exceeds the number in progress," reported Herman Goldstine in August 1944. "Requests for the preparation of new tables are being currently received at the rate of six per day."

Electromechanical assistance was not enough. In April 1943, the army initiated a crash program to build an electronic digital computer based on decimal counting circuits made from vacuum tubes. Rings of individual flip-flops (two vacuum tubes each) were formed into circular, ten-stage counters linked to each other and to an array of storage registers . . . running at about six million rpm. The ENIAC (Electronic Numerical Integrator and Computer) was constructed by a team that included John W. Mauchly, John Presper Eckert, and (Captain) Herman H. Goldstine, supervised by John G. Brainerd under a contract between the army's Ballistic Research Laboratory and the Moore School. A direct descendant of the electromechanical differential analyzers of the 1930s, the ENIAC represented a brief but fertile intersection between the otherwise diverging destinies of analog and digital computing machines. Incorporating eighteen thousand vacuum tubes operating at 100,000 pulses per second, the ENIAC consumed 150 kilowatts of power and held twenty 10-digit numbers in high-speed storage. With the addition of a magnetic-core memory of 120 numbers in 1953, the working life of the ENIAC was extended until October 1955.

Programmed by hand-configured plugboards (like a telephone switchboard) and resistor-matrix function tables (set up as read-only memory, or ROM), the ENIAC was later adapted to a crude form of stored-program control. Input and output was via standard punched-card equipment requisitioned from IBM. It was thus possible for the Los Alamos mathematicians [calculating hydrogen-bomb hydrodynamics] to show up with their own decks of cards and produce intelligible results. Rushed into existence with a single goal in mind, the ENIAC became operational in late 1945 and just missed seeing active service in the war. To celebrate its public dedication in February 1946, the ENIAC computed a shell trajectory in twenty seconds—ten seconds faster than the flight of the shell and a thousand times faster than the methods it replaced. But the ENIAC was born with time on its hands, because the backlog of firing-table calculations vanished when hostilities ceased.

Sudden leaps in biological or technological evolution occur when an existing structure or behavior is appropriated by a new function that spreads rapidly across the evolutionary landscape, taking advantage of a head start. Feathers must have had some other purpose before they were used to fly. U-boat commanders appropriated the [code-making] Enigma machine first developed for use by banks. Charles Babbage envisioned using the existing network of church steeples that rose above the chaos of London as the foundation for a packet-switched communications net. According to neurophysiologist William Calvin, the human mind appropriated areas of our brains that first evolved as buffers for rehearsing and storing the precise timing sequences required for ballistic motor control. "Throwing rocks at even stationary prey requires great precision in the timing of rock release from an overarm throw, with the 'launch window' narrowing eight-fold when the throwing distance is doubled from a beginner's throw," he observed in 1983. "Parallel timing neurons can overcome the usual neural noise limitations via the law of large numbers, suggesting that enhanced throwing skill could have produced strong selection pressures for any evolutionary trends that provided additional timing neurons. . . . This emergent property of parallel timing circuits has implications not only for brain size but for brain reorganization, as another way of increasing the numbers of timing neurons is to temporarily borrow them from elsewhere in the brain." According to Calvin's theory, surplus off-hours capacity was appropriated by such abstractions as language, consciousness, and culture, invading the neighborhood much as artists colonize a warehouse district, which then becomes the gallery district as landlords raise the rent. The same thing happened to the ENIAC: a mechanism developed for ballistics was expropriated for something else.

John Mauchly, Presper Eckert, and others involved in the design and construction of the ENIAC had every intention of making wider use of their computer, but it was von Neumann who carried enough clout to preempt the scheduled ballistics calculations and proceed directly with a numerical simulation of the super bomb.

Solving the Problems
of War

Board of Consultants to the
Secretary of State's Committee
on Atomic Energy, 1946

*Agreement between the United States, Great Britain and Canada in late 1945
"to prevent the use of atomic energy for destructive purposes" and to promote
its use "for peaceful and humanitarian ends" led U.S. Under-Secretary of State
Dean Acheson to appoint a board of consultants to consider an initiative to
internationalize nuclear arms control. Early in 1946, Tennessee Valley Authority
head David Lilienthal, Los Alamos Laboratory director Robert Oppenheimer and
three Manhattan Project industrialists—Chester I. Barnard, Charles A. Thomas and
Harry A. Winne—produced a remarkable document that came to be called the
Acheson-Lilienthal Report. It found a way to prevent a clandestine nuclear arms
race without resorting to armed world government. Suspicious of such innovative
diplomacy, the man whom President Harry Truman chose to present the plan to
the United Nations, Bernard Baruch, insisted on burdening the proposal with
conventional sanctions. It would almost certainly have been rejected by Stalinist
Russia anyway, but Baruch's modifications inevitably doomed it. Its vision of a
world where distributed nuclear capability deters nuclear adventurism implicitly
predicted the future that occurred, with the tragic difference that the Report's
alternative future contained no stockpiles of weapons. The plan remains a
practical model for a future without nuclear weapons where deterrence continues
to limit national ambitions but delivery time of weapons to targets is measured in
months from the factory rather than minutes or hours from the missile silo or
strategic bomber.*

To "outlaw" atomic energy in all of its forms and enforce such a prohibition by an army of inspectors roaming the earth would overwhelm
the capacity and the endurance of men, and provide no security. This conclusion has a further implication in a search for a security system. While
suppression is not possible where we are dealing with the quest for knowledge, this thirst to know (that cannot be "policed" out of existence) *can* be
used, affirmatively, in the design and building of an effective system of safeguards.

Human history shows that any effort to confine the inquiring human
mind, to seek to bar the spirit of inquiry, is doomed to failure. From such efforts come subversion fraught with terrible consequences: Gestapo, inquisi-
tions, wars. The development of atomic energy is one of a long, long line of
discoveries that have their well springs in the urge of men to know more

about themselves and their world. Like the jiu jutsu wrestler whose skill consists in making his opponent disable himself with his own thrusts, the designers of a system of safeguards for security should and can utilize for enforcement measures that driving force toward knowledge that is part of man's very nature.

If atomic energy had only one conceivable use—its horrible powers of mass destruction—then the incentive to follow the course of complete prohibition and suppression might be very great. Indeed, it has been responsibly suggested that however attractive may be the potentialities for benefit from atomic energy, they are so powerfully outweighed by the malevolent that our course should be to bury the whole idea, to bury it deep, to forget it, and to make it illegal for anyone to carry on further inquiries or developments in this field.

We have concluded that the beneficial possibilities . . . in the use of atomic energy should be and can be made to aid in the development of a reasonably successful system of security, and the plan we recommend is in part predicated on that idea. . . .

As a result of our thinking and discussions we have concluded that it would be unrealistic to place reliance on a simple agreement among nations to outlaw the use of atomic weapons in war. We have concluded that an attempt to give body to such a system of agreements through international inspection holds no promise of adequate security.

And so we have turned from mere policing and inspection by an international authority to a program of affirmative action, of aggressive development by such a body. This plan we believe holds hope for the solution of the problem of the atomic bomb. We are even sustained by the hope that it may contain seeds which will in time grow into that cooperation between nations which may bring an end to all war.

The program we propose will undoubtedly arouse skepticism when it is first considered. It did among us, but thought and discussion have converted us.

It may seem too idealistic. It seems time we endeavor to bring some of our expressed ideals into being.

It may seem too radical, too advanced, too much beyond human experience. All these terms apply with peculiar fitness to the atomic bomb. . . .

The proposal contemplates an international agency with exclusive jurisdiction to conduct all intrinsically dangerous operations in the field. This means all activities relating to raw materials, the construction and operation of production plants, and the conduct of research in explosives. The large field of non-dangerous and relatively non-dangerous activities would be left in national hands. These would consist of all activities in the field of research (except on explosives) and the construction and operation of non-dangerous power-producing piles. National activities in these fields would be subject to moderate controls by the international agency, exercised through licensing, rules and regulations, collaboration on design, and the like. The international agency would also maintain inspection facilities to assure that illicit operations were not occurring, primarily in the exploitation of raw materials. . . .

The Authority will be aided in the detection of illegal operations by the fact that it is not the motive but the operation which is illegal. . . . The total effort needed to carry through from the mine to the bomb, a surreptitious program of atomic armament on a scale sufficient to make it a threat or to make it a temptation to evasion, is so vast, and the number of separate difficult undertakings so great, and the special character of many of these undertakings so hard to conceal, that the fact of this effort should be impossible to hide. . . .

In strengthening security, one of the primary considerations will relate to the geographical location of the operations of the Authority and its property. For it can never be forgotten that it is a primary purpose of the Atomic Development Authority to guard against the danger that our hopes for peace may fail, and that adventures of aggression may again be attempted. It will probably be necessary to write into the charter itself a systematic plan governing the location of the operations and property of the Authority so that a strategic balance may be maintained among nations. In this way, protection will be afforded against such eventualities as the complete or partial collapse of the United Nations or the Atomic Development Authority, protection will be afforded against the eventuality of sudden seizure by any one nation of the stockpiles, reduction, refining, and separation plants, and reactors of all types belonging to the Authority.

This will have to be quite a different situation from the one that now prevails. At present with Hanford, Oak Ridge, and Los Alamos situated in the United States, other nations can find no security against atomic warfare except the security that resides in our own peaceful purposes or the attempt at security that is seen in developing secret atomic enterprises of their own. Other nations which, according to their own outlook, may fear us, can develop a greater sense of security only as the Atomic Development Authority locates similar dangerous operations within their borders. Once such operations and facilities have been established by the Atomic Development Authority and are being operated by that agency within other nations as well as our own, a balance will have been established. It is not thought that the Atomic Development Authority could protect its plants by military force from the overwhelming power of the nation in which they are situated. Some United Nations military guard may be desirable. But at most, it could be little more than a token. The real protection will lie in the fact that if any nation seizes the plants or the stockpiles that are situated in its territory, other nations will have similar facilities and materials situated within their own borders so that the act of seizure need not place them at a disadvantage. . . .

With appropriate world-wide distribution of stockpiles and facilities; with design rendered as little dangerous as possible; with stockpiles of dangerous materials kept at the lowest level consistent with good economics and engineering; there will be no need for a sense of insecurity on the part of any of the major powers. Seizures will afford no immediate tactical advantage. They would in fact be an instantaneous dramatic danger signal, and they would permit, under the conditions stated, a substantial period of time for other nations to take all possible measures of defense. For it should be borne in mind that even if facilities are seized, a year or more would be required after seizure before atomic weapons could be produced in quantities sufficient to have an important influence on the outcome of war. Considering the psychological factors in public opinion, the fixing of danger signals that are clear, simple, and vivid seems to us of utmost importance. . . .

The security which we see in the realization of this plan lies in the fact that it averts the danger of the surprise use of atomic weapons. The seizure by one nation of installations necessary for making atomic weapons would be not only a clear signal of warlike intent, but it would leave other nations in a position—either alone or in concert—to take counter-actions. The plan, of course, has other security purposes, less tangible but none the less important. For in the very fact of cooperative effort among the nations of the world rests the hope we rightly hold for solving the problem of war itself. . . .

When the plan is in full operation there will no longer be secrets about atomic energy. We believe that this is the firmest basis of security; for in the long term there can be no international control and no international cooperation which does not presuppose an international community of knowledge.

I am thinking about something much more important than bombs. I am thinking about computers.

JOHN VON NEUMANN, 1946

NAMING OF PARTS

HENRY REED, 1947

Today we have naming of parts. Yesterday,
We had daily cleaning. And tomorrow morning,
We shall have what to do after firing. But today,
Today we have naming of parts. Japonica
Glistens like coral in all of the neighboring gardens,
And today we have naming of parts.

This is the lower sling swivel. And this
Is the upper sling swivel, whose use you will see,
When you are given your slings. And this is the piling swivel,
Which in your case you have not got. The branches
Hold in the gardens their silent, eloquent gestures,
    Which in our case we have not got.

This is the safety-catch, which is always released
With an easy flick of the thumb. And please do not let me
See anyone using his finger. You can do it quite easy
If you have any strength in your thumb. The blossoms
Are fragile and motionless, never letting anyone see
    Any of them using their finger.

And this you can see is the bolt. The purpose of this
Is to open the breech, as you see. We can slide it
Rapidly backwards and forwards: we call this
Easing the spring. And rapidly backwards and forwards
The early bees are assaulting and fumbling the flowers:
    They call it easing the Spring.

They call it easing the Spring: it is perfectly easy
If you have any strength in your thumb; like the bolt,
And the breech, and the Cocking-piece, and the point of balance,
Which in our case we have not got; and the almond-blossom
Silent in all of the gardens and the bees going backwards and forwards,
    For today we have naming of parts.

In some sort of crude sense which no vulgarity, no humor, no
overstatement can quite extinguish, the physicists have known sin; and
this is a knowledge which they cannot lose.

                    J. ROBERT OPPENHEIMER, 1948

T-R-A-N-S-I-S-T-O-R

RALPH BOWN, 1948

*At a press conference on July 1, 1948, the director of research for Bell Telephone
Laboratories announced a new invention.*

We have called it the Transistor, T-R-A-N-S-I-S-T-O-R, because it is
a resistor or semiconductor device which can amplify electrical sig-

nals as they are transferred through it from input to output terminals. It is, if you will, the electrical equivalent of a vacuum tube amplifier. But there the similarly ceases. It has no vacuum, no filament, no glass tube. It is composed entirely of cold solid substances.

This tiny cylindrical object which I am holding up is a Transistor. Although it is a "little bitty" thing it can, up to a power output of about 100 milliwatts and up to a frequency of about 10 megacycles, do just about everything a vacuum tube can do and some unique things which a vacuum tube cannot do. . . .

The battery power required to operate this amplifier is less than one-tenth that used by an ordinary flashlight bulb and it is of about the same voltage as that ordinarily supplied over the telephone lines from the central office to the telephone instrument at the subscriber station. It will be evident to you immediately that there are possibilities of employing these devices to do new and interesting things in telephony.

In a radio interview the following year, transistor coinventor William Shockley commented:

There has recently been a great deal of thought spent on electronic brains or computing machines. Many of the problems of these machines are similar to those of an automatic telephone exchange which is also a sort of electronic brain. For applications of this sort there are difficulties in applying vacuum tubes because of their size and the heat which they produce. It seems to me that in these robot brains the transistor is the ideal nerve cell.

BOTTLED SUNSHINE

SUZANNE WHITE

The Florida "gold rush" in orange juice, launched in 1948 when Bing Crosby signed his first Minute Maid contract, rendered all other [frozen food] industry successes pale by comparison. Initially advertised as being cheaper than whole oranges and better for one's health, frozen orange juice was an immediate success, and more than any other single product it helped establish and stabilize the market for frozen foods. . . . Orange juice became one of the first truly national convenience foods.

MURPHY'S LAW, 1949

ROBERT A. J. MATTHEWS

In a 1997 report in Scientific American, *British physicist and science journalist Matthews reveals that Murphy's Law—"If something can go wrong, it will"— demonstrated its essential truth at its own unveiling.*

The familiar version of Murphy's Law is not quite 50 years old, but the essential idea behind it has been around for centuries. In 1786 the Scottish poet Robert Burns observed that

> The best laid schemes o' mice an' men
> Gang aft agley ("Are prone to go awry").

In 1884 the Victorian satirist James Payn described perhaps the most famous example of Murphy's Law:

> I had never had a piece of toast
> Particularly long and wide
> But fell upon the sanded floor
> And always on the buttered side.

The modern version of Murphy's Law has its roots in U.S. Air Force studies performed in 1949 on the effects of rapid deceleration on pilots. Volunteers were strapped on a rocket-propelled sled, and their condition was monitored as the sled was brought to an abrupt halt. The monitoring was done by electrodes fitted to a harness designed by Captain Edward A. Murphy.

After what had seemed to be a flawless test run one day, the harness's failure to record any data puzzled technicians. Murphy discovered that every one of its electrodes had been wired incorrectly, prompting him to declare: "If there are two or more ways of doing something, and one of them can lead to catastrophe, then someone will do it."

At a subsequent press conference, Murphy's rueful observation was presented by the project engineers as an excellent working assumption in safety-critical engineering. But before long—and to Murphy's chagrin—his principle had been transformed into an apparently flippant statement about the cussedness of everyday events. Ironically, by losing control over his original meaning, Murphy thus became the first victim of his eponymous law.

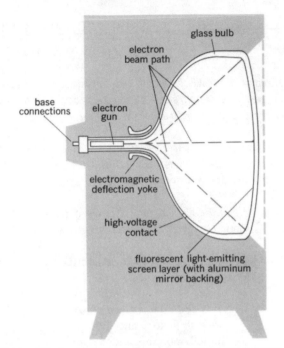

electron
beam path

glass bulb

base
connections

electron
gun

electromagnetic
deflection yoke

high-voltage
contact

fluorescent light-emitting
screen layer (with aluminum
mirror backing)

▶ Television, developed before World War II, had to wait until the war was over.

A DIME A DOZEN

VANNEVAR BUSH, 1949

In the field of complex mechanics, inventions are a dime a dozen. The question of whether a device will come into being depends upon three things: first, whether there is a practical use for it that warrants its development and manufacturing costs; second, whether the laws of physics applying to the elements available for its design allow the attainment of the needed ranges, sensitivities, or the like; and third, whether the pertinent art of manufacture has advanced sufficiently to allow a useful embodiment to be built successfully.

> Science, as well as technology, will in the near and in the farther future increasingly turn from problems of intensity, substance, and energy, to problems of structure, organization, information, and control.
>
> JOHN VON NEUMANN, 1949

THE REAL POWER

GEORGE ORWELL, 1949

1984, George Orwell's dystopian masterpiece, predicted a totalitarian world of terror and pain. Dark corners of that world were already emerging in the Soviet Union and the People's Republic of China, the latter founded the year 1984 was published. I had not noticed before that a species of antiscientific deconstructionism skulked in the shadows of Orwell's book, embodied in this crucial argument between Winston and his government torturer, O'Brien. The climactic scene mirrors the debate between the Savage and Mustapha Mond in Brave New World. Is the grandiose delusion that even science and technology are social constructions, subordinate to utopian ideology—demonstrated in the Soviet experiment with Lysenkoism, in the criminal disregard of safety measures that resulted in the Chernobyl disaster and in Mao Zedong's experiment with backyard iron production that led to widespread famine—a major reason the Communist experiment failed?

O'Brien leaned over him, deliberately bringing the worn face nearer. "You are thinking," he said, "that my face is old and tired. You are thinking that I talk of power, and yet I am not even able to prevent the decay of my own body. Can you not understand, Winston, that the individual is only a cell? The weariness of the cell is the vigor of the organism. Do you die when you cut your fingernails?"

He turned away from the bed and began strolling up and down again, one hand in his pocket.

"We are the priests of power," he said. "God is power. But at present power is only a word so far as you are concerned. It is time for you to gather some idea of what power means. The first thing you must realize is that power is collective. The individual only has power in so far as he ceases to be an individual. You know the Party slogan 'Freedom is Slavery.' Has it ever occurred to you that it is reversible? Slavery is freedom. Alone—free—the human being is always defeated. It must be so, because every human being is doomed to die, which is the greatest of all failures. But if he can make complete, utter submission, if he can escape from his identity, if he can merge himself in the Party so that he *is* the Party, then he is all-powerful and immortal. The second thing for you to realize is that power is power over human beings. Over the body—but, above all, over the mind. Power over matter—external reality, as you would call it—is not important. Already our control over matter is absolute."

For a moment Winston ignored the dial [of the torture machine]. He made a violent effort to raise himself into a sitting position, and merely succeeded in wrenching his body painfully.

"But how can you control matter?" he burst out. "You don't even control the climate or the law of gravity. And there are disease, pain, death—"

O'Brien silenced him by a movement of the hand. "We control matter because we control the mind. Reality is inside the skull. You will learn by degrees, Winston. There is nothing that we could not do. Invisibility, levitation—anything. I could float off this floor like a soap bubble if I wished to. I do not wish to, because the Party does not wish it. You must get rid of those nineteenth-century ideas about the laws of nature. We make the laws of nature."

"But you do not! You are not even masters of this planet. What about Eurasia and Eastasia? You have not conquered them yet."

"Unimportant. We shall conquer them when it suits us. And if we did not, what difference would it make? We can shut them out of existence. Oceania is the world."

"But the world itself is only a speck of dust. And man is tiny—helpless! How long has he been in existence? For millions of years the earth was uninhabited."

"Nonsense. The earth is as old as we are, no older. How could it be older? Nothing exists except through human consciousness."

"But the rocks are full of the bones of extinct animals—mammoths and mastodons and enormous reptiles which lived here long before man was ever heard of."

"Have you ever seen those bones, Winston? Of course not. Nineteenth-century biologists invented them. Before man there was nothing. After man, if he could come to an end, there would be nothing. Outside man there is nothing."

"But the whole universe is outside us. Look at the stars! Some of them are a million light-years away. They are out of our reach forever."

"What are the stars?" said O'Brien indifferently. "They are bits of fire a few kilometers away. We could reach them if we wanted to. Or we could blot them out. The earth is the center of the universe. The sun and the stars go round it."

Winston made another convulsive movement. This time he did not say anything. O'Brien continued as though answering a spoken objection:

"For certain purposes, of course, that is not true. When we navigate the ocean, or when we predict an eclipse, we often find it convenient to assume that the earth goes round the sun and that the stars are millions upon millions of kilometers away. But what of it? Do you suppose it is beyond us to produce a dual system of astronomy? The stars can be near or distant, according as we need them. Do you suppose our mathematicians are unequal to that? Have you forgotten doublethink?"

Winston shrank back upon the bed. Whatever he said, the swift answer crushed him like a bludgeon. And yet he knew, he *knew*, that he was in the right. The belief that nothing exists outside your own mind—surely there must be some way of demonstrating that it was false. Had it not been exposed long ago as a fallacy? There was even a name for it, which he had forgotten. A faint smile twitched the corners of O'Brien's mouth as he looked down at him.

"I told you, Winston," he said, "that metaphysics is not your strong point. The word you are trying to think of is solipsism. But you are mistaken. This is not solipsism. Collective solipsism, if you like. But that is a different thing; in fact, the opposite thing. All this is a digression," he added in a different tone. "The real power, the power we have to fight for night and day, is not power over things, but over men." He paused, and for a moment assumed again his air of a schoolmaster questioning a promising pupil: "How does one man assert his power over another, Winston?"

Winston thought. "By making him suffer," he said.

"Exactly. By making him suffer. Obedience is not enough. Unless he is suffering, how can you be sure that he is obeying your will and not his own? Power is inflicting pain and humiliation. Power is in tearing human minds to pieces and putting them together again in new shapes of your own choosing. Do you begin to see, then, what kind of world we are creating? It is the exact opposite of the stupid hedonistic Utopias that the old reformers imagined. A world of fear and treachery and torment, a world of trampling and being trampled upon, a world which will grow not less but *more* merciless as it refines itself. Progress in our world will be progress toward more pain. The old civilizations claimed that they were founded on love and justice. Ours is founded upon hatred. In our world there will be no emotions except fear, rage, triumph and self-abasement. Everything else we shall destroy—everything. Already we are breaking down the habits of thought which have survived from before the Revolution. We have cut the links between child and parent, and between man and man, and between man and woman. No one dares trust a wife or a child or a friend any longer. But in the future there will be no wives and no friends. Children will be taken from their mothers at birth, as one takes eggs from a hen. The sex instinct will be eradicated. Procreation will be an annual formality like the renewal of a ration card. We shall abolish the orgasm. Our neurologists are at work upon it now. There will be no loyalty, except loyalty toward the Party. There will be no love, except the love of Big Brother. There will be no laughter, except the laugh of triumph over a defeated enemy. There will be no art, no literature, no science. When we are omnipotent we shall have no more need of science. There will be no distinction between beauty and ugliness. There will be no curiosity, no employment of the process of life. All competing pleasures will be destroyed. But always—do not forget this, Winston—always there will be the intoxication of power, constantly increasing and constantly growing subtler. Always, at every moment, there will be the thrill of victory, the sensation of trampling on an enemy who is helpless. If you want a picture of the future, imagine a boot stamping on a human face—forever."

VARIETIES

DAVID RIESMAN, 1950

In his seminal work The Lonely Crowd, *the sociologist identified technology as liberating.*

The more advanced the technology, on the whole, the more possible it is for a considerable number of human beings to imagine being somebody else. In the first place, the technology spurs the division of labor, which, in turn, creates a greater variety of life experiences and human types. In the second place, the improvement in technology permits sufficient leisure to contemplate change—a kind of capital reserve in men's self-adaptation to nature—not on the part of a ruling few but on the part of many. In the third place, the combination of technology and leisure helps to acquaint people with other historical solutions—to provide them, that is, not only with more goods and more experiences but also with an increased variety of personal and social models.

THREE LAWS OF ROBOTICS

ISAAC ASIMOV, 1950

The prolific science-fiction writer published these basic ethical principles in his novel I, Robot. *They are designed to counter what might be called technology's sorcerer's-apprentice problem.*

1. A robot may not injure a human being, or, through inaction, allow a human being to come to harm.
2. A robot must obey the orders given it by human beings, except where such orders would conflict with the First Law.
3. A robot must protect its own existence, as long as such protection does not conflict with the First or Second Law.

RICHARD RHODES

DETROIT VERSUS THE HOT-ROD

GENE BALSLEY, 1950

Hot-rods—automobiles modified by dedicated amateur mechanics for high power and speed—flourished in mid-century America, when engines were still simple enough to modify at home. A student at the University of Chicago (surely a hot-rodder himself) discovers divergent interpretations of the hot-rodder's intentions and offers his own. The criticism of Detroit that emerges anticipates the import-driven design and engineering revolution of the 1980s.

The following statement by Thomas W. Ryan, director of the New York Division of Safety, presents the typical image of the hot rodder in the mediums of mass communication. He is shown as a deliberate and premeditated lawbreaker: "Possession of the 'hot rod' car is presumptive evidence of an intent to speed. Speed is Public Enemy No. 1 of the highways. It is obvious that a driver of a 'hot rod' car has an irresistible temptation to 'step on it' and accordingly operate the vehicle in a reckless manner endangering human life. It also shows a deliberate and premeditated idea to violate the law. These vehicles are largely improvised by home mechanics and are capable of high speed and dangerous maneuverability. They have therefore become a serious menace to the safe movement of traffic. The operators of these cars are confused into believing that driving is a competitive sport. They have a feeling of superiority in recklessly darting in and out of traffic in their attempt to outspeed other cars on the road.". . .

The hot rodder's picture of himself is somewhat different, though obliquely cognizant of the medium's image. Thus . . . a hot-rod organization wrote as follows: "A hot rod accident or incident is newsworthy, while an accident involving ordinary cars is so common that it is usually not newsworthy. We wonder whether you appreciate the very real contribution that the hot rod industry, for it is an industry, has made to automotive transportation. The automotive industry has the equivalent of a million dollar experimental laboratory in the hot rod industry from which they can get valuable technical information free of any expense or risk of reputation. . . ."

. . . The hot rodder and his circle are highly articulate in their objections to the Detroit product as an automobile, and the reason is that they have little respect for the Detroit solution of a problem in transportation, engineering, and esthetics. The hot rodder says that the production car is uneconomical, unsafe at modern road speeds, and uglier than it has any right to be. What is more, it is too costly, too heavy, and too complicated by class and status symbolism to be a good car. Designed in ignorance of the hot rodder's credo that driving should not be so effortless that one forgets one is driving until after the crash, this car appears to the hot rodder to be a sort of high-speed parlor sofa. In general, the hot rodder protests against the automobile production and merchandising which fail to give the public a sufficiently wide range of models to permit judgments of value. . . .

While the hot-rod culture can be seen largely as an engineering protest against Detroit, some students of American culture have looked for other sociological and psychological meanings in the activities of the hot rodder. The hot-rod culture has been called an attack on the existing channels of expression—channels which grant success and acclaim only to those who fulfill certain occupational roles. It is said that the hot rod is especially suited to this search for spheres in which it is possible to obtain nonoccupational status because its prestige points are exhibited freely and personally in the very act of driving it. David Riesman and Reuel Denney have emphasized that the hot rod is more than an isolated phenomenon, that it is rather a single example of a process that may go on in the consumption of all mass-audience products. "As the hot rodder visibly breaks down the car as Detroit made it, and builds it up again with his own tools and energies, so the allegedly passive recipient of movies or radio, less visibly but just as surely, builds up his own amalgam of what he reads, sees, and hears; and in this, far from being manipulated, he is often the manipulator." Thus, apparently, the consumer can attain a measure of autonomy despite the attempts of mass-producers to channel consumption in "respectable" directions.

PREDICTIONS:
CENTURY IN THE BALANCE

JAMES BRYANT CONANT, 1951

A distinguished chemist and president of Harvard University, Conant supervised the Manhattan Project under franchise to Vannevar Bush. At the American Chemical Society's Diamond Jubilee meeting in 1951 he essayed to predict the "outlines of the balance of the 20th century."

I see in my crystal ball—to be sure a plastic one, as befits a chemical age—I see in this instrument of prophecy neither an atomic holocaust nor the golden abundance of an atomic age. On the contrary, I see worried humanity endeavoring by one political device after another to find a way out of the atomic age. And by the end of the century this appears to have been accomplished, but neither through the triumph of totalitarianism nor by the advent of world government. . . .

The era of liquid fossil fuels is by the close of the century coming to an end, and the worry about future coal supplies is increasing. . . . Atomic energy has not proved to be an expedient way of lengthening the period in which man taps the sources of energy stored in the earth's crust. Solar energy, on the other hand, is already of significance by the time the American Chemical Society celebrates its 100th anniversary [in 1976], and by the end of the century is the dominating factor in the production of industrial power. The practical utilization of this inexhaustible source of energy, together with the great changes in the production of food, has already had enormous effects on the economic and hence political relations of nations. With cheap power the economical production of fresh water from the sea has become a reality. This was about 1985, and made more than one desert near a seacoast a garden spot. . . .

The use of new techniques has made the world food situation in 1999 something quite different from what it was 50 years before. These alterations coupled with the discoveries about the relation of dietary factors to the birth rate and the rapid rise in the standards of living in nations once overcrowded seems to provide the new century—the 21st—with an answer to Malthus. The problem of overpopulation, while not solved, promises to be in hand before 2050. . . .

[People] date the changed attitude towards population to the year 1951 when Nehru advocated the establishment of birth control clinics in India and 1961 when the biochemists made available cheap and harmless antifertility components to be added as one saw fit to the diet. As the decades went by the attitude of the religious leaders of the world on this subject, so they say, completely altered without any diminution of religious feeling. . . .

But how did the industrialized nations of the world avoid deindustrializing each other by atomic bombs, you may inquire? Only by the narrowest of margins, is the answer, and only because time and again when one side or the other was about to take the plunge . . . the expert military advisers could not guarantee ultimate success. Of course, the turning point was in 1950, the year when collective security became a reality. . . . According to my crystal ball, for a decade or more after the mid-50s a series of battles in different parts of the globe and economic sanctions had, time and again, nearly precipitated World War III. . . . But after years of such anxiety and with the economic power of the free world backed by a measure of critical stability, time to consider a way out of the atomic age had clearly come.

Unless my observations are in error, I find the 1960s a time when constructive steps away from war are first being taken. Fifteen or 20 years after the first atomic bomb was fired, a sober appraisal of the debits and credits of the exploitation of atomic fission had led people to decide the game was not worth the candle. Of course, experimental plants were then producing somewhat more power from controlled atomic reactions than was consumed in the operation of the complex process, but the disposal of the waste products had presented gigantic problems—problems to be lived with for generations. The capital investment was very great. But quite apart from the technical difficulties there was the overriding fact that the potential military applications of atomic energy were inherently inimical to the very nations that controlled the weapons. A self-denying ordinance seemed but common sense. Once the illusion of "prosperity for all through the splitting of the atom" vanished from people's minds, the air began to clear. . . . The success of a vast technical undertaking to make atomic weapons showed what could be done in other radical departures. Rapid progress in the utilization of solar energy is thus seen in the 80s as a consequence of atomic energy development in the 40s.

The mood 15 or 20 years from now, as I glimpse it, is conditioned by a set of technological military and political factors quite different from those operating in the year 1951. Just enough agreement is then possible in the United Nations to proceed with gradual disarmament. Just enough inspection proves to be possible to enable even the most suspicious to trust an international guarantee to the effect that there is no assemblage anywhere of vast amounts of fissionable materials and of guided missiles. The existing stocks of fissionable materials are put beyond the immediate reach of any nation. The possibility of wholesale atomic raids by a nation which treacherously repudiates the treaty is eliminated, sufficient information at least can be guaranteed to settle men's doubts about that sort of war. . . .

As I listen in on the 1960s people are saying, "Clearly, an industrialized civilization could destroy itself and thus leave the world to those peoples not yet heavily urbanized and mechanized but what sense is there in that?

. . . The date of the great settlement is not clear in my reading of the future, but sometime between 1960 and 1980, the climate of opinion alters. The rearmament of the free world has done its work. Armies, navies, planes are still on hand but the trend is towards less rather than more military power. So I see the physicists and engineers at that time relieved of a terrible responsibility and gladly turning to labors more congenial than making fission and fusion bombs or guided missiles. I see the chemists in increasing numbers continuing to crowd into fields once reserved for others. An era of peace and prosperity really begins to dawn.

> They are called computers, simply because computation is the only
> significant job that has so far been given to them. The name has
> somewhat obscured the fact that they are capable of much greater
> generality. . . . To describe its potentialities, the computer needs a
> new name. Perhaps as good a name as any is "information machine."
> LOUIS RIDENOUR, 1952

TAKING THOUGHT

B. V. BOWDEN, 1953

At the end of the century, experts aren't so sure that computers can't think. In the early days of the computer era, the assertion was a commonplace.

A t this point it is necessary to interpolate a remark to explain our apparent attribution to the machine of certain human qualities, such as *Memory, Judgment* and so on. Modern digital computers are capable of performing long and elaborate computations; they can retain numbers which have been presented to them or which they have themselves derived during the course of the computations; they are, moreover, capable of modifying their own programs in the light of results which they have already derived. All these are operations which are usually performed (much more slowly and inaccurately) by human beings; but it is important to note that we do not claim that the machines can think for themselves. This is precisely what they cannot do. All the thinking has to be done for them in advance by the mathematician who planned their program and they can do only what is demanded of them; even if he leaves the choice between two courses of action to be made by the machines, he instructs them in detail how to make their choice. The abilities and limitations of a machine were foreseen by Lady Lovelace, who wrote [of Charles Babbage's mechanical computer that anticipated modern machines] in 1842: "The Analytical Engine has no pretensions whatever to originate anything. *It can do whatever we know how to order it to perform.*"

TOO CHEAP TO METER

LEWIS L. STRAUSS, 1954

A controversial chairman of the U.S. Atomic Energy Commission, evoking a utopian future powered by thermonuclear fusion, coins a memorable phrase.

I t is not too much to expect that our children will enjoy in their homes electrical energy too cheap to meter, will know of great periodic regional famines in the world only as matters of history, will travel effortlessly over the seas and under them and through the air with a minimum of dangers and at great speeds, and will experience a life span longer than ours. . . . This is the forecast for an age of peace.

THE PILL: I

BERNARD ASBELL

Feminist Margaret Sanger and philanthropist Katherine McCormick proposed the development of a birth-control pill to Gregory Pincus, the world's foremost authority on the mammalian egg, in 1951. Pincus teamed with a Harvard Medical School fertility specialist, John Rock, to produce the revolutionary compound.

In 1954 Rock began his first, cautious tests of the new progestin as a blocker of ovulation. Although its true purpose was not mentioned, the tests were historic as the first human trials of an oral contraceptive. . . .

The new progestin was tried by Rock on fifty women volunteers. . . . McCormick chafed over waiting out the tedium of trying to master "that 'ol' devil' the female reproductive system!" She asked Sanger, "How can we get a 'cage' of ovulating females to experiment with?" and hovered over Rock's work all through the winter of 1955, "freezing in Boston for the pill."

Rock, as a practicing physician, instinctively emphasized the safety of the volunteers over the laboratory researcher's concern for the outcome of

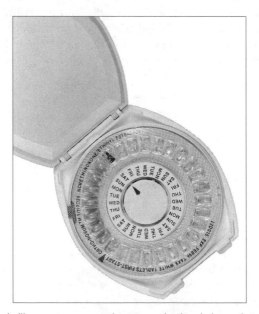

▶ The birth-control pill gave women control over reproductive choice and started an unfinished revolution.

RICHARD RHODES

the experiment. As a result, Rock's caution reached beyond normal practices of the time. A young Yale-trained obstetrician assisting Rock in the trials, Dr. Luigi Mastroianni, has recalled: "I don't really think I sensed the true significance of what was being done. The concept of informed consent that is so talked about now, and is a legal requirement of any research project involving human volunteers, didn't exist then. But Rock practiced it before it was ever defined." . . .

The work became known within the Rock team as the PPP—the Pincus Progesterone Project. But that was soon translated irreverently as the "pee-pee-pee" project to honor Mastroianni's endless task of checking daily urine samples from each of the fifty subjects.

The results were perfect. Not one of the fifty women ovulated. While requiring more proof than a trial with only fifty women, Pincus and Rock knew that they had identified an oral birth control pill.

Especially with as cautious a partner as Rock, Pincus was not about to run around shouting the triumphant news. Not yet. But he had to pour it out somewhere, to someone.

His wife, Elizabeth, who had a talent for distilling complex episodes into succinct summations, would never forget the moment her husband brought the news home. Using his intimate name for her, he said, "Lizuska, I've got it."

"What have you got?"

"I think we have a contraceptive pill."

"My God, why didn't you *tell* me?"

He replied that he was telling her now.

"Did you think you could *ever* get the pill?" she asked in awe.

Pincus replied grandly, "In science, Lizuska, everything is possible."

HAPPY AND
UNHAPPY ENDINGS

B. F. SKINNER, 1955

The master of operant conditioning asserts the long-range benefits of technologies of social control.

Those who reject the scientific conception of man must, to be logical, oppose the methods of science as well. The position is often supported by predicting a series of dire consequences which are to follow if science is not checked. A recent book by Joseph Wood Krutch, *The Measure of Man*, is in this vein. Mr. Krutch sees in the growing science of man the threat of an unexampled tyranny over men's minds. If science is permitted to have its way, he insists, "we may never be able really to think again." A controlled culture will, for example, lack some virtue inherent in disorder. We have emerged from chaos through a series of happy accidents, but in an engineered culture it will be "impossible for the unplanned to erupt again." But there is no virtue in the accidental character of an accident, and the diversity which arises from disorder can not only be duplicated by design but vastly extended. The experimental method is superior to simple observation just because it multiplies "accidents" in a systematic coverage of the possibilities. Technology offers many familiar examples. We no longer wait for immunity to disease to develop from a series of accidental exposures, nor do we wait for natural mutations in sheep and cotton to produce better fibers; but we continue to make use of such accidents when they occur, and we certainly do not prevent them. Many of the things we value have emerged from the clash of ignorant armies on darkling plains, but it is not therefore wise to encourage ignorance and darkness. . . .

To those who are stimulated by the glamorous heroism of the battlefield, a peaceful world may not be a better world. Others may reject a world without sorrow, longing or a sense of guilt because the relevance of deeply moving works of art would be lost. To many who have devoted their lives to the struggle to be wise and good, a world without confusion and evil might be an empty thing. A nostalgic concern for the decline of moral heroism has been a dominating theme in the work of Aldous Huxley. In *Brave New World* he could see in the application of science to human affairs only a travesty on the notion of the Good (just as George Orwell, in *1984*, could foresee nothing but horror). In a recent issue of *Esquire*, Huxley has expressed the point this way: "We have had religious revolutions, we have had political, industrial, economic and nationalistic revolutions. All of them, as our descendants will discover, were but ripples in an ocean of conservatism—trivial by comparison with the psychological revolution toward which we are so rapidly moving. *That* will really be a revolution. When it is over, the human race will give no further trouble." (Footnote for the reader of the future: This was not meant as a happy ending. Up to 1956 men had been admired, if at all, either for causing trouble or alleviating it. Therefore—)

It will be a long time before the world can dispense with heroes and hence with the cultural practice of admiring heroism, but we move in that direction whenever we act to prevent war, famine, pestilence and disaster. It will be a long time before man will never need to submit to a punishing environment or engage in exhausting labor, but we move in that direction whenever we make food, shelter, clothing and labor-saving devices more readily available. We may mourn the passing of heroes but not the conditions which make for heroism. We can spare the self-made saint or sage as we spare the laundress on the river's bank struggling against fearful odds to achieve cleanliness. . . .

Far from being a threat to the tradition of Western democracy, the growth of a science of man is a consistent and probably inevitable part of it. In turning to the external conditions which shape and maintain the behavior of men, while questioning the reality of inner qualities and faculties to which human achievement were once attributed, we turn from the ill-defined and remote to the observable and manipulable. Though it is a painful step, it has far-reaching consequences, for it not only sets higher standards of human welfare but shows us how to meet them. A change in a theory of human nature cannot change the facts. The achievements of man in science, art, literature, music and morals will survive any interpretation we place upon them. The uniqueness of the individual is unchallenged in the scientific view. Man, in short, will remain man.

GLOBAL EFFECTS

JOHN VON NEUMANN, 1955

A brilliant mathematician, self-taught theoretical physicist and primary contributor to the development of the digital computer, von Neumann was one of a group of Hungarian scientists, the "men from Mars," who emigrated to the United States in the 1930s to escape the Nazi persecution of the Jews and who later participated in the development of the first atomic bombs. Near the end of his life, as a member of the U.S. Atomic Energy Commission, he speculated on the technological future, identifying perspicaciously the whole-earth effects that the environmental movement would later embrace. Note his early characterization of the threat of global warming.

"The great globe itself" is in a rapidly maturing crisis—a crisis attributable to the fact that the environment in which technological progress must occur has become both undersized and underorganized. To define the crisis with any accuracy, and to explore possibilities of dealing with it, we must not only look at relevant facts, but also engage in some speculation. The process will illuminate some potential technological developments of the next quarter-century.

In the first half of this century the accelerating industrial revolution encountered an absolute limitation—not on technological progress as such but on an essential safety factor. This safety factor, which had permitted the industrial revolution to roll on from the mid-eighteenth to the early twentieth century, was essentially a matter of geographical and political *Lebens-*

raum: an ever broader geographical scope for technological activities, combined with an ever broader political integration of the world. Within this expanding framework it was possible to accommodate the major tensions created by technological progress.

Now this safety mechanism is being sharply inhibited; literally and figuratively, we are running out of room. At long last, we begin to feel the effects of the finite, actual size of the earth in a critical way.

Thus the crisis does not arise from accidental events or human errors. It is inherent in technology's relation to geography on the one hand and to political organization on the other. The crisis was developing visibly in the 1940s, and some phases can be traced back to 1914. In the years between now and 1980 the crisis will probably develop far beyond all earlier patterns. When or how it will end—or to what state of affairs it will yield—nobody can say.

In all its stages the industrial revolution consisted of making available more and cheaper energy, more and easier controls of human actions and reactions, and more and faster communications. Each development increased the effectiveness of the other two. All three factors increased the speed of performing large-scale operations—industrial, mercantile, political, and migratory. But throughout the development, increased speed did not so much shorten time requirements of processes as extend the areas of the earth affected by them. The reason is clear. Since most *time* scales are fixed by human reaction times, habits, and other physiological and psychological factors, the effect of the increased speed of technological processes was to enlarge the *size* of units—political, organizational, economic, and cultural—affected by technological operations. That is, instead of performing the same operations as before in less time, now larger-scale operations were performed in the same time. This important evolution has a natural limit, that of the earth's actual size. The limit is now being reached, or at least closely approached.

Indications of this appeared early and with dramatic force in the military sphere. By 1940 even the larger countries of continental Western Europe were inadequate as military units. Only Russia could sustain a major military reverse without collapsing. Since 1945, improved aeronautics and communications alone might have sufficed to make any geographical unit, including Russia, inadequate in a future war. The advent of nuclear weapons merely climaxes the development. Now the effectiveness of offensive weapons is such as to stultify all plausible defensive time scales. . . .

Technological evolution is still accelerating. Technologies are always constructive and beneficial, directly or indirectly. Yet their consequences tend to increase instability—a point that will get closer attention after we have had a look at certain aspects of continuing technological evolution.

First of all, there is a rapidly expanding supply of energy. It is generally agreed that even conventional, chemical fuel—coal or oil—will be available in increased quantity in the next two decades. Increasing demand tends to keep fuel prices high, yet improvements in methods of generation seem to bring the price of power down. There is little doubt that the most significant event affecting energy is the advent of nuclear power. Its only available controlled source today is the nuclear-fission reactor. Reactor techniques appear to be approaching a condition in which they will be competitive with conventional (chemical) power sources within the U.S.; however, because of generally higher fuel prices abroad, they could already be more than competitive in many important foreign areas. . . .

Yet fission is not nature's normal way of releasing nuclear energy. In the long run, systematic industrial exploitation of nuclear energy may shift reliance onto other and still more abundant modes. . . . Consequently, a few decades hence energy may be free—just like the unmetered air—with coal and oil used mainly as raw materials for organic chemical synthesis, to which, as experience has shown, their properties are best suited. . . .

Also likely to evolve fast—and quite apart from nuclear evolution— is automation. . . . Automatic control, of course, is as old as the industrial revolution, for the decisive new feature of Watt's steam engine was its automatic valve control, including speed control by a "governor." In our century, however, small electric amplifying and switching devices put automation on an entirely new footing. This development began with the electromechanical (telephone) relay, continued and unfolded with the vacuum tube, and appears to accelerate with various solid-state devices (semiconductor crystals, ferromagnetic cores, etc.). The last decade or two has also witnessed an increasing ability to control and "discipline" large numbers of such devices within one machine. . . .

Many such machines have been built to perform complicated scientific and engineering calculations and large-scale accounting and logistical surveys. There is no doubt that they will be used for elaborate industrial process control, logistical, economic, and other planning, and many other purposes heretofore lying entirely outside the compass of quantitative and automatic control and preplanning. Thanks to simplified forms of automatic or semi-automatic control, the efficiency of some important branches of industry has increased considerably during recent decades. It is therefore to be expected that the considerably elaborated newer forms, now becoming increasingly available, will effect much more along these lines.

Fundamentally, improvements in control are really improvements in communicating information within an organization or mechanism. The sum total of progress in this sphere is explosive. . . .

Such developments as free energy, greater automation, improved communications, partial or total climate control have common traits deserving special mention. First, though all are intrinsically useful, they can lend themselves to destruction. Even the most formidable tools of nuclear destruction are only extreme members of a genus that includes useful methods of energy release or element transmutation. The most constructive schemes for climate control would have to be based on insights and techniques that would also lend themselves to forms of climatic warfare as yet unimagined. Technology—like science—is neutral all through, providing only means of control applicable to any purpose, indifferent to all.

Second, there is in most of these developments a trend toward affecting the earth as a whole, or to be more exact, towards producing effects that can be projected from any one to any other point on the earth. There is an intrinsic conflict with geography—and institutions based thereon—as understood today. Of course, any technology interacts with geography, and each imposes its own geographical rules and modalities. The technology that is now developing and that will dominate the next decades seems to be in total conflict with traditional and, in the main, momentarily still valid, geographical and political units and concepts. This is the maturing crisis of technology.

What kind of action does this situation call for? *Whatever* one feels inclined to do, one decisive trait must be considered: the very techniques that create the dangers and the instabilities are in themselves useful, or closely related to the useful. In fact, the more useful they could be, the more unstabilizing their effects can also be. It is not a particular perverse destructiveness of one particular invention that creates danger. Technological power, technological efficiency as such, is an ambivalent achievement. Its danger is intrinsic.

In looking for a solution, it is well to exclude one pseudosolution at the start. The crisis will not be resolved by inhibiting this or that apparently particularly obnoxious form of technology. For one thing, the parts of technology, as well as of the underlying sciences, are so intertwined that in the long run nothing less than a total elimination of all technological progress would suffice for inhibition. Also, on a more pedestrian and immediate basis, useful and harmful techniques lie everywhere so close together that it is never possible to separate the lions from the lambs. This is known to all who have so laboriously tried to separate secret, "classified" science or technology (military) from the "open" kind; success is never more—nor intended to be more—than transient, lasting perhaps half a decade. Similarly, a separation into useful and harmful subjects in any technological sphere would probably diffuse into nothing in a decade.

Moreover, in this case successful separation would have to be enduring (unlike the case of military "classification," in which even a few years' gain may be important). Also, the proximity of useful techniques to harmful ones, and the possibility of putting the harmful ones to military use, puts a competitive premium on infringement. Hence the banning of particular technologies would have to be enforced on a worldwide basis. But the only authority that could do this effectively would have to be of such scope and perfection as to signal the *resolution* of international problems rather than the discovery of a *means* to resolve them.

Finally and, I believe, most importantly, prohibition of technology (invention and development, which are hardly separable from underlying scientific inquiry), is contrary to the whole ethos of the industrial age. It is irreconcilable with a major mode of intellectuality as our age understands it. It is hard to imagine such a restraint successfully imposed in our civilization. Only if those disasters that we fear had already occurred, only if humanity were already completely disillusioned about technological civilization, could such a step be taken. But not even the disasters of recent wars have produced that degree of disillusionment, as is proved by the phenomenal resiliency with which the industrial way of life recovered even—or particularly—in the worst-hit areas. The technological system retains enormous vitality, probably more than ever before, and the counsel of restraint is unlikely to be heeded.

A much more satisfactory solution than technological prohibition would be eliminating war as "a means of national policy." The desire to do this is as old as any part of the ethical system by which we profess to be governed. The intensity of the sentiment fluctuates, increasing greatly after major wars. . . . True, this time the danger of destruction seems to be real rather than apparent, but there is no guarantee that a real danger can control human actions better than a convincing appearance of danger.

What safeguard remains? Apparently only day-to-day—or perhaps year-to-year—opportunistic measures, a long sequence of small, correct decisions. And this is not surprising. After all, the crisis is due to the rapidity of progress, to the probable further acceleration thereof, and to the reaching of certain critical relationships. Specifically, the effects that we are now beginning to produce are of the same order of magnitude as that of "the great globe itself." Indeed, they affect the earth as an entity. Hence further acceleration can no longer be absorbed as in the past by an extension of the area of operations. Under present conditions it is unreasonable to expect a novel cure-all.

For progress there is no cure. Any attempt to find automatically safe channels for the present explosive variety of progress must lead to frustration. The only safety possible is relative, and it lies in an intelligent exercise of day-to-day judgment. . . .

The one solid fact is that the difficulties are due to an evolution that, while useful and constructive, is also dangerous. Can we produce the required adjustments with the necessary speed? The most hopeful answer is that the human species has been subjected to similar tests before and seems to have a congenital ability to come through, after varying amounts of trouble. To ask in advance for a complete recipe would be unreasonable. We can specify only the human qualities required: patience, flexibility, intelligence.

> Technology . . . the knack of so arranging the world that we don't have to experience it.
>
> MAX FRISCH, 1957

FELLOW TRAVELER

PRAVDA, 1957

The beeping of the first artificial Earth satellite—launched from Soviet Russia—shocked a complacent United States into accelerating its own satellite and ICBM programs and investing in improved science and math education. Pravda's announcement was surprisingly restrained, resorting to socialist boilerplate only in the last paragraph. "Sputnik," as the Soviets called their small satellite, means "fellow traveler," a name Senator Joseph McCarthy had recently made synonymous with Communist sympathies.

On October 4, 1957, this first satellite was successfully launched in the USSR. According to preliminary data, the carrier rocket has imparted to the satellite the required orbital velocity of about 8000 meters per second. At the present time the satellite is describing elliptical trajectories around the earth, and its flight can be observed in the rays of the rising and setting sun with the aid of very simple optical instruments (binoculars, telescopes, etc.).

According to calculations which now are being supplemented by direct observations, the satellite will travel at altitudes up to 900 kilometers above the surface of the earth; the time for a complete revolution of the satellite will be one hour and thirty-five minutes; the angle of inclination of its orbit to the equatorial plane is 65 degrees. On October 5 the satellite will pass over the Moscow area twice—at 1:46 a.m. and at 6:42 a.m. Moscow time. Reports about the subsequent movement of the first artificial satellite launched in the USSR on October 4 will be issued regularly by broadcasting stations.

The satellite has a spherical shape 58 centimeters in diameter and weighs 83.6 kilograms. It is equipped with two radio transmitters continuously emitting signals at frequencies of 20.005 and 40.002 megacycles per

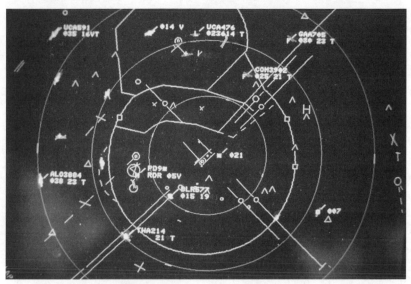

▶ World War II technology transfer: radar.

second (wave lengths of about 15 and 7.5 meters, respectively). The power of the transmitters ensures reliable reception of the signals by a broad range of radio amateurs. The signals have the form of telegraph pulses of about 0.3 second's duration with a pause of the same duration. The signal of one frequency is sent during the pause in the signal of the other frequency.

Scientific stations located at various points in the Soviet Union are tracking the satellite and determining the elements of its trajectory. Since the density of the rarefied upper layers of the atmosphere is not accurately known, there are no data at present for the precise determination of the satellite's lifetime and of the point of its entry into the dense layers of the atmosphere. Calculations have shown that owing to the tremendous velocity of the satellite, at the end of its existence it will burn up on reaching the dense layers of the atmosphere at an altitude of several tens of kilometers.

As early as the end of the nineteenth century the possibility of realizing cosmic flights by means of rockets was first scientifically substantiated in Russia by the works of the outstanding Russian scientist K[onstantin] E. Tsiolkovskii.

The successful launching of the first man-made earth satellite makes a most important contribution to the treasure house of world science and culture. The scientific experiment accomplished at such a great height is of tremendous importance for learning the properties of cosmic space and for studying the earth as a planet of our solar system. . . .

Artificial earth satellites will pave the way to interplanetary travel and apparently our contemporaries will witness how the free and conscientious labor of the people of the new socialist society makes the most daring dreams of mankind a reality.

SPACE VEGETABLES

JAMES G. FULTON, 1959

Congressman Fulton (R-Pennsylvania) was a colorful member of the Committee on Science and Astronautics (later Technology) that was formed in the U.S. Congress in the wake of Sputnik. Here, as recalled fondly by Congressman Kenneth Hechler, he pushes the space-food envelope.

In some other areas in the first few years, the . . . Committee held hearings which seemed to stretch its jurisdiction pretty far. The committee was barely a month old when [it started] some new hearings on space food by calling the Department of Agriculture over to testify. The hearings were a disaster in their lack of planning, and almost total lack of any useful information elicited. . . . The hearings would have completely collapsed had it not been for Congressman Fulton's determination "to spur you on to new ideas and new approaches. . . . We are trying to get you to raise your sights." After getting nothing but wooden responses to his questions, Fulton finally erupted with a question which literally stunned the witness and was long and fondly remembered as the greatest Fultonism of all time:

Possibly in space the approach to vegetables might be different. Did that ever strike you—because we are thinking of three-dimensional vegetables, maybe in space, where you have a lot of sunlight, you might get a two-dimensional tomato. It might be one million miles long and as thin as a sheet of paper, aimed toward the sun—a tomato.

There was a long silence, as the Department of Agriculture witness blinked, and finally blurted out softly: "It is an interesting thought." He was completely flabbergasted.

Natural Luddites

C. P. Snow, 1959

I heard this British physicist, novelist and government official lecture at Yale University in 1959, the same year he delivered his famous lectures on "the two cultures and the scientific revolution" at Cambridge University. He was a large man, bald, jowly, not handsome, with a bullet head and a high, shrill voice. Snow trained at Cambridge under the great experimental physicist Ernest Rutherford; his unusual combination of scientific and literary experience qualified him to speak to both sides of the issue of two cultures: the issue, as he puts it early in his Cambridge lectures, of "literary intellectuals at one pole—at the other scientists, and as the most representative, the physical scientists. Between the two a gulf of mutual incomprehension—sometimes (particularly among the young) hostility and dislike, but most of all lack of understanding." Early in his second lecture, he explores how such a destructive gulf came about.

The reasons for the existence of the two cultures are many, deep and complex, some rooted in social histories, some in personal histories, and some in the inner dynamic of the different kinds of mental activity themselves. But I want to isolate one which is not so much a reason as a correlative, something which winds in and out of any of these discussions. It can be said simply, and it is this. If we forget the scientific culture, then the rest of western intellectuals have never tried, wanted or been able to understand the industrial revolution, much less accept it. Intellectuals, in particular literary intellectuals, are natural Luddites.

That is specially true of this country [i.e., the United Kingdom], where the industrial revolution happened to us earlier than elsewhere, during a long spell of absentmindedness. Perhaps that helps explain our present degree of crystallization. But, with a little qualification, it is also true, and surprisingly true, of the United States.

In both countries, and indeed all over the West, the first wave of the industrial revolution crept on, without anyone noticing what was happening. It was, of course—or at least it was destined to become, under our own eyes, and in our own time—by far the biggest transformation in society since the discovery of agriculture. In fact, those two revolutions, the agricultural and the industrial-scientific, are the only qualitative changes in social living that men have ever known. But the traditional culture didn't notice: or when it did notice, didn't like what it saw. . . .

Almost none of the talent, almost none of the imaginative energy, went back into the revolution which was producing the wealth. The traditional culture became more abstracted from it as it became more wealthy, trained its young men for administration, for the Indian Empire, for the purpose of perpetuating the culture itself, but never in any circumstances to equip them to understand the revolution or take part in it. . . .

Much the same was true of the U.S. The industrial revolution, which began developing in New England fifty years or so later than ours, apparently received very little educated talent, either then or later in the nineteenth century. It had to make do with the guidance handymen could give it—sometimes, of course, handymen like Henry Ford, with a dash of genius. . . .

Almost everywhere . . . intellectual persons didn't comprehend what was happening. Certainly the writers didn't. Plenty of them shuddered away, as though the right course for a man of feeling was to contract out; some, like Ruskin and William Morris and Thoreau and Emerson and Lawrence, tried various kinds of fancies which were not in effect more than screams of horror. It is hard to think of a writer of high class who really stretched his imaginative sympathy, who could see at once the hideous back-streets, the smoking chimneys, the internal price—and also the prospects of life that were opening out for the poor, the intimations, up to now unknown except to the lucky, which were just coming within reach of the remaining 99 percent of his brother men. Some of the nineteenth-century Russian novelists might have done; their natures were broad enough; but they were living in a preindustrial society and didn't have the opportunity. The only writer of world class who seems to have had an understanding of the industrial revolution was Ibsen in his old age: and there wasn't much that old man didn't understand.

For, of course, one truth is straightforward. Industrialization is the only hope of the poor. I use the word "hope" in a crude and prosaic sense. I have not much use for the moral sensibility of anyone who is too refined to use it so. It is all very well for us, sitting pretty, to think that material standards of living don't matter all that much. It is all very well for one, as a personal choice, to reject industrialization—do a modern Walden, if you like, and if you go without much food, see most of your children die in infancy, despise the comforts of literacy, accept twenty years off your own life, then I respect you for the strength of your aesthetic revulsion. But I don't respect you in the slightest if, even passively, you try to impose the same choice on others who are not free to choose. In fact, we know what their choice would be. For, with singular unanimity, in any country where they have had the chance, the poor have walked off the land into the factories as fast as the factories could take them. . . .

The industrial revolution looked very different according to whether one saw it from above or below. It looks very different today according to whether one sees it from Chelsea or from a village in Asia. To people like my grandfather, there was no question that the industrial revolution was less bad than what had gone before. The only question was, how to make it better.

In a more sophisticated sense, that is still the question. In the advanced countries, we have realized in a rough and ready way what the old industrial revolution brought with it. A great increase in population, because applied science went hand in hand with medical science and medical care. Enough to eat, for a similar reason. Everyone able to read and write, because an industrial society can't work without. Health, food, education; nothing but the industrial revolution could have spread them right down to the very poor. Those are primary gains—there are losses too, of course, one of which is that organizing a society for industry makes it easy to organize it for all-out war. But the gains remain. They are the base of our social hope.

REINVENTING INVENTION

John Jewkes, David Sawers and
Richard Stillerman, 1959

There can be little doubt that the greater part of modern writing about invention and technical progress strongly inclines to the view that we live in a new world in which thinking of the present or the future in terms of past experience is largely irrelevant, and that our ideas must be recast and our institutions reformed to fit fresh surroundings. Social scientists are now tending to speak with more confidence about the scale on which inventions will be made and the sources from which they will arise. There have, indeed, been some odd switches of thought since the end of the First World War. In the early 1930s it was widely believed that technical progress would normally be so swift and disturbing that a high level of "technological" unemployment would be usual and inevitable. In the later thirties, due mainly to the failure of the American economy to continue to expand at an unbroken rate, the view gained currency that technical progress would usually be too sluggish to create sufficient profitable investment openings for the savings arising under full employment: secular stagnation and chronic unemployment were inevitable, in the absence of public intervention, because technical progress would never be on a large enough scale.

The period following the Second World War with its general shortage of capital, its full employment and the impact upon the public mind of the discovery of atomic energy, has brought forth a fresh crop of generalizations which, compared with the pessimistic views of the inter-war period, are oddly sanguine in tone. Unbroken and rapid technical advance, it is thought, can now be taken for granted; the causes of it, the institutions which best foster it, are understood, and understood so well that society may be within measurable distance of the power deliberately to control it or, failing that, to predict with a high degree of certainty what the future holds.

This new and fashionable doctrine, subscribed to by many scientists, technologists, economists, statesmen, businessmen and popular writers, does not seem anywhere to have been expounded in final or authoritative shape. Nor is it difficult to pick out inconsistencies between its varying formulations. The mainspring is sometimes held to be the growing power of science, sometimes the increasing skill and eagerness with which technologists pick up and use scientific knowledge. By some, inventions are now thought to be easier to make than formerly; there is a gathering momentum, an "autocatalytic process," driving things on. Others believe that inventions have now become more difficult to make because all the easy inventions have already been made. "Necessity," to some is still "the mother of invention"; to others, inventions are thought to pour out almost automatically and prodigally in an ever-increasing stream in the richer industrial communities where needs are least urgent. Some hold that there never was a time when inventions were more eagerly seized upon by industry for commercial exploitation; others that there is still a great deal of "resistance to change."

Whatever the discrepancies and dissensions, there appears to be a broad area of agreement among those who adhere to the spectacular view of modern invention. . . .

The pith of the modern view is . . . that in the nineteenth century most inventions came from the individual inventor who had little or no scientific training, and who worked largely with simple equipment and by empirical methods and unsystematic hunches. The link between science and technology was slight. Manufacturing businesses did not concern themselves with research. In the twentieth century the characteristic features of the nineteenth century are passing away. The individual inventor is becoming rare; men with the power of originating are largely absorbed into research institutions of one kind or another, where they must have expensive equipment for their work. Useful invention is to an ever-increasing degree issuing from the research laboratories of large firms which alone can afford to operate on an appropriate scale. There is increasingly close contact now between science and technology, both through the closer association of the workers in the two fields and because the borderline between the two formerly separate functions is becoming obliterated. The consequence is that invention has become more automatic, less the result of intuition or flashes of genius and more a matter of deliberate design. The growing power to invent, combined with the increased resources devoted to it, has produced a spurt of technical progress to which no obvious limit is to be seen.

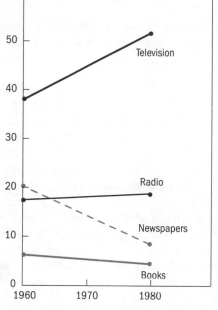

▶ Percentage of words consumed by medium, 1960–1980 (selected).

FIST FIGHTS AND FEMALES

ERNST ALEXANDERSON, 1960

A pioneer of television research at General Electric in the 1930s wrote this letter to a newspaper television critic in 1960, at eighty-two.

Forty years ago I had occasion to speculate on the impact on the world of radio and television. Now we know some of the answers and it has occurred to me that the four most popular programs could be described as follows using the classical verse measure of hexameter:

Fist fights and females and guns with horses and wagon as background
Fist fights and females and guns with a river boat as a background
Fist fights and females and guns with Indians and horses as background
Fist fights and females and guns with horses and cows as a background

This is not a sarcasm, quite the contrary. The popularity of these programs proves that the world is yearning for an escape from modern art, modern music and the Cold War, wishing to return to fundamental emotions.

I hope you will encourage such programs with themes from other ages such as the Vikings and variations as a substitute for fist fights. There is no substitute for females.

BLIND DATE: I

J. C. R. LICKLIDER, 1960

One of the pioneers of computer communication considers the possibility of cooperative interaction between humans and computers.

RICHARD RHODES

214

"Man-computer symbiosis" is a subclass of "man-machine systems." There are many man-machine systems. At present, however, there are no man-computer symbioses. . . . The hope is that, in not too many years, human brains and computing machines will be coupled together very tightly, and that the resulting partnership will think as no human brain has ever thought and process data in a way not approached by the information-handling machines we know today.

As a concept, man-computer symbiosis is different in an important way from what [J. D.] North has called "mechanically extended man." In the man-machine systems of the past, the human operator supplied the initiative, the direction, the integration, and the criterion. The mechanical parts of the system were mere extensions, first of the human arm, then of the human eye. These systems certainly did not consist of "dissimilar organisms living together" [i.e., symbiosis]. There was only one kind of organism— man—and the rest was there only to help him.

In one sense, of course, any man-made system is intended to help man, to help a man or men outside the system. If we focus upon the human operator(s) within the system, however, we see that, in some areas of technology, a fantastic change has taken place during the last few years. "Mechanical extension" has given way to replacement of men, to automation, and the men who remain are there more to help than to be helped. In some instances, particularly in large computer-centered information and control systems, the human operators are responsible mainly for functions that it proved infeasible to automate. Such systems ("humanly extended machines," North might call them) are not symbiotic systems. They are semi-automatic systems, systems that started out to be fully automatic but fell short of the goal.

Man-computer symbiosis is probably not the ultimate paradigm for complex technological systems. It seems entirely possible that, in due course, electronic or chemical "machines" will outdo the human brain in most of the functions we now consider exclusively within its province. . . . There will nevertheless be a fairly long interim during which the main intellectual advances will be made by men and computers working together in intimate association. A multidisciplinary study group, examining future research and development problems of the Air Force, estimated that it would be 1980 before developments in artificial intelligence make it possible for machines alone to do much thinking or problem-solving of military significance. That would leave, say, five years to develop man-computer symbiosis and 15 years to use it. The 15 may be 10 or 500, but those years should be intellectually the most creative and exciting in the history of mankind. . . .

It seems likely that the contributions of human operators and equipment will blend together so completely in many operations that it will be difficult to separate them neatly in analysis. That would be the case if, in gathering data on which to base a decision, for example, both the man and the computer came up with relevant precedents from experience and if the computer then suggested a course of action that agreed with the man's intuitive judgment. . . . In other operations, however, the contributions of men and equipment will be to some extent separable.

Men will set the goals and supply the motivations, of course, at least in the early years. They will formulate hypotheses. They will ask questions. They will think of mechanisms, procedures, and models. They will remember that such-and-such a person did some possibly relevant work on a topic of interest back in 1947, or at any rate shortly after World War II, and they will have an idea in what journals it might have been published. In general, they will make approximate and fallible, but leading, contributions, and they will define criteria and serve as evaluators, judging the contributions of the equipment and guiding the general line of thought.

In addition, men will handle the very-low-probability situations when such situations do actually arise. . . . Men will fill the gaps, either in the problem solution or in the computer program, when the computer has no mode or routine that is applicable in a particular circumstance.

The information-processing equipment, for its part, will convert hypotheses into testable models and then test the models against data (which the human operator may designate roughly and identify as relevant when the computer presents them for his approval). The equipment will answer questions. It will simulate the mechanisms and models, carry out the procedures, and display the results to the operator. It will transform data, plot graphs ("cutting the cake" in whatever way the human operator specifies, or in several alternative ways if the human operator is not sure what he wants). The equipment will interpolate, extrapolate, and transform. It will convert static equations or logical statements into dynamic models so the human operator can examine their behavior. In general, it will carry out the routinizable, clerical operations that fill the intervals between decisions.

In addition, the computer will serve as a statistical-inference, decision-theory, or game-theory machine to make elementary evaluations of suggested courses of action whenever there is enough basis to support a formal statistical analysis. Finally, it will do as much diagnosis, pattern matching, and relevance recognizing as it profitably can, but it will accept a clearly secondary status in these areas.

The data-processing equipment tacitly postulated in the proceeding [discussion] is not available. The computer programs have not been written. There are in fact several hurdles that stand between the nonsymbiotic present and the anticipated symbiotic future. . . .

Any present-day large-scale computer is too fast and too costly for real-time cooperative thinking with one man. Clearly, for the sake of efficiency and economy, the computer must divide its time among many users. . . .

It seems reasonable to envision, for a time 10 or 15 years hence, a "thinking center" that will incorporate the functions of present-day libraries together with anticipated advances in information storage and retrieval and the symbiotic functions suggested earlier. . . . The picture readily enlarges itself into a network of such centers, connected to one another by wideband communication lines and to individual users by leased-wire services. . . .

When we start to think of storing any appreciable fraction of a technical literature in computer memory, we run into billions of bits and, unless things change markedly, billions of dollars.

The first thing to face is that we shall not store all the technical and scientific papers in computer memory. . . .

The basic dissimilarity between human languages and computer languages may be the most serious obstacle to true symbiosis.

BLIND DATE: II

NORBERT WIENER, 1960

Norbert Wiener, professor of mathematics at MIT, coined the term cybernetics— *meaning, literally, the art of steering.*

I should like to mention a certain attitude of the man in the street toward cybernetics and automation. This attitude needs a critical discussion, and in my opinion it should be rejected in its entirety. This is the assumption that machines cannot possess any degree of originality. This frequently takes the form of a statement that nothing can come out of the machine which has not been put into it. This is often interpreted as asserting that a machine which man has made must remain continually subject to man, so that its operation is at any time open to human interference and to a change in policy. On the basis of such an attitude, many people have pooh-poohed the dangers of machine techniques, and they have flatly contradicted the early predictions of Samuel Butler that the machine might take over the control of mankind. . . .

It is my thesis that machines can and do transcend some of the limitations of their designers, and that in doing so they may be both effective and dangerous. It may well be that in principle we cannot make any machine the elements of whose behavior we cannot comprehend sooner or later. This does not mean in any way that we shall be able to comprehend these elements in substantially less time than the time required for operation of the machine, or even within any given number of years or generations.

As is now generally admitted, over a limited range of operation, machines act far more rapidly than human beings and are far more precise in performing the details of their operations. This being the case, even when machines do not in any way transcend man's intelligence, they very well may, and often do, transcend man in the performance of tasks. An intelligent understanding of their mode of performance may be delayed until long after the task which they have been set has been completed.

This means that though machines are theoretically subject to human criticism, such criticism may be ineffective until long after it is irrelevant. To be effective in warding off disastrous consequences, our understanding of our man-made machines should in general develop *pari passu* [i.e., in pace] with the performance of the machine. By the very slowness of our human actions, our effective control of our machines may be nullified. By the time we are able to react to information conveyed by our senses and stop the car we are driving, it may already have run head on into a wall. . . .

The problem, and it is a moral problem, with which we are here faced is very close to one of the great problems of slavery. Let us grant that slavery is bad because it is cruel. It is, however, self-contradictory, and for a reason which is quite different. We wish a slave to be intelligent, to be able to assist us in the carrying out of our tasks. However, we also wish him to be subservient. Complete subservience and complete intelligence do not go together. How often in ancient times the clever Greek philosopher slave of a less intelligent Roman slaveholder must have dominated the actions of his master rather than obeyed his wishes! Similarly, if the machines become more and more efficient and operate at a higher and higher psychological level, the catastrophe foreseen by Butler of the dominance of the machine comes nearer and nearer.

The human brain is a far more efficient control apparatus than is the intelligent machine when we come to the higher areas of logic. It is a self-organizing system which depends on its capacity to modify itself into a new machine rather than on ironclad accuracy and speed in problem-solving. We have already made very successful machines of the lowest logical type, with a rigid policy. We are beginning to make machines of the second logical type, where the policy itself improves with learning. In the construction of operative machines, there is no specific foreseeable limit with respect to logical type, nor is it safe to make a pronouncement about the exact level at which the brain is superior to the machine. Yet for a long time at least there will always be some level at which the brain is better than the constructed machine, even though this level may shift upwards and upwards.

It may be seen that the result of a programming technique of automatization is to remove from the mind of the designer and operator an effective understanding of many of the stages by which the machine comes to its conclusions and of what the real tactical intentions of many of its operations may be. This is highly relevant to the problem of our being able to foresee undesired consequences outside the frame of the strategy of the game while the machine is still in action and while intervention on our part may prevent the occurrence of these consequences.

Here it is necessary to realize that human action is a feedback action. To avoid a disastrous consequence, it is not enough that some action on our part should be sufficient to change the course of the machine, because it is quite possible that we lack information on which to base consideration of such an action. . . .

Disastrous results are to be expected . . . wherever two agencies essentially foreign to each other are coupled in the attempt to achieve a common purpose. If the communication between these two agencies as to the nature of this purpose is incomplete, it must only be expected that the result of this cooperation will be unsatisfactory. If we use, to achieve our purposes, a mechanical agency with whose operation we cannot efficiently interfere once we have started it, because the action is so fast and irrevocable that we have not the data to intervene before the action is complete, then we had better be quite sure that the purpose put into the machine is the purpose which we really desire and not merely a colorful imitation of it.

Up to this point I have been considering the quasi-moral problems caused by the simultaneous action of the machine and the human being in a joint enterprise. We have seen that one of the chief causes of the danger of disastrous consequences in the use of the learning machine is that man and machine operate on two distinct time scales, so that the machine is much faster than man and the two do not gear together without serious difficulties. Problems of the same sort arise whenever two control operators on very different time scales act together, irrespective of which system is the faster and which system is the slower. This leaves us the much more direct moral question: What are the moral problems when man as an individual operates in connection with the controlled process of a much slower time scale, such as a portion of political history or—our main subject of inquiry—the development of science?

Let it be noted that the development of science is a control and communication process for the longterm understanding and control of matter. In this process 50 years are as a day in the life of the individual. For this reason, the individual scientist must work as a part of a process whose time scale is so long that he himself can only contemplate a very limited sector of it. Here, too, communication between the two parts of a double machine is difficult and limited. Even when the individual believes that science contributes to the human ends which he has at heart, his belief needs continual scanning and reevaluation which is only partly possible. For the individual scientist, even the partial appraisal of this liaison between the man and the process requires an imaginative forward glance at history which is difficult, exacting, and only limitedly achievable. And if we adhere simply to the creed of the scientist, that an incomplete knowledge of the world and of ourselves is better than no knowledge, we can still by no means always justify the naive assumption that the faster we rush ahead to employ the new powers for action which are opened up to us, the better it will be. We must always exert the full strength of our imagination to examine where the full use of our new modalities may lead us.

THE FIRST LASER

Theodore H. Maiman, 1960

Although physicists Charles Townes and Arthur Schawlow won the Nobel Prize for their seminal 1958 theoretical paper "Infrared and Optical Masers," a young physicist at Hughes Research Laboratories, Ted Maiman, actually built the first working laser, a small cylinder of artificial ruby with silvered ends nested in a coiled flashlamp that pumped the ruby with energy, an assembly that fit nicely inside an aluminum reflector tube the size of a soup can. Maiman first measured strong pulses of coherent red light on May 16, 1960, nine months after he began design work.

I didn't know what material I was going to use at first, but I was thinking that I wanted it to be something fairly rugged. . . . I decided to use ruby because it had other properties that were interesting. It had a simple energy level scheme, was rugged, and in principle I wouldn't have to cool it below room temperature. To make it work, it wouldn't have to be very big. . . .

I started to get a little more interested in ruby when I found the quantum efficiency was near unity. I knew it would still take a lot of power, though. . . . I went through the calculations and came across the fact that for such a system what was really important was the brightness of the pumping source—which amounted to the equivalent of a black body of some 5,000 K [i.e., degrees Kelvin]. This was not unheard of, but it was really pushing the specifications of the common laboratory lamp. . . .

My insight regarding the high lamp temperature proved to be very important. I remember reading that the effective temperature of a xenon flashlamp was 8,000 K. Also, I didn't see any reason why I had to do this continuous-wave—pulsed mode was perfectly fine. People do a lot of things purposely with pulses, for example, radar. Besides, I was just trying to demonstrate that this could be done, not find the ultimate system. But I was determined that it not be cryogenic. . . . Another consideration that kept me hanging in with ruby was that, if it worked at all, it would be small and compact and operate at room temperature. I was also intrigued by the fact that, if it worked, ruby would emit visible light. I would be able to see it. . . .

I went through the catalogs of available laboratory lamps, and calculated the effective brightness. . . . And as I screened them and isolated them, there were only three lamps that had enough brightness to do it. All three of them were made by General Electric and all three of them were quartz helices. I bought a few versions of all three. . . . They all had enough brightness. So my first laser used the smallest of three lamps, a GE FT506 lamp. The ruby was about 1 centimeter in diameter by 2 cm long, just filling the lamp spiral. . . .

Because of their odd shape the helical lamps weren't very amenable to reflectors. In fact, I started to try to devise a focusing reflector to direct the light onto the ruby. . . . Then all of a sudden it clicked in my head. Of course! The best I could do if I put that lamp in a reflector and then collected and refocused it back on the ruby was the brightness of the lamp itself. I thought, "Instead of remotely transferring the brightness of the lamp to the ruby, why don't I put it in near proximity?" I'd still be at the same brightness. I put a reflector around the outside of the lamp to collect the radiation that would travel outward. It acts as a radiation shield and, in principle, increases the brightness of the lamp for the same input power. There's the design.

And of course it came out extremely simple in realization. . . . I evaporated silver onto the endfaces [of the ruby]. . . . At first, I wanted to put a full silver on one end and a partial silver on the other. But then I realized that had experimental problems because the half silver or partial silver was extremely thin. Any tarnishing of silver changes its transmission coefficient properties. That's when I got the idea . . . that since it was a cavity, I'll just put in a coupling hole. So I put full silver on each end and then bored a tiny coupling hole [through the silver coating] to observe what was going on. . . .

Looking back on it, I was fairly cocky, because I felt I had a sound basis for doing what I was doing. One of the nice things about simple fluorescent solids is there are only a few [electron energy] levels in there to account for. You can really keep track of what's going on. . . .

I would say that I made the commitment around August, 1959, to work on the laser. I knew there was a lot of interest. I knew what people were doing, but I felt they were going off in tough directions. Schawlow and Townes must have started their work in the spring of 1958. Here it was August, 1959, and I was aware of the fact that people all over the world were now in this race. It was a little brash for me to enter that race at that time. People with well-funded efforts had already been going for, let's say, a year. It's interesting to note that Hughes' total expenditures in our nine-month laser effort amounted to about $50,000. Contrast that with the $500,000 to $1 million that other research teams were spending. . . .

Charles Townes' comments that it turned out to be easy to make the first laser and that anything will lase if you hit it hard enough are incredible statements to me. If it was so easy, why didn't Columbia, Bell Labs, or TRG pull it off? They each had a head start, plenty of money, and heavy staffing. . . .

There was not a great deal of enthusiasm or support [at Hughes]. Of course, all hell broke loose after we got the laser working, but that's another story. I don't want to fault Hughes too much. In a lot of ways, this was just a symbol of the attitude you often find in a big company. There is a tremendous resistance to anything new and different.

People at Hughes questioned the money. Was it worth the company's investment to do this work? . . . There were several criticisms thrown at me. "Who knows if you can make a laser?" "Who knows what form it is going to take?" "There are all these other people who are ahead of you who know what they're doing." "We have limited funds and we don't know what you would do with it anyway if you got it." "There's this new field of computers—why don't you work on that." "Besides, Schawlow said that ruby can't work, so the specific approach you are working on is wrong."

It became an uphill battle. I got pretty stubborn. I had seen some of the things people had done on general research funds. I felt that, if anyone had earned the right to it, I had. I had done really well on all the government contract work and brought in some contracts for them. I felt I was entitled to use of the [general] funds. I was just carried away with that project. I knew there was something there, and I was determined to follow through on it. I had a sixth sense that it was going to work. The more I worked on it, the more convinced I became. I understood the reasons why it wasn't supposed to work. I wasn't daunted by them, though, because I felt that I knew the answers. . . .

There were a lot of doubts about how useful the laser would be. . . . Many people questioned whether it could be used outside the lab in a practical way. I had editors of magazines interview me and say, "Yes, but where is it being used?" People would announce applications on an experimental basis, but nobody was putting it on a production line. . . . The laser was, in effect, a solution looking for a problem.

People didn't appreciate that the laser offered such a radically different way of doing things that applications were bound to be slow in developing. After all, it took thirty years to get from Kitty Hawk to commercial aviation. Of course the electronics people had an easier time with the advent of transistors. For them, it was just a matter of replacing vacuum tubes in the beginning.

With one exception, that wasn't the case with the laser. The exception was repairing detached retinas. Xenon lamps were already on the market to repair detached retinas, so it was just a case of replacing them with lasers.

I interviewed Maiman for the National Geographic *at his oceanside condominium in Marina Del Ray, California, in 1988. When I asked him if he had donated his original laser to the Smithsonian, he told me the Institution had gone to Hughes for it instead of asking him. They thought they had it, he said wryly, but Hughes had unknowingly given them the second model. He still had the original; did I want to see it? I did. From a dining-room cabinet he removed an aluminum case, opened it to a coil of flashlamp and a silver-ended artificial ruby the size of a cigarette filter. I lifted the ruby between two fingers and marveled. Maiman told me:*

I set this up with my instrumentation, and I fired it, and at a certain point it went.

GLITCHES

On February 20, 1962, John Glenn became the first American to orbit the earth in space (two Russian cosmonauts had preceded him). With the other six astronauts of Project Mercury, he helped develop the equipment that supported his mission. There were frequent technical problems in those early days of spaceflight; their special character gave new meaning to a borrowed word.

We did not blame any of our problems on such things as gremlins. For one thing, these creatures belonged to another era. No matter how

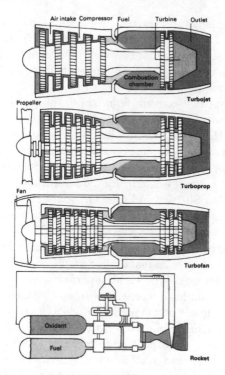

Air intake Compressor Fuel   Turbine   Outlet

Combustion chamber

Turbojet

Propeller

Turboprop

Fan

Turbofan

Oxidant

Fuel

Rocket

▶ The jet engine revolutionized military and commercial aviation after World War II.

fouled up an airplane could get, the pilot was working in an environment that was more forgiving of error and more cooperative in the sense that the pilot was not necessarily a goner just because his airplane was. In Project Mercury, however, we were preparing for a totally different environment in which the failure of a single component in the system *could* mean disastrous results for both the mission and the crew. We were also dealing with a much more sophisticated and unforgiving machine, and our approach to it had to be relentless and sophisticated, too. . . .

[The] engineering term we used to describe some of our problems was "glitch." Literally speaking, a glitch is a spike or change in voltage in an electrical circuit which takes place when the circuit suddenly has a new load put on it. You have probably noticed a dimming of the lights in your home when you turn a switch or start the dryer or the television set. Normally, these changes in voltage are protected by fuses. A glitch, however, is such a minute change in voltage that no fuse could protect against it. Glitches are a special problem when you are working with highly sensitive electronic equipment such as the many relays we had to install in the capsule to help link the complex systems together. The capsule, with its labyrinth of wiring and electronic equipment, was extremely glitch-prone and was an electronic engineer's nightmare. The crisscrossing circuits which knit all the systems into a working unit had to be plotted with great care so that a stray glitch that might arise in one circuit could not possibly find its way into other circuits and foul up another system. . . .

The most dramatic electronic failure of the program occurred at Cape Canaveral on the morning of November 21, 1960. The flight was scheduled on the board as MR-1—for Mercury Redstone 1—and it was to be the first full-scale test of a Mercury capsule on a normal ballistic flight. The capsule was empty except for instruments. It was mounted on the Redstone [rocket], however, just as it would be for the first manned [suborbital] flights by [astronauts] Alan Shepard and Gus Grissom—and before that, by the chimpanzee, Ham. The timing of these later missions depended on the success of MR-1.

I happened to be watching from inside the blockhouse that morning as the Redstone sat puffing out liquid oxygen a few hundred feet away, so I had a grandstand seat for one of the weirdest performances Cape Canaveral has ever seen. As the countdown reached "zero" and we waited for the booster to start lifting the capsule off the pad, there was a cloud of smoke and flame, and the Redstone actually budged a little. Then its engine stopped. The Redstone settled back on the pad and the capsule, which was sitting 60 feet in the air on top of it, began to act as if it were literally blowing its top. In a fraction of a second, the escape tower mounted on top of the capsule took off in a rush of flame, a parachute came shooting out of the canister where the tower had stood, and another parachute came popping out in rapid succession and plopped down over the side of the capsule like a limp rag.

When the smoke finally cleared, the pad was littered with all kinds of debris. The Redstone was still there, however, puffing out liquid oxygen as if nothing had happened. The capsule was still perched on top of it, trailing a parachute all the way to the ground and flashing a light on and off. Seeing all of this happen was like watching the fizzle of some gigantic Roman candle at a Fourth of July celebration.

It was late in the day before the technicians could get close enough to the pad to figure out what had happened. The Redstone was still full of fuel and liquid oxygen and its batteries were still full of power, so there was some danger that it might blow up on the pad if anyone tried to tamper with it. After waiting for the fuel to be drained out and for the batteries to run down, the crews finally got a good look at the bird and discovered what had happened. There were two plugs mounted at the base of the Redstone. One was a power plug which served to ground the booster's batteries until launch time. The other was a control plug which fed wires into the Redstone and allowed the technicians in the blockhouse to control its functions until just before lift-off. Both plugs were supposed to disconnect at the same time; they would be jerked out automatically as the Redstone started to pull away from the ground.

Both plugs did pull loose, for the booster managed to rise about an inch off the pad. But they did not pull out at the same instant. As it turned out, there were two reasons for this. One was that the prongs on the power plug happened to be one-eighth of an inch shorter than the prongs on the control plug. The other reason was that the Redstone carried a heavier load on this flight than it had in the past and therefore rose a little slower than it usually does. Therefore, instead of pulling both plugs free at the same instant, as it would normally have done, the Redstone disengaged them one at a time. The shorter one, the power plug, disconnected first. The lag between the two was very short—only about 20 milliseconds. But this was just enough to alert the electronic brain in the booster that something had gone wrong. Nothing like this had ever happened before. The result was that without the power plug to complete the normal circuit, an entirely new circuit was set up in the wiring. This was a classic example of a glitch, and the abort-sensing mechanism in the Redstone immediately shut off the engine. Under the circumstances, this was precisely what it should have done. The brain was there to sense trouble in the system and to stop the booster engine as soon as it spotted one.

From this point on, the capsule went to work and behaved exactly as *it* was supposed to. Its own electronic brain received the message that the Redstone's engines had shut down. This was exactly the same signal that the brain was prepared to receive when the booster had reached the altitude of 35 miles and it was time to jettison the escape tower and proceed with a normal mission. It was not the brain's fault that it behaved as it had been taught to do. It had not been wired, however, so that it could distinguish between two identical signals, one of which was caused by a small design discrepancy that had never cropped up before and was due, actually, to a change in operational procedures. There was only one conclusion for the capsule—the flight was over. So it fired off the escape tower. Then it began to tick off the entire sequence of "return-from-space" events, just as it was supposed to. Barometric pressure gauges inside the capsule sensed that the spacecraft was at a low altitude—sitting there on the pad, it certainly was—and automatically broke out the parachutes which normally come out at this time to stabilize the capsule and lower it gently to earth. The light beacon which was mounted on the capsule to attract the attention of recovery forces as soon as it hit the water started to flash on and off. Then the capsule began to transmit radio signals, just as it would at sea, to give the recovery fleet a precise fix on its location. The radio worked so well that two Navy P2V aircraft which were on recovery patrol duty 120 miles from Cape Canaveral picked up the signals, obediently turned around in mid-air and started to home in on the launching pad. Despite the comic aspects of the situation, this was nice going.

Some humorists at the Cape referred to this first test of MR-1 as "the day we launched the escape tower."

AMERICA IS PROCESS

JOHN A. KOUWENHOVEN, 1961

Asking "What's American about America?" a cultural historian assembles a list and explains what its components have in common.

Here are a dozen items to consider:

1. The Manhattan skyline
2. The gridiron town plan
3. The skyscraper
4. The Model-T Ford
5. Jazz
6. The Constitution
7. Mark Twain's writing
8. Whitman's *Leaves of Grass*
9. Comic strips
10. Soap operas
11. Assembly-line production
12. Chewing gum . . .

The central quality which all the diverse items on our list have in common . . . I would define as a concern with process rather than product—or, to re-use Mark Twain's words, a concern with the manner of handling experience or materials rather than with the experience or materials themselves. . . .

This fascination with process has possessed Americans ever since Oliver Evans in 1785 created the first wholly automatic factory: a flour mill in Delaware in which mechanical conveyors . . . are interlinked with machines in a continuous process of production. But even if there were no other visible sign of the national preoccupation with process, it would be enough to point out that it was an American who invented chewing gum (in 1869) and that it is the Americans who have spread it—in all senses of the verb—throughout the world. A non-consumable confection, its sole appeal is the process of chewing it.

THE MOST
IMPORTANT PASSION

DAVID RIESMAN, 1961

In a preface to a new edition of his seminal 1950 study The Lonely Crowd, *the sociologist assesses the impact of Western technology on the world.*

Anthropologists understandably regret the disintegration of most nonliterate cultures with the coming of the white man (or, today, the white-influenced man of any color); and we, too, feel that many of these cultures have created values our own society lacks. But a great number of nonliterates, not subject to physical coercion or dispersal, have plainly concluded that their once seemingly given culture lacked something; they have gone off, singly or in groups, to join the Big Parade—often meeting the more disenchanted Westerners going the other way. To repeat: the most important passion left in the world is not for distinctive practices, cultures, and beliefs, but for certain achievements—the technology and organization of the West—whose immediate consequence is the dissolution of all distinctive practices, cultures, and beliefs. If this is so, then it is possible that the cast of national characters is finished: men have too many to choose from to be committed to one, and as their circumstances become more similar, so will many attributes held in common, as against those unique to particular countries. Increasingly, the differences among men will operate across and within national boundaries, so that already we can see, in studies of occupational values in industrial societies, that the group character of managers or doctors—or artists—becomes more salient than the group character of Russians or Americans or Japanese, or indeed the conscious ideologies held in these societies.

A SCIENTIFIC-
TECHNOLOGICAL ELITE

DWIGHT D. EISENHOWER, 1961

Many remember President Eisenhower's warning, in his Farewell Address, *of the dangers of a "military-industrial complex." Fewer recall his concern with the influence of experts.*

Our military organization today bears little relation to that known by any of my predecessors in peacetime, or indeed by the fighting men of World War II or Korea.

Until the latest of our world conflicts, the United States had no armaments industry. American makers of plowshares could, with time and as required, make swords as well. But now we can no longer risk emergency improvisation of national defense; we have been compelled to create a permanent armaments industry of vast proportions. Added to this, three and a half million men and women are directly engaged in the defense establishment. We annually spend on military security more than the net income of all United States corporations.

The conjunction of an immense military establishment and a large arms industry is new in the American experience. The total influence—economic, political, even spiritual—is felt in every city, every State house, every office of the Federal government. We recognize the imperative need for this development. Yet we must not fail to comprehend its grave implications. Our toil, resources and livelihood are all involved; so is the very structure of our society.

In the councils of government, we must guard against the acquisition of unwarranted influence, whether sought or unsought, by the military-industrial complex. The potential for the disastrous rise of misplaced power exists and will persist.

We must never let the weight of this combination endanger our liberties or democratic processes. We should take nothing for granted. Only an alert and knowledgeable citizenry can compel the proper meshing of the huge industrial and military machinery of defense with our peaceful methods and goals, so that security and liberty may prosper together.

Akin to, and largely responsible for the sweeping changes in our industrial-military posture, has been the technological revolution during recent decades.

In this revolution, research has become central; it also becomes more formalized, complex, and costly. A steadily increasing share is conducted for, by, or at the direction of, the Federal government.

Today, the solitary inventor, tinkering in his shop, has been overshadowed by task forces of scientists in laboratories and testing fields. In the same fashion, the free university, historically the fountainhead of free ideas and scientific discovery, has experienced a revolution in the conduct of research. Partly because of the huge costs involved, a government contract becomes virtually a substitute for intellectual curiosity. For every old blackboard there are now hundreds of new electronic computers.

The prospect of domination of the nation's scholars by Federal employment, project allocations, and the power of money is ever present—and is gravely to be regarded.

Yet, in holding scientific research and discovery in respect, as we should, we must also be alert to the equal and opposite danger that public policy could itself become the captive of a scientific-technological elite.

It is the task of statesmanship to mold, to balance, and to integrate these and other forces, new and old, within the principles of our democratic system—ever aiming toward the supreme goals of our free society.

A VAST WASTELAND

NEWTON N. MINOW, 1961

Adlai Stevenson's young law partner took office as chairman of the Federal Communications Commission shortly after John F. Kennedy was inaugurated. In May 1961, Minow broke with tradition by telling the assembled membership of the National Association of Broadcasters what they didn't want to hear.

It may . . . come as a surprise to some of you, but I want you to know that you have my admiration and respect. Yours is a most honorable profession. Anyone who is in the broadcasting business has a tough row to hoe. You earn your bread by using public property. When you work in broadcasting, you volunteer for public service, public pressure and public regulation. You must compete with other attractions and other investments, and the only way you can do it is to prove to us every three years that you should have been in business in the first place. . . .

Ours has been called the jet age, the atomic age, the space age. It is also, I submit, the television age. And just as history will decide whether the leaders of today's world employed the atom to destroy the world or rebuild it for mankind's benefit, so will history decide whether today's broadcasters employed their powerful voice to enrich the people or debase them. . . .

When television is good, nothing—not the theater, not the magazines or newspapers—nothing is better.

But when television is bad, nothing is worse. I invite you to sit down in front of your television set when your station goes on the air and stay there without a book, magazine, newspaper, profit-and-loss sheet or rating book to distract you—and keep your eyes glued to that set until the station signs off. I can assure you that you will observe a vast wasteland. . . .

Gentlemen, your trust accounting with your beneficiaries is overdue.

WEIGHTLESSNESS

JOHN H. GLENN, JR., 1962

Technology projects human beings into new situations rarely or never experienced before. In the nineteenth century, some observers feared the great speeds that railroading made possible—fifteen miles per hour and more—might exceed what the human frame could bear. Orbital spaceflight—the astronaut continually falling around the earth—produces extended weightlessness, a condition not previously known on earth. John Glenn recalls the experience from his first historic orbital flight. He fared better than most; more than half the astronauts who followed him into space would experience nausea and vertigo during their first twenty-four hours of weightlessness.

Zanzibar was the next tracking station, and here the flight surgeon who was on duty came on the air to discuss how I was doing physically. I gave him a blood-pressure reading. . . . The doctor also asked me what physical reactions, if any, I had experienced so far from weightlessness. I was able to tell him that there had been none at all; I assured him that I felt fine. I had had no trouble reaching accurately for the controls and switches. There had been no tendency to get awkward and overreach them, as some people had thought there might be. I could hit directly any spot that I wanted to hit. I had an eye chart on board, a small version of the kind you find in doctors' offices, and I had no trouble reading the same line of type each time. After making a few slow movements with my head to see if this brought on a feeling of disorientation, I even tried to induce a little dizziness by nodding my head up and down and moving it from side to side. I experienced no disturbances, however. I felt no sense of vertigo, astigmatism or nausea whatever.

In fact, I found weightlessness to be extremely pleasant. I must say it is convenient for a space pilot. I was busy at one moment, for example, taking pictures, and suddenly I had to free my hands to attend to something else. Without even thinking about it, I simply left the camera in mid-air, and it stayed there as if I had laid it on a table until I was ready to pick it up again. The fact that this strange phenomenon seemed so natural at the time indicates how rapidly man can adapt to a new environment. I am sure that I could have gone for a much longer period in a weightless condition without being bothered by it at all. Being suspended in a state of zero G is much more comfortable than lying down under the pressure of 1 G on the ground, for you are not subject to any pressure points. You feel absolutely free. The state is so pleasant, as a matter of fact, that we joked that a person could probably become addicted to it without any trouble. I know that I could.

SILENT SPRING

RACHEL CARSON, 1962

A marine biologist who had already published popular books about the sea, Rachel Carson began researching her best-known work in response to a letter from a Massachusetts woman reporting that the pesticide DDT was killing birds. According to Vice-President Al Gore, who compares the book to Uncle Tom's Cabin, "the publication of Silent Spring *can properly be seen as the beginning of the modern environmental movement."*

The chemicals to which life is asked to make its adjustment are no longer merely the calcium and silica and copper and all the rest of the minerals washed out of the rocks and carried in rivers to the sea; they are the synthetic creations of man's inventive mind, brewed in his laboratories, and having no counterparts in nature.

To adjust to these chemicals would require time on the scale that is nature's; it would require not merely the years of a man's life but the life of generations. And even this, were it by some miracle possible, would be futile, for the new chemicals come from our laboratories in an endless stream; almost five hundred annually find their way into actual use in the United States alone. The figure is staggering and its implications are not easily grasped—500 new chemicals to which the bodies of men and animals are required somehow to adapt each year, chemicals totally outside the limits of biologic experience.

Among them are many that are used in man's war against nature. Since the mid-1940s over 200 basic chemicals have been created for use in killing insects, weeds, rodents and other organisms described in the modern vernacular as "pests"; and they are sold under several thousand different brand names.

These sprays, dusts, and aerosols are now applied almost universally to farms, gardens, forests, and homes—nonselective chemicals that have the power to kill every insect, the "good" and the "bad," to still the song of birds and the leaping of fish in the streams, to coat the leaves with a deadly film, and to linger on in soil—all this though the intended target may be only a few weeds or insects. Can anyone believe it is possible to lay down such a barrage of poisons on the surface of the earth without making it unfit for all life? They should not be called "insecticides," but "biocides."

. . . Along with the possibility of the extinction of mankind by nuclear war, the central problem of our age has therefore become the contamination of man's total environment with such substances of incredible potential for harm—substances that accumulate in the tissues of plants and animals and even penetrate the germ cells to shatter or alter the very material of heredity upon which the shape of the future depends.

Some would-be architects of our future look toward a time when it will be possible to alter the human germ plasm by design. But we may easily be doing so now by inadvertence, for many chemicals, like radiation, bring about gene mutations. It is ironic to think that man might determine his own future by something so seemingly trivial as the choice of an insect spray.

All this has been risked—for what? Future historians may well be amazed by our distorted sense of proportion. How could intelligent beings seek to control a few unwanted species by a method that contaminated the entire environment and brought the threat of disease and death even to their own kind? Yet this is precisely what we have done. We have done it, moreover, for reasons that collapse the moment we examine them. We are told that the enormous and expanding use of pesticides is necessary to maintain farm production. Yet is our real problem not one of *overproduction?* Our farms, despite measures to remove acreages from production and to pay farmers *not* to produce, have yielded such a staggering excess of crops that the American taxpayer in 1962 is paying out more than one billion dollars a year as the total carrying cost of the surplus-food storage program. And is the situation helped when one branch of the Agriculture Department tries to reduce production while another states, as it did in 1958, "It is believed generally that reduction of crop acreages under provisions of the Soil Bank will stimulate interest in use of chemicals to obtain maximum production on the land retained in crops."

All this is not to say there is no insect problem and no need of control. I am saying, rather, that control must be geared to realities, not to mythical situations, and that the methods employed must be such that they do not destroy us along with the insects.

HOW LIKE SAILORS THEY WERE

JOHN STEINBECK, 1962

The author of Grapes of Wrath *and* Of Mice and Men *received the Nobel Prize in Literature in 1962, and the same year published his genial report on a trip around the U.S. by camper truck,* Travels With Charley, *from which these observations on the burgeoning interstate highway system and on interstate trucking are taken.*

Trucks as long as freighters went roaring by, delivering a wind like a blow of a fist. These great roads are wonderful for moving goods but not for inspection of a countryside. You are bound to the wheel and your eyes to the car ahead and to the rear-view mirror for the car behind and the side mirror for the car or truck about to pas, and at the same time you must read all the signs for fear you may miss some instructions or orders. No roadside stands selling squash juice, no antique stores, no farm products or factory outlets. When we get these thruways across the whole country, as we will and must, it will be possible to drive from New York to California without seeing a single thing.

At intervals there are places of rest and recreation, food, fuel and oil, postcards, steam-table food, picnic tables, garbage cans all fresh and newly painted, rest rooms and lavatories so spotless, so incensed with deodorants and with detergents that it takes time to get your sense of smell back. . . .

It is life at a peak of some kind of civilization. The restaurant accommodations, great scallops of counters with simulated leather stools, are as spotless as and not unlike the lavatories. Everything that can be captured and held down is sealed in clear plastic. The food is oven-fresh, spotless and tasteless; untouched by human hands. I remembered with an ache certain dishes in France and Italy touched by innumerable human hands.

These centers for rest, food and replenishment are kept beautiful with lawns and flowers. At the front, nearest the highway, are parking places for passenger automobiles together with regiments of gasoline pumps. At the rear the trucks draw up, and there they have their services—the huge overland caravans. Being technically a truck, Rocinante [i.e., Steinbeck's camper] took her place in the rear, and I soon made acquaintance with the

▶ The Autobahn of Nazi Germany, on which he traveled while on assignment in Europe in the 1930s, inspired President Dwight D. Eisenhower to propose the Interstate Highway System.

truckers. They are a breed set apart from the life around them, the long-distance truckers. In some town or city somewhere their wives and children live while the husbands traverse the nation carrying every kind of food and product and machine. They are clannish and they stick together, speaking a specialized language. And although I was a small craft among monsters of transportation they were kind to me and helpful.

I learned that in the truck parks there are showers and soap and towels—that I could park and sleep at night if I wished. The men had little commerce with local people, but being avid radio listeners they could report news and politics from all parts of the nation. The food and fuel centers on the parkways or thruways are leased by the various states, but on other highways private enterprise has truckers' stations that offer discounts on fuel, beds, baths and places to sit and shoot the breeze. But being a specialized group, leading special lives, associating only with their own kind, they would have made it possible for me to cross the country without talking to a local town-bound man. For the truckers cruise over the surface of the nation without being a part of it. Of course in the towns where their families live they have whatever roots are possible—clubs, dances, love affairs and murders.

I liked the truckers very much, as I always like specialists. By listening to them talk I accumulated a vocabulary of the road, of tires and springs, of overweight. The truckers over long distances have stations along their routes where they know the service men and the waitresses behind the counters, and where occasionally they meet their opposite numbers in other trucks. The great get-together symbol is the cup of coffee. . . .

Quite often I sat with these men and listened to their talk and now and then I asked questions. I soon learned not to expect knowledge of the country they passed through. Except for the truck stops, they had no contact with it. It was driven home to me how like sailors they were. I remember when I first went to sea being astonished that the men who sailed over the world and touched the ports to the strange and exotic had little contact with that world. Some of the truckers on long hauls traveled in pairs and took their turns. The one off duty slept or read paperbacks. But on the roads their interests were engines, and weather, and maintaining the speed that makes the predictable schedule possible. Some of them were on regular runs back and forth while others moved over single operations. It is a whole pattern of life, little known to the settled people along the routes of the great trucks.

DOOMSDAY IN
THE WAR ROOM

STANLEY KUBRICK AND
TERRY SOUTHERN, 1963

The motion picture Dr. Strangelove, or How I Learned to Stop Worrying and Love the Bomb, introduced the general public to the notion of a Doomsday Machine. It seemed all too plausible at the height of the Cold War. Dr. Strangelove was probably modeled on the Hungarian-born theoretical physicist Edward Teller, General "Buck" O'Connor on General Thomas Power, the head of the U. S. Strategic Air Command at the height of its influence, in the 1950s. Ambassador De Sade, the first speaker, is the Soviet ambassador.

55 INTERIOR WAR ROOM

AMBASSADOR DE SADE

When it is detonated it will produce enough lethal radioactive fallout so within ten months the surface of the earth will be as dead as the moon.

GENERAL "BUCK" O'CONNOR

That's ridiculous, De Sade! Our studies show the worst fallout is down to a safe level after two weeks.

AMBASSADOR DE SADE

Have you ever heard of Cobalt-Thorium-G?

GENERAL "BUCK" O'CONNOR

What about it?

AMBASSADOR DE SADE

Cobalt-Thorium-G has a radioactive half-life of ninety-three years.

A SENIOR CIVILIAN AIDE nods grimly.

AMBASSADOR DE SADE

If you take, say, fifty H-bombs in the hundred megaton range and jacket them with Cobalt-Thorium-G, when they are exploded they will produce a Doomsday shroud, a lethal cloud of radioactivity which will encircle the earth for ninety-three years.

Murmurs and stirring.

PRESIDENT MUFFLEY

I'm afraid I don't understand something. Is the Premier threatening to explode this if our planes carry through their attack?

AMBASSADOR DE SADE

No, sir. It is not a thing a sane man would do. The Doomsday Machine is
designed to trigger itself *automatically!*

PRESIDENT MUFFLEY

But then, surely he can disarm it somehow.

AMBASSADOR DE SADE

No! It is designed to explode if any attempt is ever made to untrigger it!

GENERAL "BUCK" O'CONNOR
(aside to a Colonel)
It's an obvious commie trick, and he sits there wasting precious time.

Divided murmurs around the table.

PRESIDENT MUFFLEY

But surely, Ambassador, this is absolute madness. Why should you build
such a thing?

AMBASSADOR DE SADE

There were those of us who fought against it, but in the end we could not
keep up in the Peace Race, the Space Race and the Arms Race. Our deterrent began to lack credibility. Our people grumbled for more nylons and
lipsticks. Our Doomsday project cost us just a fraction of what we had been
spending in just a single year. But the deciding factor was when we learned
your country was working along similar lines, and we were afraid of a
Doomsday Gap.

PRESIDENT MUFFLEY

That's preposterous. I've never approved anything like that!

AMBASSADOR DE SADE

Our source was "The New York Times."

PRESIDENT MUFFLEY

Doctor Strangelove, have we anything like this in the works?

DR. STRANGELOVE

(German precision)

Mister President, under the authority granted me as Director of Weapons Research and Development, I commissioned a study last year of this project by the Bland Corporation. Based on the findings of the report, my conclusion was that this idea was not a practical deterrent for reasons which at this moment must be all too obvious.

PRESIDENT MUFFLEY

Then you mean it is unquestionably possible for them to have built this thing?

AMBASSADOR DE SADE

Mister President, the technology required is easily within the means of even the smallest nuclear power. It requires only the *will* to do so.

PRESIDENT MUFFLEY

But is it really possible for it to be triggered automatically and at the same time impossible to untrigger?

DR. STRANGELOVE

Mister President, it is not only possible, it is essential. That is the whole idea of this machine. Deterrence is the art of producing in the mind of the enemy the fear to attack. And so because of the automated and irrevocable decision-making process which rules out human meddling, the Doomsday Machine is terrifying, simple to understand and completely credible and convincing.

Murmurs around table.

SCIENCE AND TECHNOLOGY

JACQUES ELLUL, 1964

The French sociologist, in his influential book The Technological Society, *reverses a truism.*

The relation between science and technique is a standard subject for graduate theses—in all the trappings of nineteenth-century experimental science. Everyone has been taught that technique is an application of science; more particularly (science being pure speculation), technique figures as the point of contact between material reality and the scientific formula. But it also appears as the practical product, the application of the formulas to practical life.

This traditional view is radically false. It takes into account only a single category of science and only a short period of time: it is true only for the physical sciences and for the nineteenth century. . . .

In the present era, the most casual inspection reveals an entirely different relationship. In every instance, it is clear that the border between technical activity and scientific activity is not at all sharply defined. . . .

When the technical means do not exist, science does not advance. Michael Faraday was aware of the most recent discoveries concerning the constitution of matter, but was unable to formulate precise theories because techniques for the production of vacua did not yet exist. Scientific results had to await high-vacuum techniques. The medical value of penicillin was discovered in 1912 by a French physician, but he had no technical means of producing and conserving penicillin; misgivings therefore arose about the discovery and led to its eventual abandonment. . . .

It is not a question of minimizing the importance of scientific discovery, but of recognizing that in fact scientific activity has been superseded by technical activity to such a degree that we can no longer conceive of science without its technical outcome. As Charles Camichel has observed, the two are closer than ever before. The very fact that techniques advance with great rapidity demands a corresponding scientific advance, and sets off a general acceleration.

Moreover, techniques are always put to immediate use. The interval which traditionally separates a scientific discovery and its application in everyday life has been progressively shortened. As soon as a discovery is made, a concrete application is sought. Capital becomes interested, or the state, and the discovery enters the public domain before anyone has had a chance to reckon all the consequences or to recognize its full import. The scientist might act more prudently; he might even be afraid to launch his carefully calculated laboratory findings into the world. But how can he resist the pressure of the facts? How can he resist the pressure of money? How is he to resist the desire to pursue his research? Such is the dilemma of the researcher today. Either he allows his findings to be technologically applied or he is forced to break off his research. . . . The scientist is no longer able to hold out: "Even science, especially the magnificent science of our own day, has become an element of technique, a mere means" (Mauss). There we have, indeed, the final word: science has become an instrument of technique.

DOWNLOADING
(PREHISTORY OF)

NORBERT WIENER, 1964

"Downloading" in the sense I use it here isn't sucking files off the Internet but the recent fanciful notion that we might someday achieve conscious, individual immortality by storing our mental contents on computers.

This is an idea with which I have toyed before—that it is conceptually possible for a human being to be sent over a telegraph line. Let me say at once that the difficulties far exceed my ingenuity to overcome them, and that I have no intention to add to the present embarrassment of the railroads by calling in the American Telegraph and Telephone Company as a new competitor. At present, and perhaps for the whole existence of the human race, the idea is impracticable, but it is not on that account inconceivable.

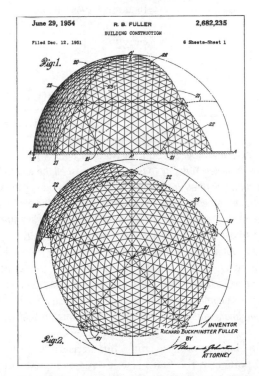

▶ New forms of construction paced twentieth-century architecture: Buckminster Fuller's original geodesic dome patent, 1954.

KING LUD

MARIO SAVIO, 1964

Savio was an influential student activist. Here he speaks from the steps of Sproul Hall at the University of California at Berkeley during a sit-in.

There is a time when the operation of the machine becomes so odious, makes you so sick at heart that you can't take part; you can't even passively take part, and you've got to put your bodies upon the gears and upon the wheels, upon the levers, upon all the apparatus and you've got to make it stop. And you've got to indicate to the people who run it, to the people who own it, that unless you're free, the machine will be prevented from working at all.

I am a human being: Do not fold, spindle, or mutilate.

> SLOGAN OF THE FREE SPEECH MOVEMENT, BERKELEY, 1964 (alluding to a warning printed on IBM punched cards used for computer data entry)

A STORY TO TELL

LOUDON WAINWRIGHT, 1965

At 5:29 P.M., November 9, 1965, the Northeast power grid failed and the lights went out from New York City to Boston and west all the way to Toronto. They stayed out all night as engineers wrestled the complex system back into sync. Attention inevitably focussed on Manhattan, where the elevators of skyscrapers stopped moving and more than a half million people were stranded in the subway. A columnist for Life *magazine muses insouciantly in the immediate aftermath, but across the page from his blithe essay Theodore H. White complained querulously that technology had been "spectacularly demonstrated" to be "too important to be left to engineers and businessmen. . . . It was required that New York come to the brink of chaos to refresh an old truth: People—men of frailty, judgment and human decision—must control machines. Not vice versa."*

It shouldn't happen every evening, but a crisis like the lights going out has its good points. In the first place it deflates human smugness about our miraculous technology, which, at least in the area of power distribution and control, now stands revealed as utterly flawed. I have never regarded as trustworthy many of the appurtenances of modern America—automatic elevators, for example, have always chilled me with their smooth mindlessness—and it is somehow delicious to contemplate the fact that all our beautiful brains and all those wonderful plans and all that marvelous equipment have combined to produce a system that is unreliable.

Even better is the fact that *something happened*. With its virtually instantaneous transmission of news from all over, modern communication leaves people with the feeling that they are spectators in a world of action. The truth is that for most people, except for whatever quality of anguish or action exists in their minds, nothing much is going on, and it is exhilarating suddenly to become a performer in a drama, even if the cast has millions. I can remember as a child the excitement that surrounded the death of one of my grandfathers; too young to grieve for him or to understand the finality of his passing, I was aware only of the solemn urgency and grave bustle brought on by the enormous event. In secret—and guiltily, because I thought I should feel sad—I hoped this strange situation would last. After a few moments of darkness the other night, holding a lighted match near the face of my watch, I forgot about those people who might be endangered by the emergency and hoped the current wouldn't come back on too soon.

It seemed to me that the blackout quite literally transformed the people of New York. Ordinarily smug and comfortable in the high hives of the city where they live and work, they are largely strangers to one another when the lights are on. In the darkness they emerged, not as shadows, but far warmer and more substantial than usual. Stripped of the anonymity that goes with full illumination, they became humans conscious of and concerned about the other humans around them. In the crowded streets businessmen, coats removed so that their light-colored shirts could be seen, became volunteer cops and directed traffic. Though the sidewalks were jammed, there was little of the rude jostling that is a part of normal, midday walking in New York. In the theatrically silver light of a perfect full moon (a must for all future power failures) people peered into the faces of passersby like children at a Halloween party trying to guess which friends hide behind which masks. In fact, the darkness made everyone more childlike. There was much laughter, and as they came down the stairs of the great office buildings in little night processions led by men with flashlights and candles, people held hands with those they could not see.

Of course, there were many people who were seriously inconvenienced by the power failure, but I was not one of those. In the midtown Manhattan restaurant where I was pleasantly marooned with about twenty-five others, I had drinks and a fine dinner by candlelight. A transistor radio was placed on the bar and we alternately listened to it and then, conversationally table-hopping, talked the news over with new friends nearby. Occasionally we looked out the window at the crowds passing (the prostitutes working that street were among the first to procure flashlights), and every hour or so I went out and walked over to Broadway, then a great dark way lit only by a stream of headlights and here and there a candle flickering in a high window. Back in the restaurant, as the news announcements made it clear that getting home would be impossible, I felt cozily snowbound, a man who had to make the best of it in ridiculously easy circumstances.

I wasn't as worried about our situation as President [Lyndon] Johnson seemed to be, but then his communications down at the ranch were better than mine were in the restaurant. Still, many were roughing it a bit, and the fact of that probably brought out the best in people. If there is any single act most Americans have to commit very seldom, it is to improvise, and quite suddenly 30 million people found that they had to improvise—right now. Walking down many flights of stairs, trudging across bridges, driving without the electronically imposed courtesy of stoplights, sleeping in chairs or in hotel lobbies—all these unusual things had to be done, and people not only did them but carried them off with a certain splendid gaiety.

When the lights came on the next morning, I saw a piece of television news which showed some people who'd been stuck all night in a subway car. A crowded mixed bag of young and old, well-dressed and shabby, they seemed absolutely overjoyed at their predicament. And when the subway policeman who had been stuck with them congratulated them on the way they had been good comrades and otherwise passed the time courageously, they cheered him wildly.

My own private fantasy of hell is a stalled elevator with me inside it, and I was astounded to hear the saga of the men trapped 25 floors up in the Empire State Building who passed their 5¼ hours of black suspension by joking, singing and not panicking.

Perhaps the best thing about such an event is that it gives all of us a story to tell. The simplest little homeward ride becomes a journey full of weird excitement; the most commonplace stranding is a spooky captivity. "What happened to *you?*" is an invitation everyone extends to another, and people can deal with their own experiences as if they were of interest and importance. We will be listening to versions of "The Night the Lights Went Out" long after a federal commission discovers that it all started when a little boy in upstate New York dropped his electric toothbrush in the toilet. It's a perfect event for the development of stories for grandchildren as yet unborn. As for my own grandchildren, though I doubt the restaurant episode provides any suspense, I ought to be able to lift enough elements here and there from other people's accounts to make my story fascinating.

▶ Microscopic transistors assembled like nerve cells on coin-size computer chips gave brains to brute machines.

MOORE'S LAW

GORDON E. MOORE, 1965

One of the founders (in 1968) of Intel Corporation, Gordon Moore formulated the principle that came to be known as Moore's Law in an article he wrote for the thirty-fifth anniversary issue of Electronics *magazine. "My assignment," he recalls, "was to predict what was going to happen in the semiconductor components industry over the next ten years—to 1975. In 1965 the integrated circuit was only a few years old and in many cases was not very well accepted. There was still a large contingent in the user community who wanted to design their own circuits and who considered the job of the semiconductor industry to be to supply them with transistors and diodes so they could get on with their jobs. I was trying to emphasize the fact that integrated circuits really did have an important role to play."*

Complexity of integrated circuits has approximately doubled every year since their introduction. Cost per function has decreased several thousand-fold, while system performance and reliability have been improved dramatically. Many aspects of processing and design technology have contributed to make the manufacture of such functions as complex single chip microprocessors or memory circuits economically feasible. It is possible to analyze the increase in complexity . . . into different factors that can, in turn, be examined to see what contributions have been important in this development and how they might be expected to continue to evolve. The expected trends can be recombined to see how long exponential growth in complexity can be expected to continue. . . .

The rate of increase of complexity can be expected to change slope in the next few years. . . . The new slope might approximate a doubling every two years, rather than every year, by the end of the decade.

Even at this reduced slope, integrated structures containing several million components can be expected within ten years. These new devices will continue to reduce the cost of electronic functions and extend the utility of digital electronics more broadly throughout society.

In 1995, speaking to the Society of Photo-Optical Instrumentation Engineers, Moore assessed the accuracy and the provenance of his predictions.

I have never been able to see beyond the next couple of generations [of integrated circuits] in any detail. Amazingly, though, the generations keep coming one after the other keeping us about on the same slope. The current prediction is that this is not going to stop soon either. . . .

The rising cost of the newer technologies is of great concern. Capital costs are rising far faster than revenue in the industry. We can no longer make up for the increasing cost by improving yields and equipment utilization. [Just as] the "cleverness" term in device complexity disappeared when there was no more room to be clever, there is little room left for manufacturing efficiency. Increasing the growth rate of the industry looks increasingly unlikely. We are becoming a large player in the world economy. . . .

In 1986 the semiconductor industry represented about 0.1 percent of the [Gross World Product]. Only ten years from now, by about 2004, if we stay on the same growth trend, we will be 1%, and by about 2025, 10%. We will be everything by the middle of the century. Clearly industry growth has to roll off.

I do not know how much of the GWP we can be, but much over one percent would certainly surprise me. I think that the information industry is clearly going to be the biggest industry in the world over this time period, but the large industries of the past, such as automobiles, did not approach anything like a percent of the GWP. Our industry growth has to moderate relatively soon. We have an inherent conflict here. Costs are rising exponentially and revenues cannot grow at a commensurate rate for long. I think that this is at least as big a problem as the technological challenge. . . .

I am increasingly of the opinion that the rate of technological progress is going to be controlled from financial realities. We just will not be able to go as fast as we would like because we cannot afford it, in spite of your best technical contributions. When you are looking at new technology, please look at how to make that technology affordable as well as functional.

Our industry has come a phenomenal distance in what historically is a very short time. I think our progress to a considerable extent is the result of two things: a fantastically elastic market with new applications that can consume a huge amount of electronics, and a technology that exploits what I have often described as an exception to Murphy's Law.

By making things smaller, everything gets better simultaneously. There is little need for tradeoffs. The speed of our products goes up, the power consumption goes down, system reliability, as we put more of the system on a chip, improves by leaps and bounds, but especially the cost of doing things electronically drops as a result of technology. . . . We have made of the order of a ten millionfold decrease in the cost of a transistor and thrown in all the interconnections free. . . . It is hard to find an industry where the cost of their basic product has dropped ten millionfold over much longer time periods.

The only one I can find that is remotely comparable is the printing industry. Carving a character into a stone tablet with a chisel probably cost quite a bit, maybe the equivalent of a few dollars today based on the time it probably took. Today people printing newspapers sell them for a price that makes the individual characters about as expensive as are individual transistors in a DRAM. Surprisingly, they sell about as many characters as we sell transistors, as near as I can estimate. Trying to estimate the number of characters printed is far more challenging than estimating the number of transistors produced. Taking into account newspapers, books, the Xerox copies that clutter up your desks, all such printed matter I estimate that it is no more than an order of magnitude greater than the number of transistors being produced.

Printing in advancing its technology over the centuries has had a revolutionary impact on society. We can now archive our knowledge and learn from the collected wisdom of the past increasing the rate of progress of mankind.

I think information technology will create its own revolution in society over a much shorter timescale, primarily because of the semiconductor technology you are driving. Semiconductor technology has made its great strides as a result of ever-increasing complexity of the products produced exploiting higher and higher density—to a considerable extent the result of progress in lithography.

As you leave this meeting I want to encourage each of you to think smaller. The barriers to staying on our exponential are really formidable, but I continue to be amazed that we can either design or build the products we are producing today.

CULTURE TO THE *Nth* POWER

HERBERT A. SIMON, 1966

Simon, who received the Nobel Prize in Economics in 1978, is widely regarded as the father of artificial intelligence.

To what extent has changing technology altered man's manner of thinking about the world and about himself? I have suggested as a measure of the change the communication distance between Pericles and ourselves. In modern man, a tragedy of Aeschylus, Sophocles or Euripides can still "arouse pity and fear, and accomplish its catharsis of such emotions." Oedipus provides to Cocteau and Stravinsky a viable text for a musical drama, and to Freud a deep insight into the springs of human action. If it be ob-

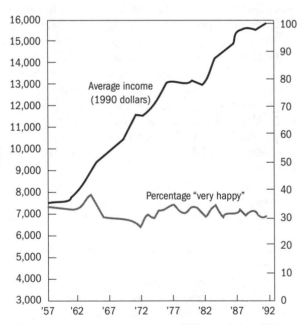

▶ Average U.S. per capita income vs. percentage of U.S. population which considers itself "very happy," 1957–1992.

jected that Oedipus, while meaningful, means something different to me than it did to Pericles, I reply that my reasons for believing that Pericles and I understand each other are as good as my reasons for believing that you and I understand each other; or that I understand Camus who expresses an existential despair that I do not feel.

It is the truth, if not the whole truth, that the focal events and the climactic emotions in my life and my neighbor's are almost identical with those in the lives of Pericles, his neighbors and his forebears. Homer would find all the materials for his third epic in this morning's newspaper: war, floods, murders, shipwrecks, negotiations, births, deaths and marriages, love, hate, curiosity, friendship, ambition, fun, pain were and are the substance of the human condition. Man is the significant part of man's environment; the nonhuman environment, whether the forest or sea designed by nature, or the farm or city by man, largely defines the rules of a particular game within which meaningful human interaction takes place.

Of course, if we take Aristotle instead of the poets as our model of the Greek mind, we arrive at a quite different measure of the communication distance that separates us from them. As far as physics, biology and mathematics are concerned, the gulf with the past has opened to limitless depth. It is not a question of whether Aristotle, Archimedes and Euclid are right or wrong. Their writings have simply become irrelevant except as data for the antiquarian or the historian of Western thought.

If we viewed science only as the foundation for our technology and our economy, we could perhaps pass the change off as unrelated to humane matters—as not touching the fundamental icons. But science is inextricably interwoven with cosmology, with theology, with epistemology, with ethics, politics and psychology, all of them crucial to exploration into the human condition. A cosmology cannot be taken seriously that is not consistent with modern mechanics and astronomy, nor an epistemology that is naive in its understanding of operationalism, quantum uncertainty or the Gödel theorems. To understand Zeno's paradoxes, we do not study the classics, but Weierstrass and Cantor.

All of this is a familiar restatement of the Two Cultures argument; it predates the computer, and would be true if the computer had never been invented. But the computer lengthens the gulf that separates our icons from those of previous cultures all the way into the sciences of man. Man has held up a mirror to nature; now he will view himself in it.

Man's words, like the clippings of his hair and nails, are divorced parts of himself, not lightly to be put in the hands of his enemies, who might use them to gain power over him. A long time ago, however, man began to entrust some of his words and numbers to clay tablets and papyrus. With the invention of printing, five hundred years ago, he became almost willing to grant them a life of their own. By freeing and cheapening their reproductive process, printing caused a great explosion in the population of words. (The explosion has been heard almost everywhere except in the lecture halls of universities, where words are still reproduced, inaccurately, by handicraft processes at a very high cost.)

Printing, however, automated only a single symbol-manipulating process—copying. Pattern recognition, information retrieval, the whole range of processes that go under the name "thinking," were untouched. Printing, for all its practical consequence, gave man little insight into his own symbol-manipulating processes, and contributed almost nothing to the science of psychology.

The computer is a different matter. It is capable, as we have seen, of a very broad range of information processes—perhaps all the kinds that man can carry out. To design, manufacture and program computers man has had to acquire an understanding of information processing, and as a result, new knowledge about the processes of his mind.

The computer is bringing about its revolution in psychology in two different ways, one general, one specific. Neither development rests on any vulgar analogy between the computer as an electronic system and the brain as a physiological one. The anatomy of a computer is visibly different from the anatomy of a brain.

The general contribution of computers to psychology has already been suggested. Airplanes have taught us a great deal about the flight of birds, although the two systems function quite differently. Likewise, when one programs a computer to play chess—even if without intent to imitate the methods of human players—one inevitably learns much about what is required of *any* information-processing system that is to perform the task successfully. Thus we have learned that in complex problem environments a very modest ability to explore selectively compensates for an enormous disadvantage of speed of search. The race is to the slow, provided that they are also thoughtful and selective.

But computers are also making a more specific and farther-reaching contribution to psychology. They are providing a language for stating psychological theories, and an effective means for generating testable implications of the theories so formulated. . . .

Through these and other lines of inquiry—research, for example, on the physiology of the brain—man's understanding of himself as a processor of information is advancing rapidly. It is still unclear how far these advances will extend into the realms of motivation and emotion, but the latter phenomena also seem amenable to statement in information-processing terms. Man is drastically reformulating his picture of himself; his distance from Greece, and even from the early twentieth century will be very much increased as a result.

What consequences will follow from these changed icons are far from obvious to me. One likely consequence should be mentioned, however, for it bears directly on the Two Cultures problem.

In reality, there are not two cultures, but n cultures, where n is a larger number. Each man shares with most of his fellows an area of common interests and concerns, and is separated from all but a few of his fellows by a multitude of special interests, skills, and bodies of knowledge. If it is obvious that scientists cannot communicate with humanists, except in the common area outside both their professional skills, it is equally obvious that musicians cannot communicate with painters, nuclear physicists with physiologists, or eighteenth-century historians with fifteenth-century historians.

Literature has faced this problem of general relevance by concerning itself almost exclusively with the "private" lives of its characters; in most cases their professions could be drastically altered without much damage. (Melville is an interesting exception; here metaphor and allegory carry the burden of making the profession of whaling humanely relevant.)

The exclusion of professional activity—whether scientific, artistic, scholarly or craft—from the area of man's common concerns is anomalous; for most creative persons the profession is the core of life and value, the "common" area almost secondary. The humane, culturally defined, fails to include a great part of what is most significant for them. If we recognize that what is most significant for the creative physicist is not the body of physical theory and data, but *doing* physics, and what is most significant for the creative musician is not the body of musical literature, but *doing* music—performing, or composing or listening—then we can see a solution to the dilemma. Doing physics and doing music are creative processes that involve thought in a rich context of motivation, evaluation and emotion. As we learn more about human thinking, we are coming to see that these processes are essentially the same, whether in physics or in music, once they are abstracted from the particular materials on which they work. They are human processes that provide common ground for all professionals in the same way that the processes of loving, hating, eating and shivering provide common ground.

Understanding our processes of thinking and searching and seeking will lead us to discover that the skills, the values, the pleasures that we share with other men are not limited to the extraprofessional hours of our lives, but extend to the very core of our daily activity, however specialized and esoteric its content may appear. Our ability to communicate about these processes will lower the barriers that separate us, as professionals, so sharply from one another.

The specialization of the disciplines that has developed since Gutenberg is a regrettable consequence of the population explosion among words.

Forbidding structures of esoteric knowledge have threatened to restrict the domain of common discourse to a narrow range of "private" matters. Perhaps the most exciting prospect of change resulting from our new technology of information processing is the likelihood that it will halt and reverse this progressive isolation of idea from idea and man from man. Mankind, in its professional as well as its nonprofessional aspects, will again become the proper study of man.

A NEW DUTY

BARRY COMMONER, 1966

Commoner, a biologist, was among the founders of the environmental movement.

RICHARD RHODES

250

Despite the dazzling successes of modern technology and the unprecedented power of modern military systems, they suffer from a common and catastrophic fault. While providing us with bountiful supply of food, with great industrial plants, with high-speed transportation, and with military weapons of unprecedented power, they threaten our very survival. Technology has not only built the magnificent material base of modern society, but also confronts us with threats to survival which cannot be corrected unless we solve very grave economic, social and political problems.

How can we explain this paradox? The answer is, I believe, that our technological society has committed a blunder familiar to us from the nineteenth century, when the dominant industries of the day, especially lumbering and mining, were successfully developed—by plundering the earth's natural resources. These industries provided cheap materials for constructing a new industrial society, but they accumulated a huge debt in destroyed and depleted resources, which had to be paid by later generations. The conservation movement was created in the United States to control these greedy assaults on our resources. The same thing is happening today, but now we are stealing from future generations not just their lumber or their coal, but the basic necessities of life: air, water and soil. A new conservation movement is needed to preserve life itself. . . .

The earlier depredations of our resources were usually made with a fair knowledge of the harmful consequences, for it is difficult to escape the fact that erosion quickly follows the deforestation of a hillside. The difficulty lay not in scientific ignorance, but in willful greed. In the present situation, the hazards of modern pollutants are generally not appreciated until after the technologies which produce them are well established in the economy. While this ignorance absolves us from the immorality of the knowingly destructive acts that characterized the nineteenth century raids on our resources, the present fault is more serious. It signifies that the capability of science to guide us in our interventions into nature has been seriously eroded—that science has, indeed, got out of hand. . . .

I believe that scientists have a responsibility in relation to the technological uses which are made of scientific developments. In my opinion, the proper duty of the scientist to the social consequence of his work cannot be fulfilled by aloofness or by an approach which arrogates to scientists alone the social and moral judgments which are the right of every citizen. I propose that scientists are now bound by a new duty which adds to and extends their older responsibility for scholarship and teaching. We have the duty to inform, and to inform in keeping with the traditional principles of science, taking into account all relevant data and interpretations. This is an involuntary obligation to society; we have no right to withhold information from our fellow citizens, or to color its meaning with our own social judgments.

Scientists alone cannot accomplish these aims, for despite its tradition of independent scholarship, science is a dependent segment of society. In this sense defense of the integrity of science is a task for every citizen. And in this sense, too, the fate of science as a system of objective inquiry, and therefore its ability safely to guide the life of man on earth, will be determined by social intent. Both awareness of the grave social issues generated by new scientific knowledge, and the policy choices which these issues require, therefore become matters of public morality. Public morality will determine whether scientific inquiry remains free. Public morality will determine at what cost we shall enjoy freedom from insect pests, the convenience of automobiles, or the high productivity of agriculture. Only public morality can determine whether we ought to intrust our national security to the catastrophic potential of nuclear war.

There is a unique relationship between the scientist's social responsibilities and the general duties of citizenship. If the scientist, directly or by inferences from his actions, lays claim to a special responsibility for the resolution of the policy issues which relate to technology, he may, in effect, prevent others from performing their own political duties. If the scientist fails in his duty to inform citizens, they are precluded from the gravest acts of citizenship and lose their right of conscience. . . .

The obligation which our technological society forces upon all of us, scientist and citizen alike, is to discover how humanity can survive the new power which science has given it. It is already clear that even our present difficulties demand far-reaching social and political actions. Solution of our pollution problems will drastically affect the economic structure of the automobile industry, the power industry, and agriculture and will require basic changes in urban organization. To remove the threat of nuclear catastrophe we will be forced at last to resolve the pervasive international conflicts that have bloodied nearly every generation with war.

Every major advance in the technological competence of man has enforced revolutionary changes in the economic and political structure of society. The present age of technology is no exception to this rule of history. We already know the enormous benefits it can bestow; we have begun to perceive its frightful threats. The political crisis generated by this knowledge is upon us.

Science can reveal the depth of this crisis, but only social action can resolve it. Science can now serve society by exposing the crisis of modern technology to the judgment of all mankind. Only this judgment can determine whether the knowledge that science has given us shall destroy humanity or advance the welfare of man.

Pax Atomica

WILLIAM G. CARLETON, 1966

A historian presciently discerns the changes forced on the nation-state system by the historic introduction into the world of an unlimited source of energy for explosives.

Some of the far-reaching consequences of the nuclear revolution in war . . . are already fairly clear. That revolution is eliminating war, banishing force from international affairs, ushering in the *Pax Atomica*. Machine technology had already rendered World War I and World War II "total." The nuclear revolution has now rendered war "totally total," and therefore obsolete. (Now, it is possible, though not probable, that man will experiment with full-scale nuclear war just once; after that, although men would survive, we may be sure that there would never be another war.)

There is still brinkmanship, but not war. The United States and the Soviet Union recoiled at Berlin and in Cuba; China recoiled at Quemoy and Matsu and in the Himalayas. The powers avoid conventional ground wars for fear these will escalate into general nuclear war. They fear even the "limited" wars which were fought in Korea and Indochina in the early 1950s, again in fear of escalation into nuclear war. The only "safe" form of force seems to be primitive guerrilla wars, and care is taken to prevent such wars from becoming "limited" wars which in turn might become nuclear ones. Guerrilla war has been going on in South Vietnam for years, yet even now, although it has been intensified, that war has not developed into "limited" war on the Korean scale.*

More than any other institution, war has been the most significant in-

*It soon did, of course, as we know. [RR]

strument of change in human affairs—creating new nations, destroying old nations, altering boundaries, stimulating and accelerating innovations within the nations themselves, making internal revolutions. It is difficult to conceive of history without war, and the full impact of its epochal elimination will not be felt for decades.

However, the nuclear revolution has already made many historical predictions and analogies obsolete. A third world war is now most unlikely. It now seems certain that the Communist revolution will have no Bonaparte. The prophecies of the cyclical historians have been rendered false, for our "time of troubles" will not degenerate into numerous and widespread wars, and a new "universal imperium" cannot be put together by a twentieth-century conquering Caesarism. A multiple balance of power in the traditional sense, based on counterbalancing alliances, cannot really revive, for, as Hans Morgenthau has pointed out, alliances can now never have what stability and reliability they formerly had; fear of nuclear destruction will make every member of a combination look sharply to its own survival.

The immediate effect of the Atomic Peace is the freezing of the international *status quo*. The actual power situations in the world seem currently sealed. Even when the balancing of power runs through countries rather than between them—as in partitioned Germany, Korea, and Vietnam—the situations cannot be altered by force and have not yielded to peaceful methods of change. The avowed purpose of American foreign policy is to prevent any territorial changes by force. American policy in Vietnam exemplifies this, and the United States has ruled out force as a means of unifying Germany and Korea. Revolutions and counter-revolutions within countries are more difficult, too, for there is fear of any significant shift in the balance of power by way of internal changes, particularly if there is interference by outside powers.

The most obvious problem posed by the Nuclear Peace is how to unfreeze the *status quo*, how to get necessary and desirable changes—the kind of changes formerly made by force, by internal revolutions and international wars. However, if the drive to machine technology has indeed become the central movement of this century's history, and if mechanized societies, however established and under whatever labels, succeed in satisfying the material aspirations of more and more peoples, then ideological and even national conflicts will recede, and the thawing of a frozen *status quo*, both internally and internationally, while still important, may in general be considerably less so in future decades than it is today, for the number of acute crises will diminish.

Advanced technology is making it easy to fool people. It would be well if technology also devoted itself to producing forms of records, photographic, printed, sound-recorded, which cannot be altered without detection, at least to the degree of a dollar bill. But it would be still more effective if the code of morals accepted generally rendered it a universally condemned sin to alter a record without notice that it is being done.

Vannevar Bush, 1967

CONSERVATIVES

F. H. Clauser, 1967

A technologist reporting on magnetohydrodynamics to a study group at the National Academy of Sciences discerns a wider truth.

RICHARD RHODES

254

In years gone by, studies aplenty have been made foretelling the future trends of speed and size of aircraft, powers and weights of engines, range and capabilities of radars, and so on. Occasionally, some devilish individual takes the trouble to go back and compare past predictions with later reality. Invariably, he finds that engineers and scientists are a conservative lot in their predictions. The immediate problems that confront them appear so formidable that they flinch from predicting ever-accelerating progress and conjure up visions of a natural barrier ahead which will cause the curve of progress to flatten off much as a biological population comes into equilibrium with its environment.

REGICIDE

Melvin Maddocks

The era of the great ocean liners came to a close in the middle of the century; 1967 marked a regicidal end.

In 1959, just one year after the *United States* set the all-time record for speed, her passenger bookings declined by 25 percent. By 1962, Cunard was reporting an annual loss of almost five million dollars and was forced to begin selling some of its real-estate properties to make ends meet.

The *United States* was faring little better. It had cost $78 million to build her, and now the United States Lines could not even afford to pay the interest on that account, much less amortize it. Except at the peak of the summer season, her voyages rarely even paid for themselves. At times, her crew members outnumbered the passengers by three or even four to one. For a while the United States government pumped a subsidy that averaged $9.7

million per year into the *United States*. Then, in November 1969, scarcely 17 years after the triumph of her maiden passage, the *United States* was mothballed at Newport News, Virginia, and to all intents and purposes, the United States retired from the transatlantic-passenger shipping lanes.

In the offices of the European shipping lines, worried men studied the rising cost curves and the plummeting passenger curves and reached some painful conclusions. The Compagnie Générale Transatlantique, the fabled French Line, had retired the 32-year-old *Ile de France* as uneconomic at the end of 1958, and had constructed a new vessel, the 66,000-ton *France*, in the hope of recapturing past glory. But it swiftly became apparent that the North Atlantic, with a steady stream of passenger jets overhead, was no longer home to the liners; just to scrape by—and that not for long—French Line ships had to spend much of their time on warm-water winter cruises to the Caribbean and elsewhere.

The Dutch, the Italians and the Scandinavians also offered transatlantic service. But their neat, small ships were equally suited for cruise duty, and that was the way they spent a good deal of their time. By the mid-1960s, only the Cunard Line's two great *Queens* still plied the North Atlantic month in, and month out, winter and summer alike. Frequently they sailed more than half empty. With a professional eye for pathos, British magazines started to liken them to deserted English holiday resorts on rainy summer days.

It was costing an average of $50,000 a day to keep the *Queen Mary* in service, leading the Cunard Line toward bankruptcy. At the end of 1965, the *Queen Elizabeth* was ready for a periodic refit. Cunard decided to refurbish her for ten more years of service, perhaps in a way that would attract cruise passengers as well. When she emerged several months later, every cabin had its own private bath, and there was air conditioning throughout.

All the desperate hopes were swiftly shattered. On one of her voyages, the *Queen Elizabeth* carried only 70 passengers one way and 130 the other. In the first-class lounge, one might take afternoon tea in the company of stewards only.

The Cunard Line's overall losses rose to nearly $20 million in 1966. The line had laid the keel for a new and much smaller vessel, the *Queen Elizabeth 2;* she would ply the Atlantic at times, but essentially she would be a cruise ship in no way to be compared with the *Queens* almost twice her size. On May 8, 1967, the captains of the *Queen Mary* and *Queen Elizabeth*, both at sea, received radio instructions to open sealed envelopes and read the contents to their crews. The letters explained in terse corporate language that each of the two liners was losing $1.8 million a year. Cunard was going to scrap them, the *Queen Mary* in the autumn, the *Queen Elizabeth* the following year.

The era of the great liners was over. The world had turned from sea to air travel as surely as a half century earlier it had turned from sail to steam. The haut monde of the great liners had become the jet set and "grand" had become a period adjective.

FOR A COMING EXTINCTION

W. S. MERWIN

Not only the great ocean liners succumbed to encroaching and changing technology in the second half of the twentieth century. Commercial hunting diminished whale populations worldwide until it seemed extinction might be looming. For some species, survival is still in question, but the gray whales of this poem have responded to an international treaty banning hunting and have begun to recover.

Gray whale
Now that we are sending you to The End
That great god
Tell him
That we who follow you invented forgiveness
And forgive nothing

I write as though you could understand
And I could say it
One must always pretend something
Among the dying
When you have left the seas nodding on their stalks
Empty of you
Tell him that we were made
On another day

The bewilderment will diminish like an echo
Winding along your inner mountains
Unheard by us
And find its way out
Leaving behind it the future
Dead
And ours

When you will not see again
The whale calves trying the light
Consider what you will find in the black garden
And its court
The sea cows the Great Auks the gorillas
The irreplaceable hosts ranged countless
And fore-ordaining as stars
Our sacrifices
Join your word to theirs
Tell him
That it is we who are important

BOOBY-TRAP TECHNOLOGY

PAUL GOODMAN, 1967

A philosopher and humanist examines the morality of scientific technology contra C. P. Snow.

The morality of technology has . . . suffered a sea-change. Historically, the main origin of technology, in the work of craftsmen, miners, navigators, etc., provided a ready check on utility, efficiency, costs, and unforeseen effect. A secondary but important origin, in the natural experiments of Medieval and Renaissance alchemists and magicians, and perhaps physicians, provided no such check; the archetypal story is *The Sorcerer's Apprentice*. But therefore these groups had a strong ethical code, to permit only white magic, and prescribing Christian virtue as the priceless ingredient of the Philosopher's Stone. The Black Magician, like our Mad Scientist, was a villain for popular tragedy. . . .

There ceases to be a morality of technique at all. A technician is hired to execute a detail of a program handed down to him. Apart from honestly trying to make his detail work, he is not entitled to criticize the program itself, in terms of its efficiency, common sense, beauty, effect on the community, or human scale. If management is not concerned with these either, a technician must often lend his wits to ludicrous contradictions. Cars are designed to go faster than it is safe to drive; food is processed to take out the nourishment; housing is expertly engineered to destroy neighborhoods; weapons are stockpiled that only a maniac would use. The ultimate of irresponsibility is that the engineer is not allowed to know what he is making, and we have had this too.

The interlocking of systems of technology without the direct check of personal acquaintance and use and political prudence creates a series of booby traps. Human scale may be quite disregarded, the time and energy that people actually have, the space they need to move in, and the rhythm or randomness with which they best operate. As the engineers design, we move, or sometimes can't move. Facilities are improved, but during the transition everybody is inconvenienced, and by the time the facility is completed it may be obsolescent. Fast trips are made possible by jet, but they prove to chop up our lives, to involve longer trips to airports and more waiting in terminals, so we have less free time. Business machines are installed and there is no longer any person from whom to get information or service for one's particular case. Cities spread so far that one can't get out of them; the country is deserted, so it is inefficient to provide means to get to it. Immense printing presses and other means of communication are devised, but to warrant such an investment of capital requires a mass audience, and it becomes hard to publish a serious book or transmit a serious message.

This sounds like chaos and modern life pretty nearly is. Apart from the cure of infectious diseases, some public services, and some household and farm equipment, there have been few recent advances in technique that have not proved to be a mixed bag in actual convenience. The great advantages, on balance, that came from universalizing basic conveniences or necessities, like electricity or water supply, do not necessarily occur when massifying comforts and luxuries. The moral advantages, of enriched opportunity, are largely delusory. New opportunities do not make time available to enjoy them, and the chance for choice works out as superficial acquaintance and confusion. The marvels of fable, like flying through the air and seeing at a distance, have not proved so beautiful in reality. . . .

A few years ago, C. P. Snow created a stir by speaking of the chasm between "two cultures," that of the scientists and that of the humanists. Since we live in times dominated by scientific technology, he castigated the humanists especially for not knowing the other language. The point of this lecture has been that Sir Charles posed the issue wrongly. There is only one culture; and probably the scientific technologists have betrayed it most. Science, the dialogue with the unknown, is itself one of the humanities; and technology, practical efficiency, is a part of moral philosophy. Scientific technology has become isolated by becoming subject to the empty system of power: excluding, expanding, controlling. The remedy is for scientists and technicians to reassert their own proper principles, and for ordinary people to stop being superstitious and to reassert their own control over their environment. Then there will be communication again.

MORE INJURY TO
MORE PEOPLE

HAROLD P. GREEN, 1967

Formerly an attorney with the U.S. Atomic Energy Commission, Green was a professor of law at George Washington University at the time he published this essay.

In 1937, a group of eminent scientists and social scientists, acting as the Science Committee of the federal government's National Resources Committee, undertook to consider the significance of science as a national resource. The Technology Subcommittee of the Science Committee issued a report, *Technological Trends and National Policy*, which evaluated the unprecedented rate of technological change during the first third of the twentieth century and discussed the social consequences of the new technologies which had come into being. The scientists also attempted to predict the course of technological progress which could be anticipated in the second third of the century, but it is clear that they did not come close to anticipating the level of technological development which would occur. Their report includes not even a hint of the possible emergence of nuclear technology, radar, computers, or the jet engine, although these came into practice within a few years. In fact, the report suggested that aviation technology had by 1937 largely run its course, and that future developments would lie in safety and comfort rather than speed.

It would be useful, as we enter the last third of the century, to consider why these eminent scientists were so wide of the mark. In 1937, resources available to science were drastically limited. Research equipment and funds to support research were scarce. Technological development depended largely on private, profit-seeking investment; government support was negligible. Since World War II, however, the game is no longer the same. In 1940, the federal government spent only $74 million for research and development. By 1950, annual federal expenditures on research and development had risen to the billion dollar level, and have steadily increased until today they are at $16 billion per year, representing about 15 percent of the federal budget. The ready availability of immense federal support has replaced the private marketplace as the principal determinant of scientific and technological progress. A large part of these federal funds is, of course, expended for national defense; but a large part—perhaps almost half—goes into development of strictly nonmilitary areas. Our society seems to be wholly committed to the pursuit of technological progress for its own sake—for reasons of national defense, the domestic economy, the general welfare, and national prestige.

VISIONS OF TECHNOLOGY

259

It is not necessary to spell out the many benefits of the mushrooming technological advance now being experienced. Instead, let us recognize that technological progress may also bring with it some undesirable consequences. At best, it frequently involves major social disruptions: unemployment, excessive leisure, and changes in social or moral standards. It may also involve hazards to the health, safety, and security of society. We already face the massive assault on human privacy represented by the computer and electronics technologies; the ill effects of pesticides, drugs, radiation, and automobile exhaust fumes; the potential consequences of atomic, biological, and chemical weapons. It is easy to visualize Washington bureaucrats determining whether, when, and where it will rain or snow; it seems inevitable that society will be living with the sonic boom, the test-tube creation of life, genetics engineering, and machines which can think and reason.

It seems clear in 1967 that continuing federal support for science and technology on the present scale is inevitable and irreversible. For one thing, the solvency of thousands of universities and business enterprises, and probably national prosperity itself, are largely dependent upon continuation of federal expenditures on research and development. It is almost certain, therefore, that technological advance in the last third of the twentieth century will dwarf even the immense advances of the past 30 years.

It is increasingly important to recognize that technological advance carries with it the very real threat of destruction of human beings and cherished human values. How much damage could one demented or evil person have inflicted on society in a single act 25 years ago? Today, such a person might inflict immense damage measured in thousands, if not millions, of human lives. If technology has not yet brought us to a crisis, it is clear that we live in peril, with a constantly shrinking margin of error or miscalculation. Nevertheless, so obsessed are we, as a matter of national policy, with technological advance as an end in itself that relatively little attention is paid to the problem of protecting society against its hazards.

Admiral Hyman Rickover, almost alone among the technologists of today, has shown an awareness of this problem. He has reminded us that technology exists to serve man, and has cautioned us against the enormous "potentialities for injury to human beings and to society . . . almost as if technology were an irrepressible force of nature to which we must meekly submit." He calls upon the legal profession, as its special civic responsibility, to protect society against rampant technology (although he does not suggest how this should be done). Admiral Rickover's summons to the legal profession squarely raises the question of the relationship between law and science and technology. . . .

Adequate protection of society against new technological hazards depends upon the speed with which the courts and the legislatures identify the new problems and create adequate rules to deal with them, and the speed with which the technology develops. Until recently, this time factor has been manageable. Only 30 years ago, the Science Committee of the National Resources Committee could say with complete accuracy [that] "from the early origins of an invention to its social effects the time interval averages about 30 years.". . .

But we know today that technological advance is so rapid that we do not have this 30-year interval. With the vast public resources available for scientific and technological advance, inventions appear suddenly, and very substantial social effects may be felt almost instantaneously. Consider, for example, the sudden emergence of atomic energy technology with its immediate major social effects.

The shorter interval between the origins of a technology and the time its social effects are felt means that the practice of a technology with intrinsic hazards will result in more injury to more people at an earlier time. There still remains the time lag, inherent in our present legal system, between the first appearance of injury and the creation of new rules of law adequate to deal with the problems. Lawyers have recognized that during this time-lag adequate justice may not be done on behalf of persons injured by the new technology, and this has been regarded as one of the costs of our form of society which places a high premium on private initiative, freedom, and progress. This price may, however, become intolerably high if the shrinkage of the time interval results in substantial injury before the law can provide adequate relief and remedy.

PREDICTIONS: ZERO POPULATION GROWTH

DONALD J. BOGUE, 1967

Recent developments in the worldwide movement to bring runaway birth rates under control are such that it now is possible to assert with considerable confidence that the prospects for success are excellent. In fact, it is quite reasonable to assume that *the world population crisis is a phenomenon of the 20th century, and will be largely if not entirely a matter of history when humanity moves into the 21st century.* No doubt there will still be problematic areas in the year 2000, but they will be confined to a few nations that were too prejudiced, too bureaucratic, or too disorganized to take action sooner, or will be confined to small regions within some nations where particular ethnic, economic, or religious groups will not yet have received adequate fertility control services and information. With the exception of such isolated remnants (which may be neutralized by other areas of growth-at-less-than-replacement), it is probable that by the year 2000 each of the major world regions will have a population growth rate that either is zero or is easily within the capacity of its expanding economy to support.

The implications of these assertions for the feeding of the human race are obvious. Given the present capacity of the earth for food production, and the potential for additional food production if modern technology were more fully employed, mankind clearly has within its grasp the capacity to abolish hunger—within a matter of a decade or two. Furthermore, it is doubtful whether a total net food shortage for the entire earth will ever develop. If such a deficit does develop, it will be mild and only of short duration. The really critical problem will continue to be one of maldistribution of food among the world's regions. . . .

This view is at variance with the established view of many population experts. For more than a century, demographers have terrorized themselves, each other, and the public at large with the essential hopelessness and inevitability of the "population explosion." Their prophecies have all been dependent upon one premise: "If recent trends continue. . . ." It is an ancient statistical fallacy to perform extrapolations upon this premise when in fact the premise is invalid. It is my major point that *recent trends have not continued, nor will they be likely to do so.* Instead, there have been some new and recent developments that make it plausible to expect a much more rapid pace in fertility control. These developments are so new and novel that *population trends before 1960 are largely irrelevant in predicting what will happen in the future.*

In times of social revolution, it often is fruitless to forecast the future on the basis of past experience. Instead, it is better to abandon time series analysis and study the phenomenon of change itself, seeking to understand it and to learn in which direction and how rapidly it is moving. If enough can be learned about the social movement that is bringing about the change, there is a hope that its eventual outcome can be roughly predicted. . . .

To summarize: wherever one looks in the underdeveloped segments of the world, one finds evidence of firmly established and flourishing family planning activity. By whatever crude estimates it is possible to make, it is quite clear that a sufficiently large share of the population already is making use of modern contraceptives to have a depressing effect on birth rate. Even conservative evaluations of the prospects suggests that *instead of a "population explosion" the world is on the threshold of a "contraception adoption explosion."* . . .

Looking at [these developments], realizing that they are only five years old or less, knowing that accomplishments in this area are cumulative and grow by exponential curves, and appreciating that new discoveries and improvements will accrue promptly along all fronts—medical, social, and psychological—both from basic research and from accumulating experience and evaluation—the following generalizations appear to be justified:

The trend of the worldwide movement toward fertility control has already reached a state where declines in death rates are being surpassed by declines in birthrates. Because progress in death control is slackening and progress in birth control is accelerating, the world has already entered a situation where the pace of population growth has begun to slacken. The exact time at which this "switch-over" took place cannot be known exactly, but we estimate it to have occurred about 1965. From 1965 onward, therefore, the rate of world population growth may be expected to decline with each passing year. The rate of growth will slacken at such a pace that it will be zero or near zero at about the year 2000, so that population growth will not be regarded as a major social problem except in isolated and small "retarded" areas.

In evaluating these conclusions, it must be kept in mind that the topic is a deadly serious one, and the penalties for misjudgment may be very great. There is one set of penalties that results from over-optimism. But there is another set of penalties that results from over-pessimism. It is quite possible that nothing has sapped the morale of family planning workers in the developing countries more than the Malthusian pessimism that has been radiated by many demographic reports. It is like assuring soldiers going into battle that they are almost certain to be defeated. If the comments made here should be so fortunate as to fall into the hands of these same family planning workers, it is hoped that those who read them will appreciate just how close they actually are to success.

The rate of world population growth had not yet declined to zero by the last years of the century, but it had declined sharply and was continuing to do so, projecting an ultimate peak world population of about 10 billion. The primary factor in its decline proved to be the education of women.

BLUE JEANS AND COCA COLA

ALVA MYRDAL, 1967

An anguished warning from the Swedish minister of disarmament.

The technical possibilities to develop the telecommunications satellite to such a point that the most isolated village in the most distant continent can be reached by one and the same powerful transmitter, faces us with the choice of utilizing this advance either for establishing a system of equitable interdependence between nations and peoples, or for a system of as yet unimaginable cultural hegemony by the technologically, industrially and economically strongest nation.

THE WHOLE EARTH

JOHN ALLEN

The chief ideologist of the Biosphere 2 project traces the history of the first photograph made public of the whole earth, aka Biosphere 1, in 1967.

In the mid-60s, Stewart Brand, editor of the *Whole Earth Catalog*, knew that the space agency NASA must have photographs of the Earth taken from spaceships, and publicly asked: "Why haven't we seen a photograph of the whole Earth yet?" Brand had buttons made which asked precisely this question and began selling them across America. NASA published the first photograph of the planet in 1967, and Brand published it on the cover of his *Whole Earth Catalog.* He felt that the image did much to shatter both the "flat earth" and "endlessly more to be used" illusions. Seeing the planet as a whole supported the idea of thinking of it as a total system and solving the problems at hand on the right scale. Like it or not, humankind had changed the face of the planet and had to acknowledge responsibility. "We are as gods," he wrote in his 1980 edition of the catalog, "so we might as well get good at it."

At the conference initiating the Biosphere 2 project, astronaut Rusty Schweickart recalled his spacewalk during the Apollo 9 flight when he looked on the Earth, miles below: "On that small blue and white planet below is everything that means anything to you; all of history and music and

▶ The whole Earth, photographed from space for the first time, became a potent symbol of environmental responsibility.

poetry and art, death and birth, love, tears, joy, games . . . all on that little spot in the cosmos. National boundaries and human artifacts no longer seem real. Only the biosphere, whole and home of life."

CONTINUITY

BRUCE MAZLISH, 1967

W e are now coming to realize that man and the machines he creates are continuous and that the same conceptual schemes, for example, that help explain the workings of his brain also explain the workings of a "thinking machine." Man's pride, and his refusal to acknowledge this continuity, is the substratum upon which the distrust of technology and an industrialized society has been reared. Ultimately, I believe, this last rests on man's refusal to understand and accept his own nature—as being continuous with the tools and machines he constructs.

TYRANNIES

EMMANUEL G. MESTHENE, 1967

After eleven years with the Rand Corporation, Mesthene directed the Harvard Program on Technology and Society from 1964 to 1974. His was the broadest perspective on technology of any I encountered in assembling this anthology.

I t might be well to start . . . by noting what is new about our age.
The fact itself that there is something new is not new. There has been something new about every age, otherwise we would not be able to distinguish them in history. What we need to examine is what in particular is new about our age, for the new is not less new just because the old was also at one time new.

The mere prominence in our age of science and technology is not strikingly new, either. A veritable explosion of industrial technology gave its name to a whole age two centuries ago, and it is doubtful that any scientific idea will ever again leave an imprint on the world so penetrating and pervasive as did Isaac Newton's a century before that.

It is not clear, finally, that what is new about our age is the rate at which it changes. What partial evidence we have, in the restricted domain of economics, for example, indicates the contrary. The curve of growth, for the hundred years or so that it can be traced, is smooth, and will not support claims of ex-

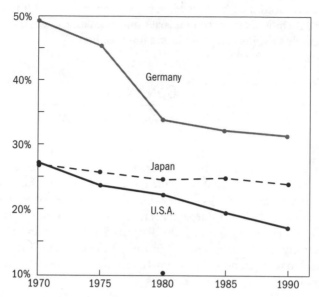

▶ Fraction of labor force employed in manufacturing, 1970–1990.

plosive change or discontinuous rise. For the rest, we lack the stability of concept, the precision of intellectual method, and the necessary data to make any reliable statements about the rate of social change in general.

I would therefore hold suspect all argument that purports to show that novelty is new with us, or that major scientific and technological influences are new with us, or that rapidity of social change is new with us. Such assertions, I think, derive more from revolutionary fervor and the wish to persuade than from tested knowledge and the desire to instruct.

Yet there is clearly something new, and its implications are important. I think our age is different from all previous ages in two major respects: first, we dispose, in absolute terms, of a staggering amount of physical power; second, and most important, we are beginning to think and act in conscious realization of that fact. We are therefore the first age who can aspire to be free of the tyranny of physical nature that has plagued man since his beginnings. . . .

Technology, in short, has come of age, not merely as technical capability, but as a social phenomenon. We have the power to create new possibilities, and the will to do so. By creating new possibilities, we give ourselves more choices. With more choices, we have more opportunities. With more opportunities, we can have more freedom, and with more freedom we can be more human. That, I think, is what is new about our age. We are recognizing that our technical prowess literally bursts with the promise of new freedom, enhanced human dignity, and unfettered aspiration.

At its best, then, technology is nothing if not liberating. Yet many fear it increasingly as enslaving, degrading, and destructive of man's most cherished values. It is important to note that this is so, and to try to understand why. I can think of four reasons.

First, we must not blink at the fact that technology does indeed destroy some values. It creates a million possibilities heretofore undreamed of, but it also makes impossible some others heretofore enjoyed. The automobile makes real the legendary foreign land, but it also makes legendary the once real values of the ancient marketplace. Mass production puts Bach and Brueghel in every home, but it also deprives the careful craftsman of a market for the skill and pride he puts into his useful artifact. Modern plumbing destroys the village pump, and modern cities are hostile to the desire to sink roots into and grow upon a piece of land. Some values are unquestionably bygone. To try to restore them is futile, and simply to deplore their loss is sterile. But it is perfectly human to regret them for a time.

Second, technology often reveals what technology has not created: the cost in brutalized human labor, for example, of the few cases of past civilization whose values only a small elite could enjoy. Communications now reveal the hidden and make the secret public. Transportation displays the better to those whose lot has been the worse. Increasing productivity buys more education, so that more people read and learn and compare and hope and are unsatisfied. Thus technology often seems the final straw, when it is only illuminating rather than adding to the human burden.

Third, technology might be deemed an evil, because evil is unquestionably potential in it. We can explore the heavens with it, or destroy the world. We can cure disease, or poison entire populations. We can free enslaved millions, or enslave millions more. Technology spells only possibility, and is in that respect neutral. The new opportunities it gives us include new opportunities to make mistakes. Its massive power can lead to massive error so efficiently perpetrated as to be well-nigh irreversible. Technology is clearly not synonymous with the good. It *can* lead to evil.

Finally, and in a sense most revealing, technology is upsetting, because it complicates the world. This is a vague concern, hard to pin down, but I think it is a real one. The new alternatives that technology creates require effort to examine, understand, and evaluate them. We are offered more choices, which makes choosing more difficult. We are faced with the need to change, which upsets routines, inhibits reliance on habit, and calls for personal readjustments to more flexible postures. We face dangers that call for constant reexamination of values and a readiness to abandon old commitments for new ones more adequate to changing experience. The whole business of living seems to become harder.

This negative face of technology is sometimes confused with the whole of it. It can then cloud the understanding in two respects that are worth noting. It can lead to a generalized distrust of the power and works of the human mind by erecting a false dichotomy between the modern scientific and technological enterprises on the one hand, and some idealized and static prescientific conception of human values on the other. It can also color discussion of some important contemporary issues that develop from the impact of technology on society, in a way that obscures rather than enhances understanding, and that therefore inhibits rather than facilitates the social action necessary to resolve them.

Because the confusions and discomfort attendant on technology are more immediate and therefore sometimes loom larger than its power and its promise, technology appears to some an alien and hostile trespasser upon the human scene. It thus seems indistinguishable from that other, older alien and hostile trespasser: the ultimate and unbreachable physical necessity of which I have spoken. Then, since habit dies hard, there occurs one of those curious inversions of the imagination that are not unknown to history. Our newfound control over nature is seen as but the latest form of the tyranny of nature. The knowledge and therefore the mastery of the physical world that we have gained, the tools that we have hewed from nature and the human wonders we are building into her, are themselves feared as rampant, uncontrollable, impersonal technique that must surely, we are told, end by robbing us of our livelihood, our freedom, and our humanity.

It is not an unfamiliar syndrome. It is reminiscent of the longtime prisoner who may shrink from the responsibility of freedom in preference for the false security of his accustomed cell. It is reminiscent even more of Socrates, who asked about that other prisoner, in the cave of ignorance, whether his eyes would not ache if he were forced to look upon the light of knowledge, "so that he would try to escape and turn back to the things which he could see distinctly, convinced that they really were clearer than these other objects now being shown to him." Is it so different a form of escapism from that, to ascribe impersonality and hostility to the knowledge and the tools that can free us finally from the age-long impersonality and hostility of a recalcitrant physical nature?

Technology has *two* faces: one that is full of promise, and one that can discourage and defeat us. The freedom that our power implies from the traditional tyranny of matter—from the evil we have known—carries with it the added responsibility and burden of learning to deal with matter and to blunt the evil, along with all the other problems we have always had to deal with. That is another way of saying that more power and more choice and more freedom require more wisdom if they are to add up to more humanity. But that, surely, is a challenge to be wise, not an invitation to despair.

CONTRA INDUSTRIAL TOURISM

EDWARD ABBEY, 1968

The irascible author of Desert Solitaire *offers a modest proposal for saving the nation's national parks from the ravages of the automobile.*

Having indulged myself in a number of harsh judgments upon the Park Service, I now feel entitled to make some constructive, practical, sensible proposals for the salvation of both parks and people.

(1) No more cars in national parks. Let the people walk. Or ride horses, bicycles, mules, wild pigs—anything—but keep the automobiles and the motorcycles and all their motorized relatives out. We have agreed not to drive our automobiles into cathedrals, concert halls, art museums, legislative assemblies, private bedrooms and the other sanctums of our culture; we should treat our national parks with the same deference, for they, too, are holy places. An increasingly pagan and hedonistic people (thank God!), we are learning finally that the forests and mountains and desert canyons are holier than our churches. Therefore let us behave accordingly.

Consider a concrete example and what could be done with it: Yosemite Valley in Yosemite National Park. At present a dusty milling confusion of motor vehicles and ponderous camping machinery, it could be returned to relative beauty and order by the simple expedient of requiring all visitors, at the park entrance, to lock up their automobiles and continue their tour on the seats of good workable bicycles supplied free of charge by the United States Government.

Let our people travel light and free on their bicycles—nothing on the back but a shirt, nothing tied to the bike but a slicker, in case of rain. Their bedrolls, their backpacks, their tents, their food and cooking kits will be trucked in for them, free of charge, to the campground of their choice in the Valley, by the Park Service. (Why not? The roads will still be there.) Once in the Valley they will find the concessioners waiting, ready to supply whatever needs might have been overlooked, or to furnish rooms and meals for those who don't want to camp out.

The same thing could be done at Grand Canyon or at Yellowstone or at any of our other shrines to the out-of-doors. There is no compelling reason, for example, why tourists need to drive their automobiles to the very brink of the Grand Canyon's south rim. They could *walk* that last mile. Better yet, the Park Service should build an enormous parking lot about ten miles south of Grand Canyon Village and another east of Desert View. At those points, as at Yosemite, our people could emerge from their steaming shells of steel and glass and climb upon horses or bicycles for the final leg of the journey. On the rim, as at present, the hotels and restaurants will remain to serve the physical needs of the park visitors. Trips along the rim would also be made on foot, on horseback, or—utilizing the paved road which already exists—on bicycles. . . .

They will complain of physical hardship, these sons of the pioneers. Not for long; once they rediscover the pleasures of actually operating their own limbs and senses in a varied, spontaneous, voluntary style, they will complain instead of crawling back into a car; they may even object to returning to desk and office and that dry-wall box on Mossy Brook Circle. The fires of revolt may be kindled—which means hope for us all.

(2) No more new roads in national parks. After banning private automobiles the second step should be easy. . . .

Once people are liberated from the confines of automobiles there will be a greatly increased interest in hiking, exploring, and back-country pack-trips. Fortunately the parks, by the mere elimination of motor traffic, will come to seem far bigger than they are now—there will be more room for more persons, an astonishing expansion of space. This follows from the interesting fact that a motorized vehicle, when not at rest, requires a volume of space far out of proportion to its size. . . . Distance and space are functions of speed and time. Without expending a single dollar from the United States Treasury we could, if we wanted to, multiply the area of our national parks tenfold or a hundredfold—simply by banning the private automobile. The next generation, all 250 million of them, would be grateful to us.

(3) Put the park rangers to work. Lazy scheming loafers, they've wasted too many years selling tickets at toll booths and sitting behind desks filling out charts and tables in the vain effort to appease the mania for statistics which torments the Washington office. Put them to work. They're supposed to be rangers—make the bums range; kick them out of those overheated aircondi-tioned offices, yank them out of those overstuffed patrol cars, and drive them out on the trails where they should be, leading the dudes over hill and dale, safely into and back out of the wilderness. It won't hurt them to work off a lit-tle office fat; it'll do them good, help take their minds off each other's wives, and give them a chance to get out of reach of the boss—a blessing for all concerned.

They will be needed on the trail. Once we outlaw the motors and stop the road-building and force the multitudes back on their feet, the people will need leaders. A venturesome minority will always be eager to set off on their own, and no obstacles should be placed in their paths; let them take risks, for Godsake, let them get lost, sunburnt, stranded, drowned, eaten by bears, buried alive under avalanches—that is the right and privilege of any free American. But the rest, the majority, most of them new to the out-of-doors, will need and welcome assistance, instruction and guidance. Many will not know how to saddle a horse, read a topographical map, follow a trail over slickrock, memorize landmarks, build a fire in the rain, treat snakebite, rappel down a cliff, glissade down a glacier, read a compass, find water under sand, load a burro, splint a broken bone, bury a body, patch a rubber boat, portage a waterfall, survive a blizzard, avoid lightning, cook a porcupine, comfort a girl during a thunderstorm, predict the weather, dodge falling rock, climb out of a box canyon, or pour piss out of a boot. Park rangers know these things, or should know them, or used to know them and can re-learn; they will be needed. . . .

Excluding the automobile from the heart of the great cities has been seri-ously advocated by thoughtful observers of our urban problems. It seems to me an equally proper solution to the problems besetting our national parks. Of course it would be a serious blow to Industrial Tourism and would be bit-terly resisted by those who profit from that industry. Exclusion of automobiles would also require a revolution in the thinking of Park Service officialdom and in the assumptions of most American tourists. But such a revolution, like it or not, is precisely what is needed. The only foreseeable alternative, given the current trend of things, is the gradual destruction of our national park system.

"Plastics"

JEFFREY L. MEIKLE

The word plastic entered the folklore of the baby boom generation just as its oldest members were coming of age. Thousands of moviegoers watched in 1968 as Dustin Hoffman, starring as *The Graduate*, received some advice from a family friend. Offering the secret of life, or at least of worldly success, he told the confused young man, "I just want to say one word to you. Just one word. . . . Plastics. . . . There's a great future in plas-tics." This odd pronouncement, isolated in the film's opening scene, con-

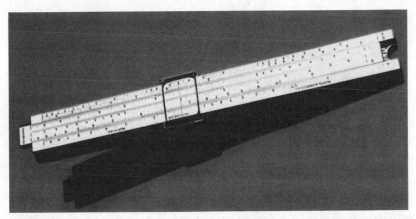

▶ The calculator that built America: the slide rule, now extinct.

vulsed audiences and became a line "repeated into classicdom by a whole generation of kids." The scene made a permanent impression—and not just among dismayed plastics executives whose trade journal could not bring itself to refer to that "tired old joke about plastics" until 1986. Most viewers would have had trouble explaining their laughter. Some perceived a comment on the banality of business, others an attack on comfortable middle-class materialism. Still others, recalling plastic's simulation of traditional materials, understood the scene as metaphoric commentary on the rhetoric of the Great Society. A few, catching an ominous note, entertained fleeting thoughts of science fiction nightmares of technology run amok. And some merely relished the absurd elevation of the commonplace. Whatever the reasons, the scene hit a nerve and entered communal memory.

SQUARE ROOTING

GENE SHALIT

The television critic recalled this University of California engineering-student cheer in a letter to The New York Times *in 1993, attributing it to "bygone days"; its reference to the slide rule locates it somewhere before the advent of the pocket electronic calculator (1971).*

> E to the X, dy! dx!
> E to the X, dx!
> Secant, cosine, tangent, sine,
> Three-point-one-four-one-five-nine;
> Square root, cube root, Q. E. D.,
> Slip stick! Slide rule!
> 'ray, U. C.!

THE RICHNESS
OF TECHNOLOGY

T. J. GORDON AND A. L. SHEF, 1968

The Sixth Goddard Memorial Symposium of the American Astronautical Society explored "the effect of technology upon human progress." Two aerospace engineers consider technology's qualitative influence.

We believe that more can be said about technology arising from a program than the mechanisms which spawn it or the means by which it is transmitted; there is a qualitative element about it, a richness, which ex-

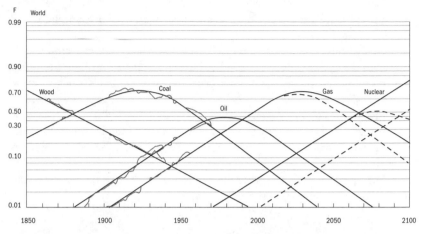

► Historical evolution of world primary energy mix (F = fraction of market in tons coal equivalent). Irregular lines = statistical data; smooth lines = calculated. Dashed lines = effect of introducing hypothetical new source of primary energy such as solar or fusion. No such source is at hand; primary energy mix in 21st century will be natural gas and nuclear power.

presses its potential for multidisciplinary application, and its eventual impact on society. Suppose we could view the future as a vast network of interrelating possibilities. The intersections of the network could represent alternative future policy choices, or the branch points of potential future developments. The technology produced by a particular enterprise, in finding transcending applications, opens some of these branch points up for alternate decisions. A rich technology finds us able to put our chips down at important intersections; a sparse technology opens relatively few to us. In other words, technology has a quality which relates to its impact on our ability to make future decisions to manipulate our future in ways we find desirable. An enterprise which may strongly affect our ability to control important aspects of the future is termed "rich."

Obviously, it is impossible to measure this factor with even moderate certainty. However, there are some attributes of technology and the programs which generate it that can be expected to improve richness. These are generality of future applications, relevance to currently perceived issues of the future, and likely impact on fundamental scientific knowledge.

Generality of future applications is a criterion which applies to the pervasiveness of the technology. Is it likely to find wide applications in many disciplines? Our current uses of the ancient skills of glassmaking, and potential future uses of nuclear energy are two examples.

Relevancy to currently perceived issues of the future is a criterion which is applied to the likely future utility of the technology. If a technology is likely to mitigate growing crime rates or food scarcity, for example, it has the component of richness.

Likely impact on fundamental scientific knowledge is a criterion which describes the opportunities opened by the technology for probing the existing disciplines of science with new precision or completely opening new disciplines to intellectual consideration. As mining enterprises provided new insight into geology, as navigation technology added to the science of geodesy, as space technology provided new models of fields and particles in interplanetary space, some transcending technology can become pivotal to science.

This quality of richness then becomes a third dimension for the comparison of technology-generating programs.

THE INTERNET PRIMEVAL

J. C. R. LICKLIDER, 1968

At the Defense Department's Advanced Research Projects Agency, computer pioneer J. C. R. Licklider commissioned the research that led to the Internet. His 1968 essay "The Computer as a Communication Device" recalls the ur-moment of its invention.

In a few years, men will be able to communicate more effectively through a machine than face to face.

That is a rather startling thing to say, but it is our conclusion. As if in confirmation of it, we participated a few weeks ago in a technical meeting held through a computer. In two days, the group accomplished with the aid of a computer what normally might have taken a week.

. . . We were all in the same room. But for all the communicating we did directly across that room, we could have been thousands of miles apart and communicated just as effectively—as people—over that distance. . . .

Creative, interactive communication requires a plastic or moldable medium that can be modeled, a dynamic medium in which premises will flow into consequences, and above all a common medium that can be contributed to and experimented with by all.

Such a medium is at hand—the programmed digital computer. Its presence can change the nature and value of communication even more profoundly than did the printing press and the picture tube, for, as we shall show, a well-programmed computer can provide direct access for both to informational resources and to the *processes* for making use of the resources. . . .

To appreciate the importance the new computer-aided communication can have, one must consider the dynamics of "critical mass," as it applies to cooperation in creative endeavor. Take any problem worthy of the name, and you find only a few people who can contribute effectively to its solution. Those people must be brought into close intellectual partnership so that their ideas can come into contact with one another. But bring these people together physically in one place to form a team, and you have trouble, for the most creative people are often not the best team players, and there are not enough top positions in a single organization to keep them all happy. Let them go their separate ways, and each creates his own empire, large or small, and devotes more time to the role of emperor than to the role of problem solver. The principals still get together at meetings. They still visit one another. But the time scale of their communication stretches out, and the correlations among mental models degenerate between meetings so that it may take a year to do a week's communicating. There has to be some way of facilitating communication among people without bringing them together in one place.

A single multiaccess computer would fill the bill if expense were no object, but there is no way, with a single computer and individual communication lines to several geographically separated consoles, to avoid paying an unwarrantedly large bill for transmission. . . . The difficulty is that the common carriers do not provide the kind of service one would like to have—a service that would let one have ad lib access to a channel for short intervals and not be charged when one is not using the channel.

It seems likely that a store-and-forward (i.e., store-for-just-a-moment-and-forward-right-away) message service would be the best for this purpose, whereas the common carriers offer, instead, service that sets up a channel for one's individual use for a period not shorter than one minute.

The problem is further complicated because interaction with a computer via a fast and flexible graphic display, which is for most purposes far superior to interaction through a slow-printing typewriter, requires markedly higher information rates. Not necessarily more information, but the same amount in faster bursts—more difficult to handle efficiently with the conventional common-carrier facilities.

It is perhaps not surprising that there are incompatibilities between the requirements of computer systems and the services supplied by the common carriers, for most of the common-carrier services were developed in support of voice rather than digital communication. Nevertheless, the incompatibilities are frustrating. It appears that the best and quickest way to overcome them—and to move forward the development of interactive *communities* of geographically separated people—is to set up an experimental network of multiaccess computers. Computers would concentrate and interleave the concurrent, intermittent messages of many users and their programs so as to utilize wide-band transmission channels continuously and efficiently, with marked reduction in overall cost. . . .

What will on-line interactive communities be like? In most fields they will consist of geographically separated members, sometimes grouped in small clusters and sometimes working individually. They will be communities not of common location, but of *common interest*. In each field, the overall community of interest will be large enough to support a comprehensive system of field-oriented programs and data.

In each geographical sector, the total number of users—summed over all the fields of interest—will be large enough to support extensive general-purpose information processing and storage facilities. All of these will be interconnected by telecommunication channels. The whole will constitute a labile network of networks—ever-changing in both content and configuration.

What will go on inside? Eventually, every informational transaction of sufficient consequence to warrant the cost. Each secretary's typewriter, each data-gathering instrument, conceivably each dictation microphone, will feed into the network.

You will not send a letter or a telegram; you will simply identify the people whose files should be linked to yours and the parts to which they should be linked—and perhaps specify a coefficient of urgency. You will seldom make a telephone call; you will ask the network to link your consoles together. . . .

Available within the network will be functions and services to which you subscribe on a regular basis and others that you call for when you need

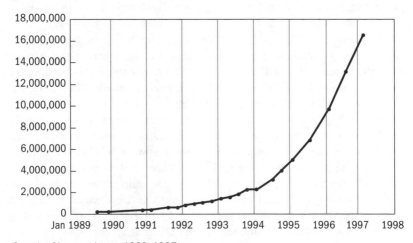

▶ Growth of Internet hosts, 1989–1997.

them. In the former group will be investment guidance, tax counseling, selective dissemination of information in your field of specialization, announcement of cultural, sport, and entertainment events that fit your interests, etc. In the latter group will be dictionaries, encyclopedias, indexes, catalogues, editing programs, teaching programs, testing programs, programming systems, data bases and—most important—communication, display, and modeling programs.

All these will be—at some late date in the history of networking—systematized and coherent; you will be able to get along in one basic [computer] language up to the point at which you choose a specialized language for its power or terseness.

When people do their informational work "at the console" and "through the network," telecommunication will be as natural an extension of individual work as face-to-face communication is now. The impact of that fact, and of the marked facilitation of the communication process, will be very great—both on the individual and on society.

First, life will be happier for the on-line individual because the people with whom one interacts most strongly will be selected more by commonality of interests and goals than by accidents of proximity. Second, communication will be more effective and productive, and therefore more enjoyable. Third, much communication and interactions will be with programs and programmed models, which will be (a) highly responsive, (b) supplementary to one's own capabilities, rather than competitive, and (c) capable of representing progressively more complex ideas without necessarily displaying all the levels of their structure at the same time—and which will therefore be both challenging and rewarding. And, fourth, there will be plenty of opportunity for everyone (who can afford a console) to find his calling, for the whole world of information, with all its fields and disciplines, will be open to him—with programs ready to guide him or to help him explore.

For the society, the impact will be good or bad, depending mainly on the question: Will "to be on line" be a privilege or a right? If only a favored segment of the population gets a chance to enjoy the advantage of "intelligence amplification," the network may exaggerate the discontinuity in the spectrum of intellectual opportunity.

On the other hand, if the network idea should prove to do for education what a few have envisioned in hope, if not in concrete detailed plan, and if all minds should prove to be responsive, surely the boon to humankind would be beyond measure.

SO IT GOES

KURT VONNEGUT, JR., 1968

As an eighteen-year-old army private in the Second World War, Kurt Vonnegut, Jr., survived the Allied firebombing of Dresden; he and other prisoners were housed fortuitously in a meat locker five stories below ground. At least 140,000 people were killed, most of them asphyxiated in basement bomb shelters. The prisoners then had the grim duty of burning the bodies without benefit of the daily ration of brandy issued to German troops. When I interviewed Vonnegut for the Paris Review *in the early 1970s he called the experience "a duty dance with death." It took him twenty years, he told me, to find a perspective from which to tell the story, his novel* Slaughterhouse Five. *In that work, his alter ego Billy Pilgrim travels back and forth in time and space, a device which effectively represents the flashbacks and nightmares of the post-traumatic stress disorder Vonnegut must have suffered. At one point, waiting for a flying saucer to pick him up, Billy Pilgrim watches a war movie. The fantasy of reversing time to reverse the baleful effects of technology will recur.*

Billy looked at the clock on the gas stove. He had an hour to kill before the saucer came. He went into the living room, swinging [a] bottle [of champagne] like a dinner bell, turned on the television. He came slightly unstuck in time, saw the late movie backwards, then forwards again. It was a movie about American bombers in the Second World War and the gallant men who flew them. Seen backwards by Billy, the story went like this:

American planes, full of holes and wounded men and corpses took off backwards from an airfield in England. Over France, a few German fighter planes flew at them backwards, sucked bullets and shell fragments from some of the planes and crewmen. They did the same for wrecked American bombers on the ground, and those planes flew up backwards to join the formation.

The formation flew backwards over a German city that was in flames. The bombers opened their bomb bay doors, exerted a miraculous magnetism which shrunk the fires, gathered them into cylindrical steel containers, and lifted the containers into the bellies of the planes. The containers were stored neatly in racks. The Germans below had miraculous devices of their own, which were long steel tubes. They used them to suck more fragments from the crewmen and planes. But there were still a few wounded Americans, though, and some of the bombers were in bad repair. Over France, though, German fighters came up again, made everything and everybody as good as new.

When the bombers got back to their base, the steel cylinders were taken from the racks and shipped back to the United States of America, where factories were operating night and day, dismantling the cylinders, separating the dangerous contents into minerals. Touchingly, it was mainly women who did this work. The minerals were then shipped to specialists in remote areas. It was their business to put them into the ground, to hide them cleverly, so they would never hurt anybody ever again.

The American fliers turned in their uniforms, became high school kids. And Hitler turned into a baby, Billy Pilgrim supposed. That wasn't in the movie. Billy was extrapolating. Everybody turned into a baby, and all humanity, without exceptions, conspired biologically to produce two perfect people named Adam and Eve, he supposed.

THE CURVE OF
TECHNOLOGICAL COMPETENCE

T. J. GORDON AND A. L. SHEF, 1968

Further exploring "the effect of technology upon human progress," aerospace engineers Gordon and Shef found a surprising regularity in technological development in the twentieth century as well as curious differences among nations. In subnational areas, California led the world, possibly because of its early involvement in aircraft production—or possibly because of the little-noticed but magnetic attraction of scientists and engineers to pleasant seaside venues.

To examine the effect of technology on societies, we have chosen nine representative, separate countries: the United States, the United Kingdom, the USSR, France, Sweden, Japan, India, China, and Brazil. In addition, we have treated the world characteristics as a bulk aggregate to form a 10th society. Also included as an 11th group is a peculiar subnational case, California, because of its special characteristics. The period extends from the beginning of the 20th century to the present time (1900 to 1965). . . .

An examination of the aggregate world society reveals a number of interesting fundamental features. The technological status of the world as a whole advances at a roughly constant exponential rate, doubling every 20 years, or in effect, every generation. Although slight temporal differences exist from an overall viewpoint, growth rate from at least the beginning of the 20th century has been relatively constant for the world as a whole. Furthermore, the present technological status of the world is roughly equivalent to the level of the United States alone at the beginning of the 20th century.

At that time, the United States and the United Kingdom were approxi-

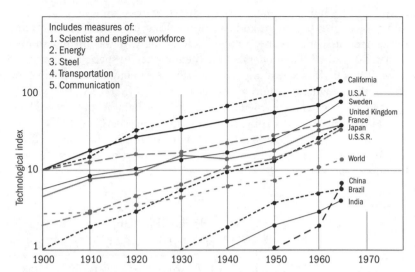

Includes measures of:
1. Scientist and engineer workforce
2. Energy
3. Steel
4. Transportation
5. Communication

▸ Technological index as a function of time for selected societies, 1900–1965 (1965 U.S. = 100).

mately equivalent. For the first two decades of the 20th century, the United States rapidly pulled ahead of the United Kingdom and, since that time, has maintained approximately a 20- to 30-year lead.

At the beginning of the 20th century, France and Sweden were approximately equivalent in technological status and this appears intuitively correct. Note, however, that commencing with the critical period following the turmoil of the "great depression" and the international conflict period of the 1930s, Sweden pulled consistently ahead of France and had surpassed the United Kingdom by the end of World War II. Sweden is now moving into a position close to that of the United States, while France continues to develop at a rate of progress close to that of the world average.

The curve depicting the world's growth of technological competence separates nations generally acknowledged as advanced from those nations recognized as "underdeveloped." It is significant that the technological growth rates of both groups of countries are approximately the same; furthermore, Japan serves as an example of a country which has crossed over the boundary from the underdeveloped technological group to the advanced group.

A point of particular interest is the technological index of China in the year 1965. This point is based on extrapolated data, because the appropriate information relating to China's technology is not readily available; therefore, this point should be viewed with caution. If this growth is true, China is exhibiting the most rapid expansion of any of the countries examined within this century.

The subnational society of the United States, California, has been included in [the graph] as an example of the extreme rapidity of growth of technological status which may occur in subsets of individual nations. California has been able to exceed the general technological status of the United States, not just recently as one would intuitively judge, but ever since World War I. From an overall standpoint, California has led the rest of the United States in technological status by approximately 15 years, ever since 1920. What caused this is a matter worthy of further study. However, it is clear that the space program alone, an invention of the late 1950s, has not been responsible for California's preeminence. Rather, the space program may have benefitted from the state's technological level and initiated the flow of technological knowledge back into the rest of the United States as a whole.

One might ordinarily expect that the curve of technological competence would vary in accordance with some characteristic growth pattern such as a logistic curve shape, that is, rising slowly at first as experience is gathered, then proceeding at an exponential rate while the reinvestment process is in progress, and finally tapering off as technological limits are reached. Within the limits of this study, it is not possible to ascertain which portion of this curve is presented in [the graph]. Intuitively, we feel it is the growth phase, that societies are beyond the beginning, and the limits of technological growth lie far in the future.

What might eventually limit growth? In terms of the definition of technology treated here, it would be when societies know how to do all things, a state of omniscience too distant to contemplate or one which may not exist. On the other hand, perhaps the practical growth limit may arrive when societies simply lose interest in the reinvestment process, or become nontechnological—like the porpoise, intelligent but without machines. This state would seem to be almost equally remote.

However, other limiting conditions may conceivably arise in the future, perhaps earlier for some societies than for others. . . . Combined technological and social advancement of a society brings social ills that sometimes require even more technology for their cure. The technological growth curve could reach a limit when society has progressed to the point that all further technological advancement is devoted to the maintenance of its status quo.

Finally, catastrophic or subtly eroding war or conflict could not only halt growth but cause a regression by destroying the minds, books, and tapes which hold the information, as surely as burning of the library of Alexandria destroyed much of the written history of man's early civilization. Other than these limits, there appears to be no upper bound to technology. . . .

In summary, our analysis of technological status has shown the following for the time span and the countries examined:

1. Technology is growing exponentially.
2. The technological index approximately doubles every 20 years.
3. The technological competence of a nation can be defined in terms of its position with respect to the aggregate world competence.
4. The rate of growth of technology appears to be generally accelerating.
5. National programs control *content* of the technology, not its rate of production.

Predicting the Unpredictable

Robert A. Nisbet, 1968

A sociologist swats at a plague of prophecies already emerging three decades before the millennium.

The approach of the year 2000 is certain to be attended by a greater fanfare of predictions, prophecies, surmises, and forewarnings than any millennial year in history. In the past twelve months, at least four books on this subject have appeared, all of them concerned with the probable shape of American and world society in the year 2000. How many articles have appeared I cannot even guess. But books and articles are in any event only the exposed part of the iceberg. There are today centers, institutes, and bureaus, not to mention specific commissions, whose principal business is to forecast or predict the future. There is with us, in short, as part of the already huge knowledge industry, the historical-prediction business; and this business is certain to become ever larger, ever more ramified. Through every conceivable means—game theory, linear programming, systems analysis, cybernetics, even old-fashioned intuition or hunch—individuals and organizations are working systematically on what lies ahead during the next thirty-two years, and indeed during the century or two after that. . . .

Why the fascination with the future at a time when so many of us are preoccupied with the roots of our not always clear cultural identity? Daniel Bell, whose knowledge of what is going on in this matter is vast, suggests two rather different forces operating in human consciousness today. There is, first the magic of the millennial number. Men have always been attracted, Bell reminds us, to the mystical allure of the *chiloi*, the Greek word for a thousand, from which we acquire our religious word *chiliasm*, the belief in a coming life free of imperfections. . . .

The second reason Daniel Bell gives for the recent upsurge of interest in the forecasting of the future is technology. As he writes, there is in our society a kind of "bewitchment of technology." So many other matters have been settled by technology, so much of space—including planetary—has been conquered by technology, why should not the marvelous skills of the computer conquer time by providing us with increasingly accurate glimpses of the future? As Bell puts it, "the possibility of prediction, the promise of technological wizardry, and the idea of a millennial turning point make an irresistible combination." There is no question of this.

What is in question, however, is whether the marvels of electronic technology, cybernetics, linear programming, systems analysis, game theory, and the like, do add anything, *can* add anything to our success in an enterprise that is at least two hundred years old in the West: the serious, conscious business of predicting the future by observation of real or imagined continuities of event, change, and circumstance in time.

The wizardry of contemporary technology notwithstanding, the essential and lasting methodology of future-predicting was set forth in the early 18th century by the great Leibniz. One sentence, taken from his "Principles of Nature and of Grace," will suffice to express the crucial elements of Leibniz's law of continuity: *The present is big with the future, the future might be read in the past, the distant is expressed in the near. . . .*

Either the future *does* lie in the present, and hence is subject to observation through dissection, or it does not. And if it does not, all the computers and systems analysis and linear programming in the world will not help us. For it is sheer delusion to suppose that anything short of H. G. Wells's Time Machine can in fact get us into the future, as technology gets us across space to the moon. . . .

Let me now state two equally important points: (1) contemporary forays into the future are no better, and generally worse, *ceteris paribus* [i.e., other things being equal], than the forays into the future that our great grandfathers—Tocqueville, Comte, Marx, et al.—made; (2) the only real utility of these fast accumulating reports and books on the future is the often enlightening, generally informative, sometimes brilliant perceptions they contain about the *present*. No doubt this point alone makes "future-predicting" worthwhile, for there is nothing like an assignment to gaze into the future for sharpening one's awareness of what lies around him in the present. . . .

What we have in [this recent literature] is a great deal that is important to know about *present* conditions, *present* structure, and *present* rates and their apparent relation to the recent past. We have much speculation about what the future might be, we have a good deal about historical "trends," and of course vast amounts of material on rates: birth rates, death rates, production rates,

rates of air miles flown, entries into the national parks, investment rates, rates of just about everything in any way amenable to quantity-statement.

Reading all of [this], we can thrill with repressed horror at the thought of the mantle of too too solid flesh that will one day cover the earth (a mantle that the physicist-population expert, Sir Charles Darwin, once told an audience, much in the manner of the old-fashioned temperance lecture, would reach, present rates continuing, one mile in height by the year 3500, or was it 2500?). One can feel his toes trampled on as he reads that by the year 2000 there will be two people for every foot of waterline in the U.S. One can hypnotize himself into a state of driver-fury by merely reading about the 250 million automobiles (we now have about 59 million) on American streets and highways. The thought of 225 billion passenger miles to be flown by the airlines in the year 2000, in contrast to a 1960 figure of 35 billion, is enough to keep everyone home, which would indeed be a change. But change is not, alas, what these books are predicting; they are only extrapolating present rates, many of which remind one of a mad physiologist predicting giants at age twenty on the basis of growth rates at age ten.

Only the unwary will be deluded into thinking that any of this is in fact the future.

That's one small step for [a] man, one giant leap for mankind.

ASTRONAUT NEIL ARMSTRONG, ON FIRST
SETTING FOOT ON THE MOON, JULY 20, 1969

▶ A human footprint on the alien moon: July 20, 1969.

BENEFIT VERSUS RISK

Chauncey Starr, 1969

The dean of the UCLA School of Engineering and a nuclear-power pioneer explores how to improve assessment of technological risk.

The broad societal benefits of advances in technology exceed the associated costs sufficiently to make technological growth inexorable. . . . Technological growth has been generally exponential in this century, doubling every 20 years in nations having advanced technology. Such technological growth has apparently stimulated a parallel growth in socioeconomic benefits and a slower associated growth in social costs.

The conventional socioeconomic benefits—health, education, income—are presumably indicative of an improvement in the "quality of life." The cost of this socioeconomic progress shows up in all the negative indicators of our society—urban and environmental problems, technological unemployment, poor physical and mental health, and so on. If we understood quantitatively the causal relationships between specific technological developments and societal values, both positive and negative, we might deliberately guide and regulate technological developments so as to achieve maximum social benefit at minimum social cost. Unfortunately, we have not as yet developed such a predictive system analysis. As a result, our society historically has arrived at acceptable balances of technological benefit and social cost empirically—by trial, error, and subsequent corrective steps.

In advanced societies today, this historical empirical approach creates an increasingly critical situation, for two basic reasons. The first is the well-known difficulty in changing a technical subsystem in our society once it has been woven into the economic, political, and cultural structures. For example, many of our environmental pollution problems have known engineering solutions, but the problems of economic readjustment, political jurisdiction, and social behavior loom very large. It will take many decades to put into effect the technical solutions we know today. . . .

In order to minimize these difficulties, it would be desirable to try out new developments in the smallest social groups that would permit adequate assessment. This is a common practice in market-testing a new product or in field-testing a new drug. In both these cases, however, the experiment is completely under the control of a single company or agency, and the test information can be fed back to the controlling group in a time that is short relative to the anticipated commercial lifetime of the product. This makes it possible to achieve essentially optimum use of the product in an acceptably short time. Unfortunately, this is rarely the case with new technologies. Engineering developments involving new technology are likely to appear in

VISIONS OF TECHNOLOGY

285

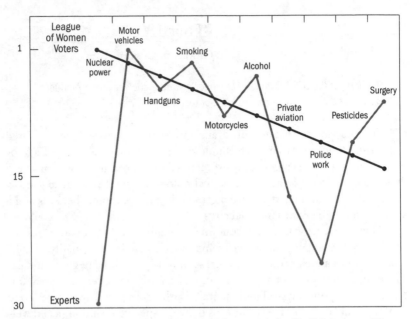

▶ Ranking of risks for ten activities and technologies as preceived by U.S. League of Women Voters vs. risk experts (1 = maximum risk).

many places simultaneously and to become deeply integrated into the systems of our society before their impact is evident or measurable.

This brings us to the second reason for the increasing severity of the problem of obtaining maximum benefits at minimum costs. It has often been stated that the time required from the conception of a technical idea to its first application in society has been drastically shortened by modern engineering organization and management. In fact, the history of technology does not support this conclusion. The bulk of the evidence indicates that the time from conception to first application (or demonstration) has been roughly unchanged by modern management, and depends chiefly on the complexity of the development.

However, what *has* been reduced substantially in the past century is the time from first use to widespread integration into our social system. The techniques for *societal diffusion* of a new technology and its subsequent exploitation are now highly developed. Our ability to organize resources of money, men, and materials to focus on new technological programs has reduced the diffusion-exploitation time by roughly an order of magnitude in the past century.

Thus, we now face a general situation in which widespread use of a new technological development may occur before its social impact can be properly assessed, and before any empirical adjustment of the benefit-versus-cost relation is obviously indicated. . . .

Although this study is only exploratory, it reveals several interesting points. (i) The indications are that the public is willing to accept "voluntary" risks roughly 1,000 times greater than "involuntary" risks. (ii) The statistical risk of death from disease appears to be a psychological yardstick for establishing the level of acceptability of other risks. (iii) The acceptability of risk appears to be crudely proportional to the third power of the benefits (real or imagined). (iv) The social acceptance of risk is directly influenced by public awareness of the benefits of an activity, as determined by advertising, usefulness, and the number of people participating.

EMERGENT JAPAN

NIGEL CALDER, 1970

For the Japanese, an American toy locomotive brought by Matthew Perry to Yokohama began a developmental experience that culminated in the *Little Boy* A-bomb carried by Superfortress to Hiroshima. By the late 1960s, their country of 100 million souls was firmly established as an outstanding industrial nation that could easily become the third "superpower" if they chose that role. Lacking the world-embracing zeal of the Americans or Russians, or the nostalgic habits of industrial enterprise in Europe, the Japanese seemed able to outpace them in economic growth without destroying or abandoning their national customs.

The success of Japan makes it, in some people's judgment, a model of development for the poor countries of the world. So perhaps it does, but it has taken Japan one century to catch up. On that timescale, projected from now into the 21st century, one may doubt whether world conditions will be comparable, or even whether the aims of development will remain the same. Nevertheless, Japan in the 1960s was a fascinating case of a country that had been a technological follower for a hundred years, seeking at last to develop the scientific base that would enable it to be more of an originator. The Japanese having to change was a sign that the character of industrial technology was changing, as its dependence on research became more intimate. . . .

The policy of parasitism for countries and companies, that would attempt no original research but would make only whatever scientific effort was necessary for exploiting innovations from outside, had ceased to be effective. The reason was more than a matter of self-respect, although that was a good motive, too. Research had become an essential part of the self-education of an industry, as it was for the self-education of university teachers. While it might not matter much in the end whether an invention originated with yourself or someone else, you had to be well enough informed by your own research to respond quickly but sensibly when an opportunity arose. The important word was "quickly," because by the 1960s it was apparent that, in the science-based industries, the profitable market life of a particular kind of product could often be shorter than the time taken to develop it.

In retrospect, Japan nevertheless illustrates the separability of three functions in the cultivation of technology: research, innovation and exploitation. During the 1950s, the Japanese were doing far less research than were the Americans or British, but their economic growth was much faster. Nor were they particularly strong in being the first to bring a novel product to the market. Their great strength was exploitation, using their abundant capital and labor to make the product more profitably. Such was the pattern from shipbuilding to transistor radios.

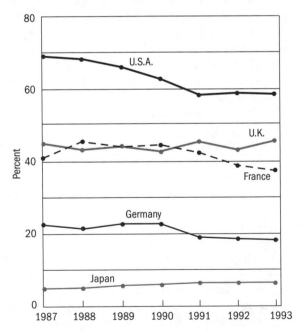

▶ Military share of science budgets in five countries, fiscal years 1987–1993.

PRIMARY INSTINCTS

VANNEVAR BUSH, 1970

Bush had informed knowledge of the development of new weapons and their influence on national policy.

We are not at the end of war. But we may indeed be at the end of world wars, with A-bombs, chemical warfare, biological warfare. This is by no means sure, of course, and there is little doubt what such a war would mean. Yet it is not just wishful thinking to believe we may escape. It is based on one of the primary instincts of the race, the central urge for self-preservation. Nations, rulers, do not commit suicide knowing that they are doing so. There is not much doubt, with modern communication, that they will know. There will be secondary wars, wars with conventional weapons. They will be disastrous and absurd. But they will not hold the population in check. We again replace one problem with another. It is a good exchange.

The advent of the A-bomb is generally regarded as a catastrophe for civilization. I am not convinced that it was. With the pace of science in this present century it was inevitable that means of mass destruction should appear. Since the concept of one world under law is far in the future, it was also inevitable that great states should face one another thus armed. If there were no A-bombs the confrontation would still have occurred and the means might well have been to spread among a people a disease or a chemical that would kill or render impotent the whole population. History may well conclude, if history is written a century from now, that it was well that the inevitable confrontation came in a spectacular way that all could recognize, rather than in a subtle form which might tempt aggression through ignorance. At least we all know, we and the rest of the world, that there are A-bombs and what they can do. And other devilish forms of assault are pushed into the background, for one does not pick up a rock in the presence of an antagonist who carries a gun.

SETTING STANDARDS:
TWO LISTS

NIGEL CALDER, 1970

I. Twelve Standard Horrors

* Thalidomide disaster: malformed babies follow use of new drug.
* Abuse of antibiotics: resistant strains of bacteria appear.

- Abuse of pesticides: killing of wildlife and environmental contamination.
- *Lucky Dragon* incident: fishermen caught in unexpectedly far-reaching H-bomb fallout, one dies.
- *Starfish* high-altitude H-bomb [test]: "wrecking" of Earth's radiation belts despite protests of astronomers.
- Water pollution: especially (at one time) from detergents not susceptible to natural breakdown.
- Air pollution: especially from power stations and automobile exhaust.
- Modification of atmosphere: by carbon dioxide from fuel, and by exhausts of jets and rockets, with unknown effects on the Earth's climate.
- Overfishing: technological aids cause depletion of fishing grounds, near extinction of whales.
- Bugging devices: electronic invasion of privacy.
- Sonic boom of supersonic airliners: pending.
- The great power blackout: northeastern USA, 9 November 1965, a failure of automatic systems.

II. Twelve Recent Boons

- Nuclear power: relieves the human species of fears of rapid exhaustion of energy sources.
- Microbiology: infectious disease can now be largely eliminated.
- Space technology: much improved weather observation and telecommunications.
- Desalination of water: makes deserts habitable, perhaps one day cultivable.
- Computer: eliminates much clerical drudgery and aids in mastery of complex systems.
- Plant and animal breeding: vastly increased yields of food per acre.
- Technology in general: rising productivity means rising living standards, increased leisure.
- Technology in general: machinery, detergents, synthetic fibers, etc., greatly reduce domestic drudgery.
- Air transport: safer, faster, cheaper.
- Television: medium of entertainment and instruction of unprecedented potency.
- Psychiatry: much mental illness now susceptible to therapy.

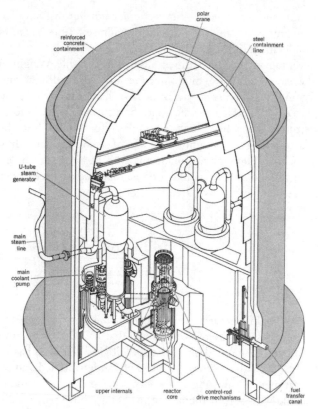

▶ Nuclear power, the first major energy source not derived from sunlight, expanded to account for 20 percent of U.S. energy supply.

labels: polar crane; reinforced concrete containment; steel containment liner; U-tube steam generator; main steam line; main coolant pump; upper internals; reactor core; control-rod drive mechanisms; fuel transfer canal

TRUE DEVOTION

VANNEVAR BUSH, 1970

An invention has some of the characteristics of a poem. Standing alone, by itself, it has no value; that is, no value of a financial sort. This does not mean that inventions—or poems—have no value. It is said that a poet may derive real joy out of making a poem, even if it is never published, even if he does not recite it to his friends, even if it is not a very good poem. No doubt one has to be a poet to understand this. In the same way an inventor can derive real satisfaction out of making an invention, even if he never expects to make a nickel out of it, even if he knows it is a bit foolish, provided he feels it involves ingenuity and insight. An inventor invents because he cannot help it, and also because he gets quiet fun out of doing so. Sometimes he even makes money at it, but not by himself. One has to be an inventor to understand this.

One evening in Dayton I dined alone with Orville Wright. We had been on the way to a dinner which was canceled. During dinner, and during a long evening, we discussed inventions we had made which had never amounted to anything. He took me up to the attic and showed me models of various weird gadgets. I had plenty of similar efforts to tell him about, and we enjoyed ourselves thoroughly. Neither of us would have thus spilled things except to a fellow practitioner, one who had enjoyed the elation of creation and who knew that such elation is, to a true devotee, independent of practical results. So it is also, I understand, with poets.

MUGS AND ZEALOTS

NIGEL CALDER, 1970

The clash of opinions can now be clarified which characterizes the political struggle about the uses of science that has already begun and will intensify. The tough-minded and the technological opportunists I propose to call Zealots, for self-evident reasons. The tender-minded and the scientific conservationists I call Mugs, this term serving as a shorthand for mugwumps (people who regard themselves as being above party politics) and also having an appropriate slang connotation of passivity.

. . . In practice Zealots disagree with one another because their beliefs are incompatible; Mugs because they are disinclined to believe anything. Zealots may be muggish, and Mugs zealous, in their own professional activities. Neither side has a monopoly of virtue. A man may be zealous in a good cause or a bad one; a Mug may be a saint or a cynic. Zealots dominate traditional centers of political power, because they are more forceful. The Mugs dominate the arts and sciences for an equally simple reason: parroting of existing ideas is boring and professionally unadmired. The Mug's weakness is an aversion from action, whether political fighting or major technological enterprise; he may be too "reasonable." On the other hand, the most potent new ideas, social and scientific, tend to originate from Mugs. . . .

Technology itself has a built-in bias towards zealotry, especially when embodied in huge industries or elaborate nationalistic projects like sending men to the Moon. The concept of the technological fix may be Zealot-like, even when applied to muggish ends. . . .

The Mugs will gain power because, as Winston Churchill said of science, the new empires are empires of the mind. That is perhaps the chief reason why those who know all the reasons for being pessimistic about our technological future can yet remain optimistic. It is scarcely credible that the well-informed, research-trained, interconnected communities of the foreseeable future should tolerate men in power who want power for its own sake, who have simple-minded theories of society, or who seek to profit from the differences between men. The wind of learning and self-knowledge will topple them. Their fall will not end the search for new ways of life; it will merely allow it to begin in earnest.

SOME CHARACTERISTICS	MUGS	ZEALOTS
Admired virtues	skepticism	conviction
	prudence	boldness
	iconoclasm	authority
	nonconformity	loyalty
Political connotations	internationalism	patriotism
	"doves"	"hawks"
	nostalgic conservatism	economic conservatism
	ideal communism	communism in practice
	liberalism	fascism
	democracy	revolution
Common descriptions	concerned	decisive
	woolly-minded	fanatical
Some heroes	Mahatma Gandhi	V. I. Lenin
	Robert Kennedy	Winston Churchill
	Martin Luther King	Moshe Dayan
	Albert Einstein	Henry Ford
	Charlie Brown	Superman

PARTICULAR ATTITUDES TO TECHNOLOGIES

Military use of H-bombs is	indefensible	conceivably justifiable
Development of biological and chemical weapons is	indefensible	prudent
Manned spaceflight is	wasting resources	tremendously exciting
Space exploration is	for cooperation	for competition
Satellite broadcasting is nationally for	nonprofit agency	government or private enterprise
internationally for	international agency	national or private enterprise

Ocean rights assigned	under international control	as rewards for enterprise
Wildlife conservation is	desperately important	okay within limits
Top priorities in medical research are	family planning preventive medicine	cure for cancer heart replacements
The automobile is	dangerous and obsolete	a man's best friend
Monitoring devices may	enslave us	eliminate crime
Super-intelligent computers may	enslave us	make us smarter
Mood-controlling drugs may	enslave us	make us more sociable
Use of psychedelic drugs is	tolerable, or may be discouraged	salvation, or to be prohibited
Attitudes to eugenics are that	human beings are okay	human beings should be improved

THE TECHNOLOGICAL IMPERATIVE

LEWIS MUMFORD, 1970

Western society has accepted as unquestionable a technological imperative that is quite as arbitrary as the most primitive taboo: not merely the duty to foster invention and constantly to create technological novelties, but equally the duty to surrender to these novelties unconditionally just because they are offered, without respect to their human consequences. One may without exaggeration now speak of technological compulsiveness: a condition under which society meekly submits to every new technological demand and utilizes without question every new product, whether it is an actual improvement or not; since under this dispensation the fact that the proffered product is the result of a new scientific discovery or a new technological process, or offers new opportunities for investment, constitutes the sole proof required of its value.

This situation was well characterized by the mathematician John von Neumann: "Technological possibilities are irresistible to man. If man *can* go to the moon, he will. If he can control the climate, he will." Though von Neumann himself was properly alarmed by this condition, he himself attributed too glibly to "man" characteristics that belong only to this particular moment of Western culture, which has so concentrated its energies and its hopes for salvation on the machine that it has stripped itself of all the ideas, institutions and habits that have in the past enabled other civilizations to overcome these obsessions and compulsions. Earlier communities, in contrast, strenuously resisted technological innovations—sometimes quite unreasonably—and either delayed them until they conformed to other human requirements and proved their worth, or rejected them altogether.

Now there is no doubt that the "irresistible" impulse that von Neumann has described actually pervades the present-day scientific and technological world. Hermann Muller, the American geneticist, used von Neumann's dictum as a clincher in his argument for establishing genetic controls by scientists over the breeding of human populations. Speaking of the possibility of using banks of frozen human sperm cells taken from "geniuses," as one now preserves similar cells from prize bulls, Muller said, with alarming naïveté, "Their mere existence will finally result in an irresistible incentive to use them." Psychologists know this "irresistible incentive" in many forms: for at the moment any impulse, however normal, becomes irresistible in its own right and for no other reason than that it exists, it becomes pathological, and the unawareness of this pathology among scientists whose discipline supposedly serves as a safeguard against irrational conclusions or actions is just a further evidence of that pathology.

There is a simple way of establishing the downright absurdity—or more accurately the menacing irrationality—of accepting such technological compulsiveness; and that is to carry von Neumann's dictum to its logical conclusion: *If man has the power to exterminate all life on earth, he will.* Since we know that the governments of the United States and Soviet Russia have already created nuclear, chemical, and bacterial agents in the massive quantities needed to wipe out the human race, what prospects are there of human survival, if this practice of submitting to extravagant and dehumanized technological imperatives is "irresistibly" carried to its final stage?

IV. YESTERDAY, TODAY AND TOMORROW: 1970–

AT THE DAM

JOAN DIDION, 1970

The native Californian novelist and essayist excels at articulating the quality of anxiety boiling off disjunctions between natural and artificial, internal or external structures—whether large federal dams or the human heart.

Since the afternoon in 1967 when I first saw Hoover Dam, its image has never been entirely absent from my inner eye. I will be talking to someone in Los Angeles, say, or New York, and suddenly the dam will materialize, its pristine concave face gleaming white against the harsh rusts and taupes and mauves of that rock canyon hundreds or thousands of miles from where I am. I will be driving down Sunset Boulevard, or about to enter a freeway, and abruptly those power transmission towers will appear before me, canted vertiginously over the tailrace. Sometimes I am confronted by the intakes and sometimes by the shadow of the heavy cable that spans the canyon and sometimes by the ominous outlets to unused spillways, black in the lunar clarity of the desert light. Quite often I hear the turbines. Frequently I wonder what is happening at the dam this instant, at this precise intersection of time and space, how much water is being released to fill downstream orders and what lights are flashing and which generators are in full use and which just spinning free.

I used to wonder what it was about the dam that made me think of it at times and in places where I once thought of the Mindanao Trench, or of the stars wheeling in their courses, or of the words *As it was in the beginning, is now and ever shall be, world without end, amen.* Dams, after all, are commonplace: we have all seen one. This particular dam had existed as an idea in the world's mind for almost forty years before I saw it. Hoover Dam, showpiece of the Boulder Canyon project, the several million tons of concrete that made the Southwest plausible, the *fait accompli* that was to convey, in the innocent time of its construction, the notion that mankind's brightest promise lay in American engineering.

Of course the dam derives some of its emotional effect from precisely that aspect, that sense of being a monument to a faith since misplaced. "They died to make the desert bloom," reads a plaque dedicated to the 96 men who died building this first of the great high dams, and in context the worn phrase touches, suggests all of that trust in harnessing resources, in the meliorative power of the dynamo, so central to the early Thirties. Boulder City, built in 1931 as the construction town for the dam, retains the ambience of a model city, a new town, a toy triangular grid of green lawns and trim bungalows, all fanning out from the Reclamation building. The bronze sculptures at the dam itself evoke muscular citizens of a tomorrow that never came, sheaves of wheat clutched heavenward, thunderbolts defied. Winged Victories guard the flagpole. The flag whips in the canyon wind. An empty Pepsi-Cola can clatters across the terrazzo. The place is perfectly frozen in time.

But history does not explain it all, does not entirely suggest what makes that dam so affecting. Nor, even, does energy, the massive involvement with power and pressure and the transparent sexual overtones to that involvement. Once when I revisited the dam I walked through it with a man from the Bureau of Reclamation. For a while we trailed behind a guided tour, and then we went on, went into parts of the dam where visitors do not generally go. Once in a while he would explain something, usually in that recondite language having to do with "peaking power," with "outages" and "dewatering," but on the whole we spent the afternoon in a world so alien, so complete and so beautiful unto itself that it was scarcely necessary to speak at all. We saw almost no one. Cranes moved above us as if under their own volition. Generators roared. Transformers hummed. The gratings on which we stood vibrated. We watched a hundred-ton steel shaft plunging down to that place where the water was. And finally we got down to that place where the water was, where the water sucked out of Lake Mead roared through thirty-foot penstocks and then into thirteen-foot penstocks and finally into the turbines themselves. "Touch it," the Reclamation man said, and I did, and for a long time I just stood there with my hands on the turbine. It was a peculiar moment, but so explicit as to suggest nothing beyond itself.

There was something beyond all that, something beyond energy, beyond history, something I could not fix in my mind. When I came up from the dam that day the wind was blowing harder, through the canyon and all across the Mojave. Later, toward Henderson and Las Vegas, there would be dust blowing, blowing past the Country-Western Casino FRI & SAT NITES and blowing past the Shrine of Our Lady of Safe Journey STOP & PRAY, but out at the dam there was no dust, only the rock and the dam and a little greasewood and a few garbage cans, their tops chained, banging against a fence. I walked across the marble star map that traces a sidereal revolution of the equinox and fixes forever, the Reclamation man had told me, for all time and for all people who can read the stars, the date the dam was dedicated. The star map was, he had said, for when we were all gone and the dam was left. I had not thought much of it when he said it, but I thought of it then, with the wind whining and the sun dropping behind a mesa with the finality of a sunset in space. Of course that was the image I had seen always, seen it without quite realizing what I saw, a dynamo finally free of man, splendid at last in its absolute isolation, transmitting power and releasing water to a world where no one is.

FOLLOWING ORDERS

DANIEL C. DRUCKER, 1971

What degree of responsibility do engineers bear for technological problems? Not much, a dean of engineering argues.

The attackers of science and engineering really have far too much faith in the ability of engineers to solve problems. They have so high an opinion of our technical ability that they are convinced our errors of omission as well as commission are purposeful. In their childlike faith they believe that all our engineering-related societal problems—housing, transportation, pollution—would be solved, or never would have been created, were we not venal or amoral or lazy. After all, we did put a man on the moon according to plan. . . .

Engineers view themselves, on the whole, as servants of society. Those who view society or the Establishment as inherently evil then have cause for transferring the label of evil to engineers. When society, as represented by those who govern the setting of priorities, votes for military weapons, weapons are developed. When laws permit pollution by industry and tax incentives attract polluting industries, industrial plants which pollute will be built by engineers.

A split personality sometimes develops with an engineer or scientist serving the demands of society during his workday and protesting the decisions of society in his leisure hours. Engineers may be criticized for this subservience to the will of the majority, but their behavior really is more appropriate than the technocracy concept of engineers governing society. If we believe in democracy, we must behave in accord with that belief. Anarchy is not a viable alternative in the interdependent world of today. It is necessary to work within the system to change the system. . . .

Engineering has served society extremely well by the standards society has set for itself. The basic criticism is that engineering did not lead in the political movement for automobile safety, or for more stringent laws against air pollution. This is a serious charge against a professional group, and its partial validity is now acknowledged by the engineering societies. However, it should not be confused with the charge that engineers caused vehicles to be unsafe, or caused pollution, or failed in the role society as a whole had assigned to the profession. It is the ground rules of society which have changed in the last 25 years and have undergone especially rapid change in the last decade. Translated to its simplest and possibly unacceptable extreme form, the engineer (really the political or economic policymaker) no longer is to give the public what it thinks and says it wants. This historic turning point may be here or just around the corner. It will be a turning for the better only when we agree that we know enough about how to solve our problems.

DAILY DELIBERATE PROTECTION

GIL ELLIOT, 1972

I believe this Scottish writer's little-known Twentieth Century Book of the Dead *to be one of the most important books of the century. It counts how many lives have been lost to man-made death—particularly war and war's privations—in this bloodiest of all centuries in human history and identifies public health ("public death," in Elliot's terms) as a methodology adaptable to preventing or limiting violence at every scale.*

Our societies are dedicated to the preservation and care of life. Official concern ceases at death, the rest is private. Public death was first recognized as a matter of civilized concern in the nineteenth century, when some health workers decided that untimely death was a question between men and society, not between men and God. Infant mortality and endemic disease became matters of social responsibility. Since then, and for that reason, millions of lives have been saved. They are not saved by accident or goodwill. Human life is daily deliberately protected from nature by accepted practices of hygiene and medical care, by the control of living conditions and the guidance of human relationships. Mortality statistics are constantly examined to see if the causes of death reveal any areas needing special attention. Because of the success of these practices, the area of public death has, in advanced societies, been taken over by man-made death—once an insignificant or "merged" part of the spectrum, now almost the whole.

When politicians, in tones of grave wonder, characterize our age as one of vast effort in saving human life, and enormous vigor in destroying it, they seem to feel they are indicating some mysterious paradox of the human spirit. There is no paradox and no mystery. The difference is that one area of public death has been tackled and secured by the forces of reason; the other has not. The pioneers of public health did not change nature, or men, but adjusted the active relationship of men to certain aspects of nature so that the relationship became one of watchful and healthy respect. In doing so they had to contend with and struggle against the suspicious opposition of those who believed that to interfere with nature was sinful, and even that disease and plague were the result of something sinful in the nature of man himself. To this day there survives, amongst some of the well fed and cared for, a nostalgia for the slums of disease. . . .

The manner in which people die reflects more than any other fact the value of a society.

REVOLUTIONARY TOOLMAKING

GERARD PIEL, 1972

Technology is concerned not alone with the means but also with the ends of life. The toolmaking propensity of man, from his beginnings, has expanded the possibilities of his existence, exciting as well as implementing his aspirations. It is only in the milieu of industrial society that we can speak of equality without the silent discount of eight out of ten of our brothers. It is by toolmaking that men have at least freed themselves from physical bondage to toil and to one another. One of the aspirations that now becomes feasible, therefore, is self-government. With the open acknowledgment that our toolmaking has brought a revolutionary change in man's relationship to nature, we can proceed to the necessary revisions in our relationships with one another.

> Whereas it was once believed that every new technological possibility was automatically and inevitably beneficial, the great achievements in outer space [among others] have helped to dim the light once cast by technological progress. . . . Science, engineering and technology have all become amalgamated into a single entity which is conceived as a source of damage and costly waste. The research workers, engineers, military men, industrialists and politicians are seen as homogeneous groups with each section pursuing its own advantage at the expense of the rest of society.
> EDWARD SHILS, 1972

HOW TO SOLVE AMERICA

L. RUST HILLS, 1972

For many years the fiction editor of Esquire, *Hills is also a humorist whose special gift is preserving the intimacies of a past that advancing technology has nearly obliterated—including (in his essay collection* How To Do Things Right*) making and eating milk toast, organizing a family picnic, setting a mechanical alarm clock, eating an ice-cream cone and, my personal favorite, four dumb tricks you can do with a pack of Camels (my favorite because my father smoked Camel cigarettes and taught me the tricks, which I subsequently forgot). Here Hills proposes a way to maintain the national economy while fixing everything that has gone wrong. His solution recalls Kurt Vonnegut's vision of a war movie run backward, and like that elegiac fantasy, emphasizes how irreversible technological change really is.*

The problem of America, then, is one hell of a problem. But, fortunately, my solution is also one hell of a solution.

So what am I going to do about it? How are we going to solve America? You can't slow down or stop all this growth and progress without causing disaster to the economy; we know that. My solution for America takes this into account. It takes *everything* into account.

How to solve America is profoundly and beautifully simple: *just turn everything around and start going the other way.* This would keep America moving, but now we'd be moving in the right direction, backward—from the complex back toward the simple, from the new back to the old, from the ugly and shoddy back to the lovely and the sturdy.

The idea, basically, is to decide today to undo tomorrow what was done yesterday, day-after-tomorrow undo what was done day-before-yesterday, and so on—with some important variations and refinements—backward in time, until we get just the kind of America we used to have and liked so much.

This would cause no unemployment, because there'd be just as much work undoing all the things we've done recently as there was doing them; actually it will be *more* work, the way we're going to do it. The real cause of unemployment is mechanization, right? Well, not only are we going to de-mechanize America, we are going to demechanize her, in so far as possible, quietly and slowly, without using any noisy machines, and that's going to be a lot of work. No de-assembly lines: men will take computers apart using just screwdrivers and pliers. So actually this job of painstakingly dismantling modern America would require fantastic amounts of manpower—and re-building the old America would take even more, for we'd have to relearn the skills and attitude of good workmanship. For instance, the tunnels under the rivers around Manhattan would have to be undug, and the bridges un-built—both dangerous and time-consuming jobs. Then the ferry boats that the tunnels once replaced would now have to be rebuilt to replace them—and because wood is going to be at a premium in this new, older society, un-til we can get some trees to grow again—the ferries would have to be built of available scrapwood by skilled shipwrights. Taking down skyscrapers will be done carefully, too—with no noise or dust. Hardhats will stand proudly and respectfully and quietly as the mayor says a few grateful words to them at the joyous ceremonies celebrating the unlaying of a cornerstone. Taking up the asphalt and concrete would be done by friendly groups of overweight businessmen working easily together with sledgehammers and crowbars, for the sound of the air compressor and the jackhammer would pass from the land. And as the good earth is revealed again under all the crud we've put over it, how lovely it will be! Perhaps a tenth of our men and a quarter of our women will have to work as landscapists and gardeners for a while—not just as a hobby, as they do now, but as their work. Imagine finding plea-sure in work!

There would be a great need for people to raise and train horses. What a job it would be to dismantle all the automobiles, melting down all the steel and ultimately replacing it in the ground in a form as close to iron ore as possible. Everything—steel, concrete, plastics—all would go back into the earth where it came from. What planning it would require to get it all put back neatly into place! Archivists would search the records to determine what existed previously in each area; long-range planners would speculate about what existed even earlier; and implementers would plan older, slower ways to achieve it.

The intention wouldn't be to take the nation as a whole back to an abo-riginal state, living in caves and tepees and hunting from ponies, or anything like that—although that style of life would be available for those who wanted it. What's to be achieved is a return to some sort of approximation of the American ideal, having the best of each era, without anybody being ex-ploited.

Thus, most Americans would live in small towns, but there would be isolated rural areas and there would be moderate-sized, tenable, livable cities for those that wanted them. The economy would be basically agricul-tural—small farms, unmechanized, family-owned and family-worked—but there would also be small, local industries manufacturing things *well*. There would be traveling companies of repertory players, putting on plays in ren-ovated movie houses of each village. Books would be published in each city, sold in the bookstores in each town. Everything would be much cheaper, the way it was before we got mass-production. Saturday afternoons there would be baseball games, and one would play—or watch men one knew play—against the teams of other towns. People would once again know where they belonged, who they were.

We would take something of the best from each era, but very little of the very modern world would endure. There could, for instance, be a few quiet old-fashioned airplanes, kept as museum pieces for those that wanted them; but no jetports, nothing remotely resembling a jetport. There would be old, grand, luxurious trans-Atlantic liners and romantic tramp steamers, as well as lots of sailboats, not just yachts, but fast clipper ships, and working schooners. An adequate train service might remain, or be re-created—with elegant dining cars, as in days gone by. All the elevated highways and the throughways and the parking lots would be replaced with trees and grass, of course; but there might be a few hundred cars left, for the nuts. That disturbing nuisance, the telephone, would go, as soon as we didn't need it. Electric alarm clocks would go right away—in fact, alarm clocks won't be necessary, for people will go to bed early, sleep well in the quiet, and arise refreshed, the way they once did. All the noisy inconvenient "conveniences" of modern life will disappear more or less in reverse order of their appearance—the snowmobile taking precedence over all others, however, no matter what newer horrors they've thought of by the time we start this.

The family will assume a vital role again, as demechanization creates chores inside and outside the house for older people and for children. In coastal areas children may be able to help with boat chores as soon as they can walk. "Community" will come to mean something good again, too, as the process of urbanization is reversed.

There'll be a need to subvert and eventually obvert our national mania that "growth" is good, that it is "vital." A nation that is really vital, we can say, is one that endures in its best phases, that refuses to grow toward its own extinction. We must make our Gross National Product *less* gross, more refined. Those in charge of our bureaus of government must show a substantial decrease in all aspects of their organizations at year's end, or be replaced by can-undo men who will. Advertising men must devise campaigns that imply that owning a car will make you feel, not young and sexy and rebellious as now, but old, worn out and subservient. Eventually advertising will disappear, too, except for useful announcements of events and services and perhaps a few of those nice old Burma Shave signs by the side of the road. Conglomerates need to be unconglomerated; mutual funds need to be unmutualized; everything *mass* must be made as individual as possible as soon as possible.

Those who have difficulty dismantling their own modern appliances can pay others to do it for them, earning money themselves in some kind of demerchandising, like arranging for the giant chain stores in the shopping centers to be divided up into village general stores, or starting old-fashioned bakeries with delicious nutritious fresh-baked bread, or local butcher shops with tender local beef placidly raised cropping some local hillside into a lovely meadow instead of being toughened by being whooped at and "git along"-ed at up and down some trail out West and then packed and processed in Chicago. But by and large, men and women would want to do as much of their own work as they could. Independence would once again become a virtue, which is good, and also become possible, which is even better. With despecialization, demechanization and a general decomplication of everything, ingenuity and resourcefulness and real know-how would gradually again become part of the Yankee character. And gradually, too, as more and more of the complexity has been removed and buried and grown over, Americans would once again be faced with work that they could handle and understand, that had real meaning for them, that told them something about themselves.

When America, the leading industrial power in the world, thus rejects industrialism, we can be sure that the Germans and Japanese, great emulators of The American Way, will reject industrialism too. Emerging nations will stop emerging, take a second look and go right back into the forest. In fact, it won't be many years before America will feel a real rapprochement with the other underdeveloped countries of the world. And when they start sending us teams of advisers, think what that will do for world peace.

Doesn't it all sound nice? An ease and a peace will come over the restless, troubled people of this land, as gradually out of the noise and grime and busyness of our "civilization" appear once again the rolling plains and the farms and the wooded hillsides.

Isn't this a good plan for how to solve America?

PREDICTIONS: SUPER-SUPER SSTS

SPIRO AGNEW, 1972

The vice-president of the United States prognosticates the future of supersonic transport.

For it must be obvious to anyone with any sense of history and any awareness of human nature that there *will* be SSTs. And Super SSTs. And Super-Super SSTs. Mankind is simply not going to sit back with the Boeing 747 and say "This is as far as we go."

WHITE BREAD AND TECHNOLOGICAL APPENDAGES: I

THEODORE ROSZAK, 1972

Part One of a pickup debate on who is responsible for bad technology, from an incisive social critic; Part Two, from the founder of the magazine Scientific American, *follows next on.*

How much of what we readily identify as "progress" in urban-industrial society is really the undoing of evils inherited from the *last* round of technological innovation? In England, the cradle of industrialism, one looks in vain to find more than a handful of reforms in the period from Arkwright and Watt down to World War II (the so-called "age of improvement") which were not essentially efforts to cancel out social dislocations born of industrialism in the first place. So in our own day, we "progress" toward scientific husbandry by way of nitrate fertilizers and hyper-lethal pesticides, only to discover the noxious ecological consequences when they are well advanced. Then we moved to correct the imbalance and confidently identify this as progress too. A moribund patient flat on his back in a hospital bed can also be said to be making such progress if he is further from death today than yesterday. But he is nowhere near being a healthy person.

Since Charlie Chaplin's *Modern Times*, people as the victims of their own technical genius have been a standard subject of popular comedy. But the ironies of progress are far more than laughingstock for casual satire. Even crusading reformers like Ralph Nader or Paul Ehrlich who gamely challenge each new criminality or breakdown of the system, often fail to do justice to the depth of the problem. While each failure of the technology may allow us to pillory a few profiteering culprits for incompetence or selfishness, the plain fact is that few people in our society surrender one jot of their allegiance to the urban-industrial system as a whole. They cannot afford to. The thousand devices and organizational structures on which our daily survival depends are far more than an accumulation of technical appendages that can be scaled down by simple subtraction. They are an interlocking whole from which nothing can easily be dropped. How many of us could tolerate the condition of our lives if but a single "convenience" were taken away . . . the telephone . . . automobile . . . air conditioner . . . refrigerator . . . computerized checking account and credit card . . . ? Each element is wedded to the total pattern of our existence; remove any one and chaos seems to impend. As a society, we are addicted to the increase of environmental artificiality; the agonies of even partial withdrawal are more than most of us dare contemplate.

From this fact springs the great paradox of the technological mystique: its remarkable ability to grow strong by virtue of chronic failure. While the treachery of our technology may provide many occasions for disenchantment, the sum total of failures has the effect of increasing our dependence on technical expertise. After all, who is there better suited to repair the technology than the technicians? We may, indeed, begin to value them and defer to them more as repairmen than inventors—just as we are apt to appreciate an airplane pilot's skill more when we are riding out a patch of severe turbulence than when we are smoothly under way. Thus, when the technology freezes up, our society resorts to the only cure that seems available: the hair of the dog. Where one technique has failed, another is called to its rescue; where one engineer has goofed, another—or several more—are summoned to pick up the pieces. What other choice have we? If modern society originally embraced industrialism with hope and pride, we seem to have little alternative at this advanced stage but to cling on with desperation. So, by imperceptible degrees, we license technical intelligence, in its pursuit of all the factors it must control, to move well beyond the sphere of "hardware" engineering—until it begins to orchestrate the entire surrounding social context. The result is a proliferation of what Jacques Ellul has called "human techniques": behavioral and management sciences, simulation and gaming processes, information control, personnel administration, market and motivational research, etc.—the highest stage of technological integration.

Once social policy becomes so determined to make us dance to the rhythm of technology, it is inevitable that the entire intellectual and moral context of our lives should be transformed. As our collective concern for the stability of urban-industrialism mounts, science—or shall we say the scientized temperament?—begins its militant march through the whole of culture. For who are the wizards of the artificial environment? Who, but the scientists and their entourage of technicians? Modern technology is, after all, the scientist's conception of nature harnessed and put to work for us. It is the practical social embodiment of the scientific worldview; and through the clutter of our technology, we gain little more than an occasional and distorted view of any other world. People may still nostalgically honor prescientific faiths, but no one—no priest or prophet—any longer speaks with authority to us about the nature of things except the scientists.

White Bread and
Technological
Appendages: II

Gerard Piel, 1972

Economic questions are not so easily avoided today [as they were in Classical-era discussions of democratic institutions]. Scarcity has given way to abundance. The compulsions prescribed by natural law are weakening. Our lives are cast, however, in the uncomfortable transition period. The advent of abundance is not yet comprehended in the theory and practice of our economy. On the contrary, our abundance is dodged, minimized, and concealed as well as squandered and burned. To the degree that we have failed to come to terms with our historic attainment of abundance, we have botched and corrupted its distribution and realization.

The failure to comprehend this failure has sadly misled the students of our plight. They blame the machine for our willful mismanagement of it. This compound failure diverts public discussion from what ought to be the issues before the electorate and reduces our politics to trivia. The stress on our economic arrangements has been transferred to our democratic institutions.

Consider the affluence that yields no satisfaction, the convergence on mediocrity and indistinguishability that characterizes so many of the common articles of commerce—white bread, light beer, and colorless vodka, to mention a few correspondingly tasteless products. This looks so much like the universal process of entropy that it is taken without question to be a bane laid on by technology. The real truth is that technology frees the task of design and production from technical limitations. By doing so, it permits other considerations to hold sway. It is these other considerations, I submit, that assert the common denominator of mediocrity. . . .

The money and the social cost of this way of managing abundance is plainly demonstrated in the American automobile. Here is, above all, a social not a technological artifact, and one symbolic of the pathology to which I am urging your attention. Technological considerations of function, efficiency, and utilization of resources might dictate one or more designs for the automobile. None of the American automobiles is any one of these. The gifts of abundance are bodied forth in waste: of materials, fuel, and space, to name the principal categories. Here also we see the relentless convergence on the same design, the pursuit of identical oscillations in design change, and the packaging that is supposed to make all the difference in the world.

▶ Precision aerospace technologies produce 300 million aluminum beverage cans per day—more cans annually than nails or paper clips.

Against the argument that mass production must necessarily make the autos all alike, it can be shown again that the sales function integrated in the virtuosity conferred by technology throws away the economies of mass production on junk. As a supposedly durable good, what is more, the automobile does no credit to American technology. According to the standard practice of our durable goods industries—always with the aim of perpetuating scarcity in the face of abundance—the automobile is designed for 1,000 hours of service, to be traded in at 40,000 miles or less.

Now, it is plain that all of this has to do with the maintenance of prosperity, with the creation of jobs, with the struggle for profit margins and other vital objectives. But the compulsions that dictate these objectives arise from the economics of scarcity. They are not determinants of technology. In the place of abundance, proffered by technology, we get affluence.

The Bell Telephone Company, responding to different economic compulsions, builds to far higher standards of service. The telephone handset, the cheapest thing of its kind in the world, is built for amortization over twenty years. During that time, of course, the telephone company sees to it that you put the instrument to a great many more than 1,000 hours of service.*

*Urban archaeology: There was once only one telephone company, which supplied telephone handsets as part of basic service; they were plain, simple and rugged. [RR]

SMALL IS BEAUTIFUL

E. F. SCHUMACHER, 1973

*The title of Schumacher's influential book became the rallying cry of the
alternative technology movement. The argument, summarized here—a familiar
one, as we have seen—urged a return to craft rather than mass production but
justified it on ecological as well as humanistic grounds.*

Suddenly, if not altogether surprisingly, the modern world, shaped by
modern technology, finds itself involved in three crises simultaneously.
First, human nature revolts against inhuman technological, organizational
and political patterns, which it experiences as suffocating and debilitating;
second, the living environment which supports human life aches and groans
and gives signs of partial breakdown; and, third, it is clear to anyone fully
knowledgeable in the subject matter that the inroads being made into the
world's non-renewable resources, particularly those of fossil fuels, are such
that serious bottlenecks and virtual exhaustion loom ahead in the quite foreseeable future.

Any one of these three crises or illnesses can turn out to be deadly. I do
not know which of the three is the most likely to be the direct cause of collapse. What is quite clear is that a way of life that bases itself on materialism,
i.e. on a permanent, limitless expansionism in a finite environment, cannot
last long, and that its life expectation is the shorter the more successfully it
pursues its expansionist objectives.

If we ask where the tempestuous developments of world industry during the last quarter-century have taken us, the answer is somewhat discour-
aging. Everywhere the problems seem to be growing faster than the
solutions. This seems to apply to the rich countries just as much as to the
poor. There is nothing in the experience of the last twenty-five years to suggest that modern technology, as we know it, can really help us to alleviate
world poverty, not to mention the problem of unemployment which already
reaches levels like thirty percent in many so-called developing countries,
and now threatens to become endemic also in many of the rich countries. In
any case, the apparent yet illusory successes of the last twenty-five years
cannot be repeated: the threefold crisis of which I have spoken will see to
that. So we had better face the question of technology—what does it do and
what should it do? Can we develop a technology which really helps us to
solve our problems—a technology with a human face?

... We may say ... that modern technology has deprived man of the
kind of work that he enjoys most, creative, useful work with hands and
brains, and given him plenty of work of a fragmented kind, most of which he

does not enjoy at all. It has multiplied the number of people who are exceedingly busy doing kinds of work which, if it is productive at all, is so only in an indirect or "roundabout" way, and much of which would not be necessary at all if technology were rather less modern. . . . All this confirms our suspicion that modern technology, the way it has developed, is developing, and promises further to develop, is showing an increasingly inhuman face, and that we might do well to take stock and reconsider our goals.

Taking stock, we can say that we possess a vast accumulation of new knowledge, splendid scientific techniques to increase it further, and immense experience in its application. All this is truth of a kind. This truthful knowledge, as such, does *not* commit us to a technology of giantism, supersonic speed, violence and the destruction of human work-enjoyment. The use we have made of our knowledge is only one of its possible uses and, as is now becoming ever more apparent, often an unwise and destructive use.

. . . Directly productive time in our society has already been reduced to about 3½ percent of total social time, and the whole drift of modern technological development is to reduce it further, asymptotically to zero. Imagine we set ourselves a goal in the opposite direction—to increase it sixfold, to about twenty percent, so that twenty percent of total social time would be used for actually producing things, employing hands and brains and, naturally, excellent tools. An incredible thought! Even children would be allowed to make themselves useful, even old people. At one-sixth of present-day productivity, we should be producing as much as at present. There would be six times as much time for any piece of work we chose to undertake—enough to make a really good job of it, to enjoy oneself, to produce real quality, even to make things beautiful. Think of the therapeutic value of real work; think of its educational value. No one would then want to raise the school-leaving age or to lower the retirement age, so as to keep people off the labor market. Everybody would be welcome to lend a hand. Everybody would be admitted to what is now the rarest privilege, the opportunity of working usefully, creatively, with his own hands and brains, in his own time, at his own pace—and with excellent tools. Would this mean an enormous extension of working hours? No, people who work in this way do not know the difference between work and leisure. Unless they sleep or eat or occasionally choose to do nothing at all, they are always agreeably, productively engaged. Many of the "on-cost jobs" would simply disappear; I leave it to the reader's imagination to identify them. There would be little need for mindless entertainment or other drugs, and unquestionably much less illness.

Now, it might be said that this is a romantic, a utopian, vision. True enough. What we have today, in modern industrial society, is not romantic and certainly not utopian, as we have it right here. But it is in very deep trouble and holds no promise of survival. We jolly well have to have the courage to dream if we want to survive and give our children a chance of survival. The threefold crisis of which I have spoken will not go away if we simply carry on as before. It will become worse and end in disaster, until or unless we develop a new life-style which is compatible with the real needs of human nature, with the health of living nature around us, and with the resource endowment of the world.

SOLVING SOCIAL PROBLEMS

AMITAI ETZIONI, 1973

A Columbia University sociologist defends technology against flower power.

In cocktail party sociology, where slogans serve as substitutes for thinking, technology is often depicted as anathema to a humane, just, "liberated" society. The defense made by the friends of technology is equally simplistic: "Technology is a set of neutral means; whether it is used to good or evil purposes is not determined by the technology itself."

Both of these viewpoints are as valid as half-truths usually are. In fact, most technologies do have fairly specific uses; no one has yet been killed by a cable television. And, while some technological developments do promote an impersonal, efficiency-minded, mass-production society, other technologies are essential for a more humane society.

Some recent technological developments take over routine and repetitive jobs, freeing people from the drudgery of counting, calculating, remembering numerous dull details. It is also true that these same technologies, those of the computer for instance, generate such routine work as key punching. But they eliminate more drudgery than they impose. Automatic switchboards of telephones do routine work which would require several million people, while generating little menial work. And, the way to combat remaining and newly created routines is to advance technology—to create, for example, computers that understand spoken English—surely not to condemn the machines.

Beyond this, new technological developments contribute to the solution of societal problems very close to the hearts of the deriders of technology, often making progress precisely where nontechnological attempts have failed. Thus, one of the barriers to arms limitation was the demand for human, on-site inspection, a demand quite unacceptable to the U.S.S.R. and unattractive to U.S. corporations worried about their trade secrets. The development of powerful inspection satellites made this issue obsolete. Another example: a cost-effectiveness study made by the Department of Health, Education, and Welfare shows that it is much more economical to avert a death by means of seat belts, a technological innovation, than driver education. The birth control pill, a chemical technology, is much more potent in reducing family size than are efforts to educate people to have smaller families. Instructional television saves teachers the time often used to repeat exercises to their classes ad nauseam; it allows pupils to view the lesson when they choose, as often as they need to, and, soon, at the pace they wish; and it is as effective or more effective than live teaching—tune in "Sesame Street" sometime.

As for the future, pollution will be reduced through the development of less polluting, substitute technologies, not by a return to the pretechnological age. Distance and isolation will be further bridged through technological means such as two-way cable television and more suitable housing patterns. More and more people will be able to enjoy increased free time, culture, education, and each other because more of their chores will be done by machines and supervised by machines, whose excesses are corrected largely by other machines.

All of this is surely less romantic than the world depicted by the advocates of a return to nature, but it is also more likely to be realized, and it promises a *more* livable world, by practically any humane standard, than our Stone Age past. The task before us is to marshal more of technology to the service of human purposes, not to put technology into a self-destruct, reverse-thyself gear. This will not be achieved by a blind, wholistic approval of technology, but by carefully developing those tools which can be geared to advance our true values.

THE PILL: II

LORETTA LYNN, 1973

For several years I've stayed at home
    while you had all the fun,
And every year that came by
    another baby come.
There's gonna be some changes made
    right here on nursery hill.
You've set this chicken your last time
    'cause now I've got the Pill.

FIVE-DIMENSIONAL
TECHNOLOGY

DANIEL BELL, 1973

RICHARD RHODES

Technology, like art, is a soaring exercise of the human imagination. Art is the aesthetic ordering of experience to express meanings in symbolic terms, and the reordering of nature—the qualities of space and time—in new perceptual and material form. Art is an end in itself; its values are intrinsic. Technology is the instrumental ordering of human experience within a logic of efficient means, and the direction of nature to use its powers for material gain. But art and technology are not separate realms walled off from each other. Art employs *techne*, but for its own ends. *Techne*, too, is a form of art that bridges culture and social structure, and in the process reshapes both.

This is to see technology in its essence. But one may understand it better, perhaps, by looking at the dimensions of its existence. For my present purposes, I will specify five dimensions in order to see how technology transforms both culture and social structure.

1. *Function.* Technology begins with an aesthetic idea: that the shape and structure of an object—a building, a vehicle, a machine—are dictated by its function. Nature is a guide only to the extent that it is efficient. Design and form are no longer ends in themselves. Tradition is no justification for the repetition of designs. Form is not the unfolding of an immanent aesthetic logic—like the musical forms of the eighteenth to early twentieth centuries. There is no dialogue with the past. It is no accident that adherents of a machine aesthetic in the early twentieth century, in Italy and Russia, flaunted the name Futurism.

2. *Energy.* Technology is the replacement of natural sources of power by created power beyond all past artistic imagination. Leonardo made designs for submarines and air conditioning machines, but he could not imagine any sources of power beyond what his eyes could behold: human muscles or animal strength, the power generated by wind and falling water. Visionaries of the seventeenth century talked grandiosely of mechanized agriculture, but their giant combines were to be driven by windmills and thus could not work. Energy drives objects—ships, cars, planes, lathes, machines—to speeds thousands of times faster than the winds, which were the limits of the "natural" imagination; creates light, heat and cold, extending the places where people can live and the time of the diurnal cycle; lifts weights to great heights, permitting the erection of scrapers of the skies and multiplying the densities of an area. The skyscraper lighted at night is as much the technological symbol of the modern city as the cathedral was the emblem of medieval religious life.

3. *Fabrication.* In its oldest terms, technology is the craft or scientific knowledge that specifies ways of doing things in a reproducible manner. The replication of items from templates or dies is an ancient art; its most common example is coinage. It is the essence of technology that its reproduction is much cheaper than its invention or development, owing to the standardization of skills and objects. Modern technological fabrication introduces two different factors: the replacement of manual labor and artisan skills by programmed machines; and the incredible rapidity of reproduction—printing a million newspaper copies a night—which is the difference in scale.

4. *Communication and Control.* Just as no one before the eighteenth century could imagine the new kinds of energy to come, so, well into the nineteenth century, no one could imagine—even with the coding of messages into dots and dashes, in telegraphy—the locking of binary digits with electricity, or the amplification of ethereal waves, which has produced modern communication and control systems. With telephone, radio, television and satellite communication, one person can talk to another in any part of the globe, or one person can be seen by hundreds of millions of persons at the same moment. With programmed instructions—through the maze of circuits at nanosecond speed—we have control mechanisms that switch trains, guide planes, run automated machinery, compute figures, process data, simulate the movement of the stars, and correct for error both human and machine. An odd phrase sums it up; these are all done in "real time."

5. *Algorithms.* Technology is clearly more than the physical manipulation of nature. There is an "intellectual technology" as well. An algorithm is a "decision rule," a judgment of one or another alternative course to be taken, under varying conditions, to solve a problem. In this sense we have technology whenever we can substitute algorithms for human judgment.

We have here a continuum with classical technology, but it has been transposed to a new qualitative level. Physical technology—the machine—replaced human power at the *manual* level of raw muscle power or finger dexterity or repetition of tasks; the new intellectual technology—embodied in a computer program, or a numerically controlled machine tool—substitutes algorithms for human *judgments*, where these can be formalized. To this extent, the new intellectual technology marks the last half of the twentieth century, as the machine was the symbol of the first half.

Life: the expression of the surface properties of the biopolymers.
MOLECULAR BIOLOGY GROUP, ASILOMAR, C.
1970

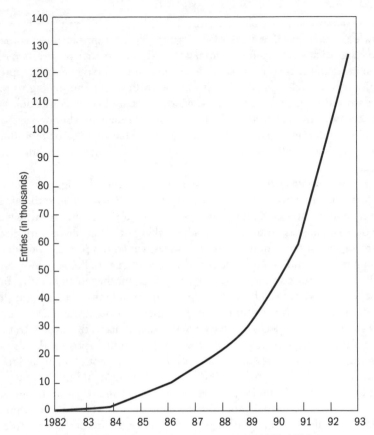

▶ Increase in identified animal and human DNA sequences, 1982–1993.

PRACTICAL VIRTUE

ISAIAH BERLIN, 1974

Not all values are compatible. It is quite clear that—this is an a priori proposition—one cannot have everything. Not everything can be done. Instead, we must sacrifice some things to obtain others. In *The Prince*, Machiavelli stated very clearly—and truly—that one cannot both lead what he called a Christian life and also be effective in public life. By that he meant that ideal virtue is incompatible with being effective in practice, and that in public affairs something, therefore, must be sacrificed. This uncomfortable truth is something that people on the whole are not prepared to face.

The only conclusion I should like to draw from this fact is that I think in general it is very desirable for those who go into dangerous professions, namely, medicine, technology, biology, physics, chemistry and the like, to be aware that their inventions and discoveries are likely to have some effects that they may regard as deleterious; and yet progress in the acquisition and dissemination of knowledge cannot, and should not, be stopped until all humankind is morally purified. The dilemmas leading to extreme anxiety, the agonies Nobel went through, and then Oppenheimer, Einstein and others, are not accidental.

SOLVING UNSUSPECTED PROBLEMS

ELTING E. MORISON, 1974

The dean of historians of technology assesses industrial research.

In the history of the incandescent lamp at the General Electric Company from 1900 to 1913 there is to be discovered a simple model for the way reciprocal connections between making things and having ideas can do useful work. . . .

For twenty-one years [i.e., 1879–1900] the incandescent lamp had been stabilized at a point of interesting, promising inefficiency. Then, in half that time [i.e, 1900–1913], the life of the lamp had been extended by 400 percent and the efficiency of its performance (reckoned in candle power per watt) improved by 700 percent [by developing the tungsten filament and argon gas filling]. As supplementary fallout there were, on one side, some general ideas everyone could continue to use in the search for further understanding of certain physical phenomena and, on the other side, a comparative advantage that amounted almost to a monopoly for the General Electric Company.

The point of this demonstration was not lost on interested parties on the American industrial scene. Companies that worked with chemicals soon adopted the model. By 1920, Dupont, Standard Oil, Kodak and U. S. Rubber had all established laboratories. In 1925, the American Telephone and Telegraph Company consolidated its various small, excellently manned and organized, research groups into the Bell Telephone Laboratories. The power of the model spread. By 1930 there were 1,200 industrial laboratories in this country and by 1950 the figure was 2,200. A good number of these, seeking merely small and temporary advantage in the market, often simply test and tinker with existing products. A somewhat smaller number confine their deeper researches to narrow channels determined by the nature of the things their parent companies make. A very significant few devote a considerable fraction of their resources (sometimes 15 to 20 percent) to the investigation of first causes. With these funds they have sought to create conditions "where men could ask whatever questions of nature they could devise." Quite often the answers discovered by these men have made fundamental contributions to our understanding of how nature works, and quite often, too, they have laid the foundation for a new product line. Which is only to suggest the efficiency of the system that has been developed to convert general ideas into goods and services. . . .

The process as now developed seeks to close [the gap that lies between a new understanding and a new use] in a controlled and systematic way. Using the idea to predict the possibility of the thing . . . the skills and energies of a trained band are mobilized to develop a thing that will work. By virtue of their understanding and training, the members of this band are prepared to drive chance out of the predetermined path that leads to the conversion of the idea into the product. By a multiplication of their numbers—reckoned in man years—the time span of that conversion is collapsed.

All this is called development and can be thought of as the place where the carefree search through nature's mysteries leaves off and purposeful dog work begins. But even the dog work falls within an intellectual scheme. The logic of the originating idea and of the family of ideas to which it belongs runs right through the whole process and controls its development. Even the pitch and position of each new blade or vane in a new kind of turbine, for instance, is designed in accordance with the dictates of recognized principles. The work of modern development is something of a miracle of intellectual sophistication, joinery and management.

It is in this kind of work, hidden among the statistics on man hours, that a good many engineers are to be discovered today. What they do is not much like what [the early General Electric engineers] did before them. For one thing they work, more often than not, as members of teams. For another, they tend to work on parts of things. And for a third, they know more about what they are doing,which tends to keep them working on the parts they know about. The structure of knowledge and administration they work within supports but also confines. With [the early engineers], it was different. The scope of their tasks and the limitation of their resources, intellectual and material, forced along such art, intuition and spirit of independence as was in them.

Of course not everyone gets lost in systematic dog work. Edwin Land has conceived and put together his series of progressively more amazing cameras. Stark Draper has brought into being the intricate systems of inertial navigation. But the norm for modern engineering is more properly to be found in that primitive model which begins with [the General Electric laboratory]. It is, in its way, as ordered and restrictive as any assembly line must be.

On that model we have created a whole system of production that can generate very powerful effects. We can now pour into the society a continuous stream of goods and services, massive in volume, almost infinite in variety and subtly perturbing in its changing character. Alfred North Whitehead used to say that the world had changed more in his lifetime than it had in all the years from the time of Christ to his own birth. That, when he said it, was generally understood to be the course of Progress—more things for more people to reduce the hardships of a forbidding nature. Today, as the capacity of the system still increases, that is not so clear. Those things Emerson put into the saddle have multiplied and are now riding off in all directions. Looking at it only a few weeks ago, a colleague of mine who is deep into the system said, "Everything we did or made up to 30 years ago seemed on the whole to be for good. Now, often as not, it seems to increase confusion, perhaps even to harm."

Maybe it is time to take a closer and more detached look at the system. One place to begin is with an observation made by [General Electric scientist] Irving Langmuir forty years ago. The forced marches of science and the close couple developed between science and engineering have "enabled us," he said, "to solve a problem where a few years previously it was not even suspected that there was a problem." "Who knew," he asked, "that we needed the telephone or the victrola?" Henry Ford carried the idea somewhat further. The object of the system, he said, was to fill "needs the public [is] not yet conscious of." In the press to solve these problems and serve these unknown needs we have introduced so many goods and services that we have not yet had the time to figure out what to do with them. The flood of loose things is, in places, bursting through the joinery of the Western culture so painfully put together in centuries past.

A second thing about the system is that it has its own dynamic. Where so much is now known, it is possible to give rather precise definition to things that are not known—the hollow spots in the body of knowledge. The urge is to fill these empty places, that is, in Langmuir's words, to problem solve. It goes like this: select something that is not yet known or cannot yet be done—a problem; solve it using what is already known from past experience and theory; in the solving obtain some new information that supports a prediction of some new thing that might be done under the sun; subject the prediction to an engineering feasibility study; then realize the prediction by making the new thing. What we have here is a marvelous machinery for the mass production of self-fulfilling prophecies and it appears to run by perpetual motion.

CLOSING THE DISTANCE

JOSEPH WEIZENBAUM, 1976

A computer pioneer assesses the disconnection that makes technological violence "rational"—in this instance, in the scientific panel that advised the U.S. Department of Defense during the Vietnam War.

These men were able to give the counsel they gave because they were operating at an enormous psychological distance from the people who would be maimed and killed by the weapons systems that would result from the ideas they communicated to their sponsors. The lesson, therefore, is that the scientist and technologist must, by acts of will and of the imagination, actively strive to reduce such psychological distances, to counter the forces that tend to remove him from the consequences of his actions.

Hitch's Rule: A new enterprise always costs from two to twenty times as much as the most careful official estimate.

GARRETT HARDIN, 1976

POWER

LANGDON WINNER, 1977

Conceptions of power and authority in technocratic writings have remained virtually unchanged since [Francis] Bacon. Power is ultimately the power of nature itself, released by the inquiries of science and made available by the inventive, organizing capacity of technics. All other sources of political power—wealth, public support, personal charisma, social standing, organized interest—are weak by comparison. They are anachronisms in a technological age and will ultimately decline as scientific technology and the people who most directly control its forces become more important to the workings of society. To say this is only to recognize the drift of history and to be realistic. Elaborate political facades of various kinds may still surround the exercise of technocratic power, but beneath the surface, the true situation will always be evident.

MURPHY'S LAW: COROLLARIES

ARTHUR BLOCH, 1977

Over the years since its formulation in 1949, Murphy's Law ("If anything can go wrong, it will") has accumulated corollaries.

1. Nothing is as easy as it looks.
2. Everything takes longer than you think.
3. If there is a possibility of several things going wrong, the one that will cause the most damage will be the one to go wrong.
4. If you perceive that there are four possible ways in which a procedure can go wrong, and circumvent these, then a fifth way will promptly develop.
5. Left to themselves, things tend to go from bad to worse.
6. Whenever you set out to do something, something else must be done first.
7. Every solution breeds new problems.
8. It is impossible to make anything foolproof because fools are so ingenious.
9. Nature always sides with the hidden flaw.
10. Mother nature is a bitch.

GOING SOLAR

JEROME MARTIN WEINGART, 1978

In the wake of the Oil Shock of the mid-1970s and at the height of the environmental movement, solar power became an attractive rallying cry for those who deplored fossil-fuel pollution on the one hand and feared nuclear power on the other. In a prizewinning paper, a systems analyst imagines a sustainable world-scale solar energy system, not neglecting to estimate as well such a system's economic constraints and environmental impact.

Energy is a central issue in present discussions of the "limits to growth." In much of the world, the growing disparity between rich and poor is closely related to a gap in the amount and thermodynamic quality of available energy and the efficiency with which it is used. One dilemma is that modern technology and abundant energy, which together could help to erase much of this disparity, constitute in their use a major source of environmental disruption. A great challenge to our technological and social ingenuity will be the navigation of the transition to a world in which we can operate well within the carrying capacity of natural systems and at the same time extend justice, equity and a first-class environment to all. . . .

It is sobering to realize that *only the [nuclear] fast breeder [reactor] and harnessing the sun are technically more or less assured and also adequate to meet even the most modest of projected world energy needs over the coming century and beyond.* . . .

Analysis suggests that sunlight could eventually be the primary and even exclusive source of heat, electricity and synthetic fuels for the entire world, continuously and eternally on a scale . . . generally regarded possible only with [thermonuclear] fusion or with [nuclear] fission via the fast breeder. . . . It appears that this can be achieved through a global network of solar-conversion facilities coupled with appropriate energy-transport and storage systems, and that this is possible within acceptable constraints on energy payback time, capital investment, and available suitable land. . . .

Naively, sunlight seems an ideal source of energy. The source itself is eternal and unchanging; the resource is globally distributed, not subject to embargo or depletion, and is of sufficient thermodynamic quality to produce at high efficiency the heat, electricity and synthetic fuels required by a technologically advanced society. On the other hand, sunlight has characteristics that make it problematic to convert and use reliably and economically. Difficulties include the diurnal and seasonal cycles, the unpredictable effects of weather, the nonstorability of the energy in its primary form (photons), and the "low" power density of the direct radiation. A further difficulty is the lack of a practical technology for truly large-scale seasonal electricity storage.

Technically, but at a price, these difficulties can be resolved by a suitable network of solar energy conversion systems. . . . This "network" could be a richly structured set of systems ranging from very small, localized units to very large complexes, producing electricity and synthetic fuels, with interconnections over thousands of kilometers. A richly articulated hierarchical structure, loosely analogous to a complex ecosystem, could provide a stability and resilience that may not be possible with other long-term options, which provide for energy conversion only at very large scales of production and system complexity. This global system would exhibit the following features:

1. Local use of solar-generated heat for space heating, water heating and industrial processes where economically and logistically suitable.
2. Local and regional use of small-scale mechanical, electrical and fuel-generating units, especially in developing countries.
3. Solar electric power plants of various sizes located throughout the world, primarily in sunny regions, interconnected through large integrated electric utility systems over distances up to several thousand kilometers.

4. Solar fuel generation units primarily in sunny regions and interconnected globally via pipeline and, for a few locations (Japan), by tanker (cryogenic and liquid fuel).

In particular, the large-scale generation of hydrogen and of liquid fuels would permit, through long-distance energy transport and seasonal energy storage, the complete decoupling in space and time of the solar source and energy needs. . . .

For solar energy to provide a substantial fraction of world energy needs, the production of electricity and synthetic fuels is essential. Solar thermal techniques, including water and space heating as well as process heat, can displace at most 5–10% of the primary energy use in industrialized countries and are likely to displace even less in much of the tropical, semitropical and arid parts of the developing world.

Second, the scale of future energy use, even in the most modest scenarios and using the most efficient of solar technologies, will require substantial land areas. Yet in spite of competing pressures for land from increasing food demands, urbanization and the needs for forests and the maintenance of ecological diversity, the arid sunny wastelands of the globe—some 20 million square kilometers—will remain essentially unused and potentially available for large-scale use, even in an ecumenopolis of 20 billion people.

Third, the price of solar-derived energy will be (approximately) inversely proportional to the magnitude of the available solar resource. For direct conversion technologies, this means that the least expensive secondary energy production will be in the sunniest regions; for those technologies (solar thermal electricity, solar thermochemical production of hydrogen) that respond only to direct beam sunlight, location in arid, sunny regions will be essential.

Long-distance transport permits such a siting strategy. All of Europe is within practical high-voltage transmission distances of Portugal, Spain and Turkey; in a few decades undersea cable from North Africa could also bring solar electricity to Europe. With the exception of Japan, which must be served by liquid fuels via tanker, virtually the entire world is within practical hydrogen pipeline transport distances (5000 km) of large regions of arid, sunny land. . . .

The use of indirect forms of sunlight (hydropower, ocean thermal gradients, wind and waves) appears limited. . . . Only the high-efficiency direct conversion options appear to have the potential for practical energy supply of 20–100 [terawatts] or more, comparable with the potential from the fast breeder and fusion.

A global transition to such a solar energy system, if it is possible, would require a century or more. Urbanization of the human population is expected to continue during this period and the fraction of the world population potentially served by such extensive technologically sophisticated energy networks would increase. . . .

A global solar-energy system would have important potential benefits and liabilities for mankind. The system itself would be structurally resilient to a variety of natural and sociopolitical upheavals. The enormous geographic and geopolitical diversity of similarly sunny locations would permit global dispersion of production capacity, decreasing the possibility of embargo by any one bloc of nations. Since the resource is nondepletable, stopping operation of the conversion facilities would result in loss of revenue (but not in continued amortization costs). The economic incentives associated with keeping oil and gas in the ground won't exist. This will be especially true for electricity production, where real bulk storage is not yet possible. . . . In addition, user nations such as Japan and most of Europe could develop several years of strategic stockpiles (underground hydrogen) over a period of several decades, permitting more flexibility in responding to energy production shortages than is now possible. . . .

The construction of such a system and its maintenance and operation would be the largest and most daring activity of mankind, and would not be without considerable difficulties—technical, economic, cultural and environmental. But in terms of the scale of energy production that will ultimately be required even in the most modest growth scenarios, we must be willing to consider this route since we have only two options that we can more or less count on—the fast breeder reactor and the sun. . . .

The global scenario . . . is subject to important constraints. Capital requirements may well exceed those for an equivalent scale of energy production based on the breeder. . . . Also, the problems and costs of constructing large industrial facilities in arid lands cannot be overestimated; they may be as problematical and expensive as production and transport of oil from Alaska and coal from Siberia. The large network of transmission lines, storage facilities and pipelines will also cause disruption of regions and will experience increasing social resistance in some parts of the world.

Requirements for land (and perhaps ocean) areas will be extensive. However, considering that almost 10% of the world's land mass is under cultivation and another 14% is partially used pasture, it seems reasonable to consider conversion of 1% of arid, nonproductive regions to solar-energy "farming" (5% corresponds to 50–100 terawatts primary energy conversion). Depending on the classification chosen, the arid regions of the world, including the deserts, comprise 22–30 million square kilometers of land, or roughly 15–20% of total land.

In the United States the total present energy demands could be provided from high-efficiency solar-energy conversion systems sitting on less than 1% of the land, compared with the 41% already committed to crops and grassland pasture. (For comparison, roads cover 1% of the continental U.S.) Even in Western Europe, where there is no Arizona, the present energy demand of France and Italy could be provided from high-efficiency solar conversion systems on about 1% of the land and on 3% of the land in West Germany. . . .

Similarly, requirements for steel and concrete and other materials will be enormous—construction of 50 terawatts of solar thermochemical hydrogen and solar thermal electric units plus the associated energy transport and storage elements will require 20% of identified world iron resources if the present material-intensive designs are retained. (Concrete, glass and silicon are not resource-limited). . . . Even if the average system lifetime is 50 years, annual materials requirements for replacement in a steady-state situation would be a substantial fraction of present world production. . . .

Operation of solar energy systems and the industrial infrastructure required for their construction and replacement will have environmental consequences, in spite of some widely prevailing myths that solar technologies will be relatively benign. We know that new technology, when used on a large scale, will often have unexpected and sometimes unwanted consequences. . . . Fragile desert ecosystems would be severely impacted during construction, with the fine desert crust broken, leading to erosion and dust. The habitats of burrowing animals would be destroyed and the ecology of the region permanently altered. While on the national scale, additional air pollution resulting from production of glass, concrete and steel for the solar plants would not be substantial, the *local impact* of these emissions would constitute an environmental charge against the facilities.

Because the systems will require energy transport and storage, other environmental impacts, such as the flooding of valleys to provide pumped hydrostorage units, and the aesthetic impacts of long-distance high-voltage transmission lines will arise, often at long distances from the solar conversion facilities themselves.

Potential climatic effects exist with many of the solar technologies. [They] will modify both the boundary conditions and energetics of the climatic system. Surface albedo and surface roughness will be altered, as will surface hydrology in nonarid regions. Solar radiation may be converted to latent heat through evaporative cooling. . . . Ocean thermal-electric conversion systems will decrease the surface temperature of the tropical oceans by as much as a few tenths of a degree Celsius, sufficient to cause large climatic changes on the synoptic scale and may also change the ocean/atmosphere equilibration dynamics of carbon dioxide leading to an increased atmospheric carbon-dioxide burden. All of this deserves close attention. The potential climatic effects of these physical changes, especially when modification of as much as one million square kilometers may be involved, are virtually unknown, and there has been little effort to investigate them.

Solar energy systems will also be important because of what they *do not do*. There is rapidly growing agreement that the increasing atmospheric carbon-dioxide levels associated with fossil-fuel combustion presents a potentially severe threat to mankind via massive changes in the climatic system within a century, if present trends in fossil-fuel use are not modified. It may prove necessary to consider the large-scale use of both fission and solar energy systems in order to minimize the risk associated with these severe climatic changes.

SHARING THE BLAME

HARVEY BROOKS, 1979

Science is blamed not only for the arms race, but also for environmental degradation, largely because of its indirect role as the basis for technology, especially in the area of the chemical industry. Yet, ironically enough, many of the environmental issues which cause such concern today would never have been suspected were it not for the instruments, techniques and theories provided by recent advances in science. The notorious Los Angeles smog was attributed to backyard incinerators until a California Institute of Technology chemist, Haagen-Smit, proposed in the early 1950s a complicated mechanism for the catalytic formation of smog through the action of sunlight on the invisible and inodorous constituents of automobile exhausts, and was able to prove his hypothesis by laboratory experiments. Unlike the belching smokestacks of the 19th century, today's pollution consists mostly of substances undetectable by the human senses.

That which *is* detectable is usually the product of secondary chemical or biological action on the primary effluents which produces new substances. Invisible auto exhausts produce smog. Invisible sulfur dioxide emissions from power plants and industry interact with subvisible smoke particles to produce sulfate aerosols, the primary health hazard from air pollution. Apparently innocuous phosphates dissolved in municipal waste waters become nutrients for the growth of algae, which in turn die and are decomposed by oxygen-consuming bacteria, depriving water bodies of life-sustaining dissolved oxygen, and thus forming stinking streams and eutrophied lakes. Invisible carbon dioxide from the burning of fossil fuels may be a potential source of dangerous worldwide climatic changes in the future.

Such effects can only be predicted or understood through the use of elaborate computer models of the atmospheric radiation balance and the global atmospheric circulation patterns. Little of the pollution which is a public issue today would have produced health or environmental effects that could have been detected by the means available to our ancestors, beset as they were by often-endemic infectious disease in the environment, especially in drinking water. . . .

These global effects are threats of future (rather than contemporary) consequences of the continuation of current trends rather than immediate threats, which makes their regulation much more problematical. They could never have been detected by human means in the absence of scientific theory.

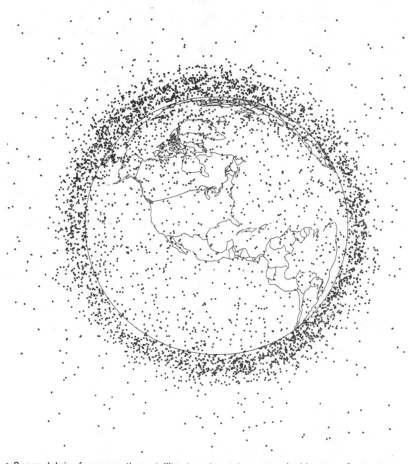

▶ Space debris—from operating satellites to astronaut sewage—doubles every five years. Military computers track every piece larger than 10 centimeters.

Catch-22

David Collingridge, 1980

Two things are necessary for the avoidance of the harmful social consequences of technology; it must be known that a technology has, or will
have, harmful effects, and it must be possible to change the technology in
some way to avoid the effects. In the early days of a technology's development it is usually very easy to change the technology. Its rate of development and diffusion can be reduced, or stimulated, it can be hedged around
with all kinds of control, and it may be possible to ban the technology altogether. But such is the poverty of our understanding of the interaction of
technology and society that the social consequences of the fully developed
technology cannot be predicted during its infancy, at least not with sufficient confidence to justify the imposition of disruptive controls. The British
Royal Commission on the Motor Car of 1908 saw the most serious problem
of this infant technology to be dust thrown up from untarred roads. With
hindsight we smile, but only with hindsight. Dust was a recognized problem
at the time, and so one which could be tackled. The much more serious social consequences of the motor car with which we are now all too familiar
could not then have been predicted with any certainty. Controls were soon
placed on the problem of dust, but controls to avoid the later unwanted social consequences were impossible because these consequences could not be
foreseen with sufficient confidence.

Our position as regards the new technology of microelectronics mirrors
that of the Royal Commissioners in 1908. This technology is in its infancy,
and it is now possible to place all kinds of controls and restrictions on its development, even to the point of deciding to do without it altogether. But
this is a freedom which we cannot exploit because the social effects of the
fully developed technology cannot be predicted with enough confidence to
justify applying controls now.

THE GROWTH OF
ANYTHING WHATEVER

CYRIL STANLEY SMITH, 1981

A metallurgist by training, Smith directed the work at wartime Los Alamos of purifying, alloying, casting and plating uranium and plutonium for the first atomic bombs. After the war he taught at MIT, enlarging his range of study from materials science to a deepening understanding of structure at every scale. I met him in Cambridge, Massachusetts, in the late 1980s, by which time he had evolved into a ruddy, cheerful, white-haired wizard who reminded me of Shakespeare's Prospero. After dinner at his house one evening , he brought out for inspection a flat gold disk about four inches in diameter with a hole in the middle like a Chinese pi that I recognized to be a spare shim prepared to improve the fit of the two hemispheres of the plutonium core of the Fat Man device tested in the New Mexican desert on July 16, 1945. The second of the two graphs reproduced here is one Smith inked by hand into my copy of his book A Search for Structure, *from which this meditation on the relationships between art, invention and technology is taken.*

N early everyone believes, falsely, that technology is applied science. It is becoming so, and rapidly, but through most of history science has arisen from problems posed for intellectual solution by the technicians's more intimate experience of the behavior of matter and mechanisms. Technology is more closely related to art than to science—not only materially, because art must somehow involve the selection and manipulation of matter, but conceptually as well, because the technologist, like the artist, must work with many unanalyzable complexities. Another popular misunderstanding today is the belief that technology is inherently ugly and unpleasant, whereas a moment's reflection will show that technology underlies innumerable delightful experiences as well as the greatest art, whether expressed in object, word, sound or environment.

Even less widely known, but important for what it tells of man and novelty, is the fact that historically the first discovery of useful materials, machines or processes has almost always been in the decorative arts, and was not done for a perceived practical purpose. Necessity is *not* the mother of invention—only of improvement. A man desperately in search of a weapon or food is in no mood for discovery; he can only exploit what is already known to exist. Discovery requires aesthetically motivated curiosity, not logic, for new things can acquire validity only by interaction in an environment that has yet to be. Their origin is unpredictable. A new thing of any kind whatsoever begins as a local anomaly, a region of misfit within the pre-

331

VISIONS OF TECHNOLOGY

existing structure. This first nucleus is indistinguishable from the few fluctuations whose time has not yet come and the innumerable fluctuations which the future will merely erase. Once growth from an effective nucleus is well under way, however, it is then driven by the very type of interlock that at first opposed it: it has become the new orthodoxy. In crystals undergoing transformation, a region having an interaction pattern suggesting the new structure, once it is big enough, grows by demanding and rewarding conformity. With ideas or with technical or social inventions, people eventually come to accept the new as unthinkingly as they had at first opposed it, and they modify their lives, interactions and investments accordingly. But growth too has its limits. Eventually the new structure will have grown to its proper size in relation to the things with which it interacts, and a new balance must be established. The end of growth, like its beginning, is within a structure that is unpredictable in advance.

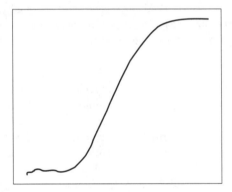

▶ Curve depicting the beginning, growth and maturity of anything whatever. . . . Both the beginning and the end depend on highly localized conditions and are unpredictable in detail. [C.S.S.]

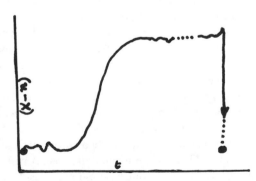

▶ Cyril Smith's hand-drawn improvement, headed "or, better?"

The "S" curve [previous page] (adapted from a paper on the transformations of microstructure responsible for the hardening of steel) can be used to apply to the nucleation and growth of anything, *really* any "thing" that has recognizable identity and properties depending on the coherence of its parts. It reflects the underlying structural conflicts and balance between local and larger order, and the movement of interfaces in response to new conditions of components, communication, cooperation and conflict.

Applied to the growth of either individual technologies or to the development of whole civilizations based upon interactive technologies, the "S" curve reflects origin in art, growth in social acceptance and the eventual limitation of growth by interactions within a larger structure which is itself nucleated in the process. The conditions of beginning, development and maturity are very different from each other.

Though a computer program can duplicate such curves, it is only by looking at the whole hierarchical substructure and superstructure that intuitive understanding can be gained. All stages involve a balance between local structure and overarching regional restraints. All change involves a catastrophic change of connections at some level while topological continuity is maintained, though perhaps with strain, at levels both above and below. Human history follows the same general principles of structural rearrangement as a phase change in a chemical system, though most teachers of history ignore the nucleating role of technology and concentrate on the social changes that are engendered by it.

The transition from individual discovery and rare use of techniques to the point where they affect the environment of Everyman and the content and means of communication between people and peoples underlies virtually every fundamental change in man's view of the world. Few general histories reflect this. An understanding of the proper place of technology within the whole human experience is desperately needed in order that society can wisely decide what to develop and what to discourage. Technology needs to be seen in the perspective that humanists have traditionally applied to man's other activities. . . .

Everything complicated must have had a history, and its internal structural features arise from its history and provide a specific record of it. One might call these structural details of memory "funeous," after the unfortunate character in [Jorge Luis] Borges's story "Funes the Memorious" who remembered everything. The aim of respectable science throughout most of history has been to study afuneous details; it has been analytic, seeking the parts or ideally simplified wholes. Analysis is, of course, absolutely essential for understanding, but no synthesis based upon it can reproduce the funeous structures that provoked interest in the first place unless the essentials of their individual histories are repeated. History selects and biases statistics. The particular structures that do exist, however improbable they may be, must be given priority in man's studies. . . .

Practical metallurgy is seen to have begun with the making of necklace beads and ornaments in hammered native copper long before "useful" knives and weapons were made. The improvement of metals by alloying and heat treatment and most methods of shaping them started in jewelry and sculpture. Casting in complicated molds began in making statuettes. Welding was first used to join parts of bronze sculpture together; none but the smallest bronze statues of Greece or the ceremonial vessels of Shang China would exist without it, and neither would most of today's structures or machines. Ceramics began with the fire-hardening of fertility figurines molded of clay; glass came from attempts to prettily glaze beads of quartz and steatite. Most minerals and many organic and inorganic compounds were discovered for use as pigments; indeed, the first record that man knew of iron and manganese ores is in cave paintings where they make the glorious reds, browns and blacks, while the medieval painter controllably used pH-sensitive color changes long before the chemist saw their significance. In other fields, archaeologists have shown that the transplanting and cultivation of flowers for enjoyment long preceded useful agriculture, while playing with pets probably gave the knowledge that was needed for purposeful animal husbandry. To go back even earlier, it is hardly possible that human beings could have decided logically that they needed to develop language in order to communicate with each other before they had experienced pleasurable interactive communal activities like singing and dancing. . . .

In all these cases, and many more that could be cited, it was aesthetic curiosity that led to initial discovery of some useful property of matter or some manner of shaping it for use. Although the maker of weapons was quick to follow, it was nearly always the desire for beauty or the urge to make art available to the masses (or, if you will, the desire to exploit mass desire for pretty things in order to make profit) that led to advances in production techniques. The desire to beautify the utilitarian has always stretched the ingenuity of the mechanic, who made drawbenches, stamps and screwpresses to shape trinkets before automobile parts or weapons. It is the same in building construction: temples and churches, greenhouses and Crystal Palaces, not necessarily shelter, led to imaginative new structural methods. Even railroad rails and the steel girders for today's skyscrapers needed a precursor in the form of the little mill that rolled lead *cames* to be used in medieval stained-glass windows, and it was a French gardener who invented reinforced concrete because he wanted a larger flower pot for more magnificent display.

In the nineteenth century the milieu of discovery began to expand. Science created a new environment in which imaginative curiosity could operate. Though the discovery of voltaic electricity could have come from metal-replacement reactions used in the arts or from the delightful philosopher's toy, the *Arbor Dainae* (an electrolytic tree of crystalline silver), it actually came from an unaesthetic experiment on a frog's leg. It remained unused until 1837, when the electric telegraph and electrotyping were both seen to be useful. The utility of the latter, however, at first lay only in the arts: it was used as a process for electrolytically duplicating coins, plaques, statuary and engraved or etched plates for the graphic artist. All the great illustrated newspapers stem from this—the *Illustrated London News*, the *Scientific American*, *L'Illustration*, and *Harper's Weekly*. Soon electrolytic baths were giving rise to monumental sculpture, some weighing over 7500 pounds. . . .

Almost immediately an even larger use for "galvanism" developed—the production for middle-class tables of metalware having all the glitter of the rich man's silver and gold. Within a decade, Sheffield plate was supplanted by electrodeposited silver plate, with a not always felicitous relaxation of restraint on design.

At first the electric current for these applications came from banks of small batteries (Daniell cells) in which nearly three pounds of zinc and acid were consumed for every pound of copper deposited. The larger uses of electricity could develop only after a steam-driven generator had grown out of an 1832 lecture-demonstration device made to intrigue physics students with the realities of Oersted's electromagnetic interaction. The first commercial electrical generator was constructed in 1844 to the 1842 design of J. S. Woolrich (whose patent includes also a plating bath); it was used in the shops of the Elkington Company for several decades before it was donated to the Birmingham City Museum, where it now stands. The giant electric power industry of today thus did not begin with a preconceived desire for its utility; the first suggestion came from the arts. Once power generation had been demonstrated, however, it was ripe for development and use by men of a different cast of mind; soon came lighting (beginning with arc lights for lighthouses), and then motive power. All big things grow from little things, but new little things will be destroyed by their environment unless they are cherished for reasons more like love than purpose. . . .

The simple picture of origins outlined above, which applies so well to the early stages of many early technologies, seems hardly applicable to the twentieth century. The experience of discovery in the laboratory is still an essentially aesthetic one (a fact rather thoroughly disguised by the accepted style of reporting the results), but the motivation is rarely a desire to create beauty. Why is this? Is it just that the patronage for creation has changed, or is it that most of what we notice today is not creation but merely a natural or unnatural refinement of the old, while the really new is around unnoticed, awaiting an environment that does not exist?

Information wants to be free.

HACKERS' CREDO, C. 1983

All 3 billion of us are being connected by telephones, radios, television sets, airplanes, satellites, harangues on public-address systems, newspapers, magazines, leaflets dropped from great heights, words got in edgewise. We are becoming a grid, a circuitry around the earth. If we keep at it, we will become a computer to end all computers, capable of fusing all the thoughts of the world.

LEWIS THOMAS, 1983

336

GIRLS JUST WANT TO HAVE COMPUTERS

NANCY KREINBERG AND ELIZABETH K. STAGE, 1983

Exploring equal access to computer technology, the authors arranged for grade- and high-school teachers to ask their students to "imagine they were thirty years old and describe how they would be using a computer." Three girls responded.

When I am 30 I'll have a computer that has long arms that can clean the house and cook meals. And another computer that has a little slot that money comes out to pay for groceries and stuff. (6th-grader)

When I have a computer I hope it will be a computer that tells you how to do certain things and answers you. I hope for my computer to also give me reports on the news so I won't have to buy a paper. I also hope my computer could create anything I want. Or do anything I wish or please, but my computer would not be programmed for evil deeds, such as stealing, cheating, or lying. I would like my machine to be loyal and trustworthy. So my machine would be the greatest machine around. (6th-grader)

I would ask my computer to find me a perfect husband. I would tell it to tell me where to find the prettiest clothes and tell me where to get my hair done. Tell me where the most famous restaurant is in the world. Then I would meet my future husband, get married, have children—two, named Ben and Angie—then I would get a divorce and raise my children on my own till they're grown. And then it's just me and my computer. All alone again. (7th-grader)

I want to be a machine.

ANDY WARHOL, 1984

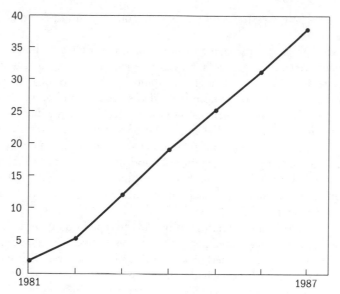

▶ U.S. domestic personal computers in use, 1981–1987 (in millions).

TAKING A MEETING

MICHAEL H. JORDAN, 1984

The president and CEO of Frito-Lay explores the humanization of office management.

A year ago we thought that we would be the ideal company to go to a totally automated office environment. Because of our rapid growth we were a relatively young company. We had a lot of aggressive, intelligent employees. Most of our individual professionals were "computer friendly." We had a large concentration of people in numbers-oriented jobs like Marketing and Accounting and we were moving into a new headquarters that could be tailored to exact specifications. We were going to have the whole operation wired and ready to go by mid- to late-1985. With electronic mail, computer scheduling, central databases—in short, a paperless world.

Wrong.

What we found is that it's going to take a lot longer to make people comfortable with these kinds of systems than we thought a year ago. The same feeling of losing control of your immediate environment is as probable for a marketing manager as it is for a factory worker, or a clerical worker. Maybe even more so. Although they are familiar with computers for specialized tasks, the white-collar executive has *never* had to face a machine on a daily basis for every job aspect. And for many, it could be frightening.

We took people through a proposal on what life would be like with a computer-generated, paperless office. The executives would "hook onto" the system to start their day. They would be able to receive all of their memos and respond to them on the video screen by pulling up data from electronic files, read the mail, call up their calendar for the day's work and look up meeting times and agendas. All of this without paper. Everyone agreed that would be nice, but they didn't want all of that. They still want paper and files and something to hold and make notes on whenever and wherever they want. We found that even something as seemingly harmless as computer scheduling of an appointment calendar could become a problem if used incorrectly.

Current software technology allows me to type into my terminal that I would like to arrange a meeting with three executives, for about an hour, sometime next week, on a certain subject. The computer can then set up the meeting, adjust everyone's schedule and report to me when and where my meeting will take place. The office computer makes this possible—but I believe strongly that using this would be a mistake. The feeling of the Orwellian Society, of marching from one spot to another as the computer dictates, would never be accepted.

The alternative is to use all of the devices for putting schedules on the computer, but stop with the *access* to information. It will be possible to display each of the three executive's calendars to see what they have scheduled, but the actual setting up of a meeting will still come by communicating directly.

Despite all the surveys and studies, when it comes down to practical application, people don't want all the bells and whistles of our "brave new world," at least not yet. What people want are systems to do the drudgery work. They want a system to handle accounts payable invoices and capital expenditures paperwork and all the other processes that are nothing more than time-consuming.

Currently it takes nineteen signatures before a capital expenditure can be instituted at Frito-Lay. That means one or more visits to nineteen individual offices to follow up on the proposal. If a computer can take care of circulating the paperwork instead of a manager physically doing it, then we have come up with something worthwhile.

People want to run their computers and have *them* do the things that *they* really don't want or have time to do themselves. So we've reevaluated our priorities as to what the real impact of our electronic office will be. We will still have some 80 percent of the work areas computerized. But they will still have file cabinets and paper notes too. And we probably will integrate all the bells and whistles eventually, but we'll never do it at the expense of personal interaction and communication. In fact, communication may be the key to humanism in the end.

OBJECTIFIED HUMAN
COMPASSION

ELAINE SCARRY, 1985

In a remarkable book, The Body in Pain, *this eloquent philosopher and literary critic explores the effect on the world of human making.*

It is the benign, almost certainly heroic, and in any case absolute intention of all human making to distribute the facts of sentience outward onto the created realm of artifice, and it is only by doing so that men and women are themselves relieved of the privacy and problems of that sentience. . . .

A chair, as though it were itself put in pain, as though it knew from the inside the problem of body weight, will only then accommodate and eliminate the problem. A woven blanket or solid wall internalize within their design the recognition of the instability of body temperature and the precariousness of nakedness, and only by absorbing the knowledge of these conditions into themselves (by, as it were, being themselves subject to these forms of distress), absorb them out of the human body. A city, as though it incorporates into its unbroken surfaces of sand and stone a sentient uneasiness in the presence of organic growth and decay (the tyranny of green things that has more than once led people to the desert whose mineral expanse is now mimed in every modern urban oasis) will only then divest human beings of that uneasiness, divest them to such an extent that they may even come to celebrate and champion that green world, reintroducing it into their midst in the delicate spray of an asparagus fern or in a breathtaking framed photograph of the Andes. A clock or watch, as though it were itself sentient, as though it knew from the inside the tendency of individual sentient creatures to become engulfed in their own private bodily rhythms, and simultaneously knew of their acute and frustrated desire to be on a shared rhythm with other sentient creatures, will only then empower them to coordinate their activities, to meet for a meal, to meet to be schooled, to meet to be healed, after which the clock can be turned to the wall and the watch can be taken off, for these objects also incorporate into their (set-asidable) designs an awareness of sentient distress at having to live exclusively on shared time.

The naturally existing external world—whose staggering powers and

339

beauty need not be rehearsed here—is wholly ignorant of the "hurtability" of human beings. Immune, inanimate, inhuman, it indifferently manifests itself in the thunderbolt and hailstorm, rabid bat, smallpox microbe and ice crystal. The human imagination reconceives the external world, divesting it of its immunity and irresponsibility not by literally putting it in pain or making it animate but by, quite *literally, "making it" as knowledgeable about human pain as if it were itself animate and in pain.* . . .

The general distribution of material objects to a population means that a certain *minimum* level of objectified human compassion is built into the revised structure of the external world, and does not depend on the day-by-day generosity of other inhabitants which itself cannot be legislated. . . . It is almost universally the case in everyday life that the most cherished object is one that has been hand-made by a friend; there is no mystery about this, for the object's material attributes themselves record and memorialize the intensely personal, extraordinary because exclusive, interior feelings of the maker for just this person—This is for you. But anonymous, mass-produced objects contain a collective and equally extraordinary message: Whoever you are, and whether or not I personally like or even know you, in at least this small way, be well.

Karl Taylor Compton, the famous physicist, told of his sister, who lived in India, hiring an electrician to make improvements in her home. Unable to explain herself, she said, "Oh, you know what's needed here, just use common sense and do it." To which he replied, "Alas, madam, common sense is a great gift of God. I am a humble soul with only a technical education."

ALLAN H. CLARK, 1985

The number of computer specialists in the [U. S.] labor force rose from 167,000 in 1974 to 576,000 in 1986—an increase of 345 percent in twelve years. . . . The number of [computer and information sciences] doctorates more than doubled during that interval, while the number of master's degrees increased by 508 percent and of bachelor's degrees by 1754 percent!

Recent Social Trends in the United States
1960–1990

In general, older technologies tend to serve recreational or aesthetic roles once the replacement [by newer technologies] is complete. This was the case with fuel wood, sailing ships, horses, convertible (open) cars, and many other examples.

NEBOJSA NAKICENOVIC, 1986

As with all good things, technology sometimes misfires; and as it progresses, its turbulence often shakes people and institutions. Yet in the United States, as only one of the capable runners in today's international technology race, our real standard of living has about doubled every forty years. Technology and innovation cannot take all the credit, but they have made a crucial difference.

DOUGLAS E. OLESEN, 1986

SMOKE AND MIRRORS

EDWIN T. LAYTON, JR., 1986

A parable from a professor of the history of technology at the University of Minnesota, delivered as part of a presidential address to the Society for the History of Technology.

I wish to report on a rare variant of [Lewis] Carroll's classic [*Through the Looking Glass*] published by an obscure technological press. In the edition I used, the rooms on both sides of the looking glass were filled with people at work: one side was labeled "science" and the other side "technology."

In this version Alice had difficulty telling one side from the other. The people seemed to be doing the same sorts of things; they fiddled with computers, manipulated various pieces of apparatus, made drawings and calculations, and argued with one other. On both sides some of the people performed experiments and developed theories while others designed technological devices. Though the proportion of people engaged in these tasks varied somewhat, Alice found it very confusing, especially considering that the initial output in both cases appeared to be information.

Not only were the goings-on similar, but Alice also found that now she had to run as fast as possible just to stay in one place on both sides of the looking glass. In addition, the Jabberwocky was present in both worlds, and the people in both seemed about equally eager to supply him with still sharper and more deadly teeth and claws. There were differences, but they were often rather subtle. When Alice held up the book titled "Thermodynamics" it changed to "Engineering Thermodynamics" in the looking glass. Alice became a bit confused and asked whether it really made any important difference what side of the looking glass people were on.

At this point in the variant edition, an alien from another galaxy appeared with a time machine. . . . When they traveled in the time machine, the images in the mirror became different. The further they traveled backward in time, the more unlike the images became, until Alice could see little similarity between the two.

On the science side, the big particle accelerators and giant telescopes disappeared, as did nearly all of the scientists engaged in industrial research and design. Instead, Alice saw a Cambridge University professor examining light with a prism and an Italian professor experimenting with balls of different weight; further back there was a long line of Greek philosophers, Babylonian priests, and so forth. They seemed to be preoccupied with building theories or simply with "knowing."

On the technology side the changes were even more dramatic. The theories and experimentation diminished, but they never quite disappeared completely. The engineers shrank from a multitude almost filling the room to a mere handful. After a long time they too faded away. In their place was an even longer line of hunters, gatherers, farmers, artisans, and artists. They appeared to be working people making or doing things.

But while the differences grew greater with time, Alice noticed that at certain times the same people tended to appear on both sides of the looking glass. There were, for example, a number of 18th-century philosophers: she thought she recognized James Watt and Benjamin Franklin. Also present on both sides were a number of Renaissance artist-engineers, including Leonardo da Vinci. At an earlier period there were cathedral-building masons and, still further in the past, Hellenistic engineers.

The probability of someone being on both sides of the looking glass appeared to increase as she approached the present. This was particularly the case in industrial research laboratories. In fact, the mirror became so permeable and even difficult to see that many people denied that it existed. Things also got more complex in every way. Alice concluded that the difference was sometimes important and sometimes not, depending on the questions asked. She noticed that some people on both sides seemed to be obsessed with determining who got credit for what. She concluded that this was not, by itself, a very important question.

As in the original version, Alice passed through the looking glass. She found that what could be seen from the old room was the same. But the parts she could not see from the old room were quite different in unexpected and delightful ways. Workers, farmers, artists—all kinds of people of both sexes—were still there. They were just as numerous and hardworking as they had been in the distant past revealed by the time machine. But when viewed from the science side in the present time, they had been pushed down out of sight by the weight of the giants standing on their shoulders.

From the science side it had appeared to be a masculine affair no matter which way she looked. However, from the technology side of the mirror she saw that both sexes were well represented in a variety of roles. . . .

Now, in this version of *Through the Looking Glass*, Alice decided that the technology side was much more fun, more dynamic, more open to new ideas, to novel methodologies, to women, and to younger people. So she stayed on the technology side of the looking glass, founded an organization with the acronym WITH, and lived happily ever after. (As noted, this edition was produced by a technological press.)

CHRONICLE OF A
DISASTER FORETOLD

ROGER M. BOISJOLY, 1986

The failure of two O-rings—rubber gaskets sealed with putty—because they lost resiliency at low temperature led to the explosion that killed seven astronauts and destroyed the space shuttle Challenger *on January 28, 1986. Morton Thiokol, Inc. (MTI), manufactured the solid-fuel booster rocket whose interlocking upper and lower casings the O-rings sealed. Roger Boisjoly was an engineer at MTI expert in the O-rings' performance.*

Morton Thiokol, Inc.
Wasatch Division
Interoffice Memo

31 July 1985
TO: R. K. Lund, Vice President, Engineering
FROM: R. M. Boisjoly, Applied Mechanics
SUBJECT: SRM [i.e., solid rocket motor] O-Ring Erosion/Potential
            Failure Criticality

This letter is written to insure that management is fully aware of the seriousness of the current O-Ring erosion problem in the SRM joints from an engineering standpoint.

The mistakenly accepted position on the joint problem was to fly without fear of failure and to run a series of design evaluations which would ultimately lead to a solution or at least a significant reduction of the erosion problem. This position is now drastically changed as a result of the SRM 16A nozzle joint erosion which eroded a secondary O-Ring with the primary O-Ring never sealing.

If the same scenario should occur in a field joint (and it could), then it is a jump ball as to the success or failure of the joint because the secondary O-Ring cannot respond to the clevis opening rate and may not be capable of pressurization. The result would be a catastrophe of the highest order—loss of human life.

An unofficial team (a memo defining the team and its purpose was never published) with leader was formed on 19 July 1985 and was tasked with solving the problem for both the short and long term. This unofficial team is essentially nonexistent at this time. In my opinion, the team must be officially given responsibility and the authority to execute the work that needs to be done on a non-interference basis (full time assignment until completed).

It is my honest and very real fear that if we do not take immediate action to dedicate a team to solve the problem, with the field joint having the number one priority, then we stand in jeopardy of losing a flight along with all the launch pad facilities.

(signed) Roger M. Boisjoly

The team was duly reconstituted, met regularly, and reported its findings to NASA, which had a counterpart team. When the time came to decide whether to launch the Challenger, *on January 27, the night before the flight, the fact that the O-rings sometimes failed was fully known. Boisjoly and other MTI engineers had in fact opposed the launch because of predicted low temperatures at the Florida launch site—the only launch MTI had opposed in 12 years. That was the reason for the January 27 meeting, a teleconference between MTI and NASA officials at the Marshall Space Flight Center. Boisjoly kept notes of the meeting.*

Summary notes for January 27 and 28, 1986, written 2/13/86

1. The first I heard of a cold temperature at Kennedy Space Center prior to the launch was at 1:00 p.m. on January 27, 1986. As time went on, I heard that it had been cold for several preceding days also.
2. I and others spent much of the afternoon trying to convince management not to fly in weather so cold. By now I heard that it was predicted to get to 18° F. overnight.
3. We were successful in convincing engineering management (Bob Lund) that we should not fly. This fact is evident from the last two charts [of 13 total] in the [January 27 evening] presentation. Bob Lund prepared these two charts during the presentation preparation. The last two charts were conclusions and recommendations.
4. A telecon was set up with Kennedy Space Center and Marshall Space Flight Center in mid-evening (approximately 8:00 p.m.) to give our presentation.
5. The portions of the presentation were given by the person who was tasked with preparing that chart.

6. I gave my portion and explained that I was deeply concerned about launching due to the low temperature and its effect on the O-rings. I was asked why and I responded that the timing function of the seals to create a seal into the extrusion gap would be affected but that I was unable to quantify the threshold of acceptability, except to say that it was away from goodness and our experience base. I reminded everyone about the soot blow-by Solid Rocket Motor-15 (launched January 24, 1985 at 1:50 p.m. EST) in two field joints [i.e., O-rings] and how it was the coldest launch at that time and it was the first time that we had blow-by of a primary seal in a field joint. A comment was then made [that] the Solid Rocket Motor-22 (launched October 31, 1985 at 11:00 EST) also had soot past the primary seal and it was essentially a warm launch temperature. A conclusion was made that on this basis, temperature was *not* a discriminator.

6a. I was asked if I had any data to support my claim. I said I have been trying to get lab data since last October but was not successful due to problems with the instron machine.* To this answer Mr. Mason [Jerry Mason, senior vice-president, Wasatch Operations, MTI] looked very unpleased.

7. The presentation continued, then Bob Lund gave his conclusions and recommendations. The listeners on the other lines [i.e., NASA officials at Marshall and Kennedy] were not pleased and began to question the rationale behind such a set of conclusions and recommendations. Bob Lund explained why we should stay within our data base [of bad experience with O-rings at low temperature] as recommended.

8. George Hardy [deputy director, Science and Engineering, Marshall Space Flight Center] was then asked by someone at either Kennedy Space Center or Marshall Space Flight Center what he thought. He said he was appalled at Morton Thiokol Inc.'s recommendation. He was then asked if he would fly and he said not if the contractor is recommending not to fly.

9. A short discussion followed by various people. Morton Thiokol, Inc. management asked for a short five minute caucus offline and said they would get right back to them. The caucus lasted approximately 20–25 minutes.

10. The caucus started by Mr. Mason stating that a management decision was necessary.

11. Those of us who opposed the launch were still trying to convince management that temperature was indeed a very important parameter.

12. Arnie Thompson went up and placed a pad of paper right in front of Mr. Mason on the table and attempted once again to explain the effect of a lower tempera-

*Boisjoly explained this jargon during his appearance before the presidential commission: "I had been trying to get data since October on O-ring resiliency, and I did not have it in my hand. We have had tremendous problems in trying to get a function generator and a machine to actually operate and characterize this particular pressurization function rate." [R.R.]

ture. Arnie stopped when it became apparent that he wasn't getting through. I then spoke up again and this time made a big point about the photo evidence between Solid Rocket Motor-15 (launched on January 24, 1985, at 1:50 p.m. EST with a 53° F O-ring temperature) and Solid Rocket Motor-22 (launched October 31, 1985 at 11:00 a.m. EST with 75° F O-ring temperature). I tried to explain that Solid Rocket Motor-15 had a lot (approximately 110 degree arc) of black soot while Solid Rocket Motor-22 had a small (local grayish spots) amount of soot. I too stopped when it was apparent to me that no one wanted to hear what I had to say.

13. Sometime during the caucus, Mr. Mason asked if he was the only one that was taking the position to fly. I really think that many people didn't hear him due to all the discussion that was taking place.

14. After Arnie and I had our last say, Mr. Mason turned to Bob Lund and asked him to take off his engineering hat and put on his management hat.

15. Shortly after this we went back on the telecon and made the recommendation to launch Solid Rocket Motor-25 without any temperature restriction.

16. The decision was basically based on the items that were listed on a chart signed by Joe Kilminster [vice-president, Space Booster Systems, MTI]. This signed sheet was requested by NASA.

17. I was not a party to the contents on the chart and in fact did not see the chart until the next day.

18. I left the room feeling badly defeated but felt I did all I could short of being fired. I knew that management had made a tough decision but I didn't agree with it at all.

19. One of my colleagues put it best to sum it up. "This was a meeting where the determination was to launch and it was up to us to prove beyond a shadow of doubt that it was not safe to do so." I might comment that this is *not* the usual tone of pre-flight meetings. We usually have to *prove* beyond a shadow of doubt that it is okay to launch.

20. Refer to notebook #2 pages 9 and 10 for real time entries about feeling prior to and after launch.

21. My feelings are well documented by memo and internal activity reports long before this incident. I couldn't stand the thought of a disaster any longer so I wrote memo 2870:FY 86:073 sometime in July-August of 1985. A seal task team was subsequently formed for real as a result of my memo—refer to notebook #1 page 1 for entry note.

22. I hope and pray that I have not risked my job and family security for being honest in my conviction, to stand up for what is right and honorable.

(Signed) Roger M. Boisjoly 2/13/86
4:50 p.m.
EST

BAD CHARTS

Edward Tufte, 1997

Many theories evolved to explain the Challenger *disaster. Edward Tufte, a statistician and information designer at Yale, explores them in his book* Visual Explanations, *and offers a simpler explanation of his own.*

The immediate cause of the accident—an O-ring failure—was quickly obvious. . . . But what are the general causes, the lessons of the accident? And what is the meaning of *Challenger?* Here we encounter diverse and divergent interpretations, as the facts of the accident are reworked into moral narratives. These allegories regularly advance claims for the special relevance of a distinct analytic approach or school of thought: if only the engineers and managers had the skills of field X, the argument implies, this terrible thing would not have happened. Or, further, the insights of X identify the deep causes of the failure. Thus, in management schools, the accident serves as a case study for reflections on groupthink, technical decision-making in the face of political pressure, and bureaucratic failures to communicate. For the authors of engineering textbooks and for the physicist Richard Feynman, the *Challenger* accident simply confirmed what they already knew: awful consequences result when heroic engineers are ignored by villainous administrators. In the field of statistics, the accident is evoked to demonstrate the importance of risk assessment, data graphs, fitting models to data, and requiring students of engineering to attend classes in statistics. For sociologists, the accident is a symptom of structural history, bureaucracy, and conformity to organizational norms. Taken in small doses, the assorted interpretations of the launch decision are plausible and rarely mutually exclusive. But when *all* these accounts are considered together, the accident appears thoroughly overdetermined. It is hard to reconcile the sense of inevitable disaster embodied in the cumulated literature of post-accident hindsight with the experiences of the first 24 shuttle launches, which were distinctly successful.

Regardless of the indirect cultural causes of the accident, there was a clear proximate cause: an inability to assess the link between cool temperature and O-ring damage on earlier flights. Such a pre-launch analysis would have revealed that this flight was at considerable risk.

On the day before the launch of *Challenger,* the rocket engineers and managers needed a quick, smart *analysis* of evidence about the threat of cold to the O-rings, as well as an effective *presentation* of evidence in order to convince NASA officials not to launch. Engineers at Thiokol prepared 13 charts to make the case that the *Challenger* should *not* be launched the next day, given the forecast of very chilly weather. Drawn up in a few hours, the charts were faxed to NASA and discussed in two longer telephone conferences between Thiokol and NASA on the night before the launch. The charts were unconvincing; the arguments against the launch failed; the *Challenger* blew up. . . .

These charts *defined the database for the decision:* blow-by (not erosion) and temperature for two launches, SRM 15 and SRM 22. Limited measure of effect, wrong number of cases. Left out were the other 22 previous shuttle flights and their temperature variation and O-ring performance. A careful look at such evidence would have made the dangers of a cold launch clear. Displays of evidence implicitly but powerfully define the scope of the relevant, as presented data are selected from a larger pool of material. Like magicians, chartmakers reveal what they choose to reveal. The selection of data, whether partisan, hurried, haphazard, uninformed, thoughtful, wise—can make all the difference, determining the scope of the evidence and thereby setting the analytic agenda that leads to a particular decision.

SIMPLE AND COMPLEX TECHNOLOGIES

John Manley, 1987

A pioneer participant in the Manhattan Project, active in nuclear science and technology for forty years, draws lessons from two spectacular and unsettling accidents.

In 1986 two technical disasters, *Challenger* and Chernobyl, were much in the public view. Surely people wondered whether these were examples of technology developing ahead of its time or, more generally, whether these events suggested that nuclear power and space travel are technologies unfit for human consumption. . . .

As technological complexity increases, more humans become involved, with each possessing a specialized responsibility and purview. Each has less opportunity to experience the two forms of intimacy so significant in simpler technologies: intimacies with the details of the process and with one's fellow workers. The former provides a knowledge and understanding of the nature and basics of the technology; the latter lays the foundation for confidence and trust in the expertise of one's associates. Complexity thus leads to an inherent paradox: it demands greater interaction between people and their technical systems, and among the people involved as creators, operators, and users. Yet the complexity generates an environment that makes this rapport more difficult.

In the early years of automobile technology, for example, the user of a car also had to be a mechanic. As car ownership became the norm, however, user expertise became less important because of two concurrent developments. One was the improved reliability of all parts of the machine; the other was the growth of a service infrastructure: mechanics and other service technicians on whom the user could rely, even if with justified reluctance. Such developments take time, generations in this case. They require a transfer of familiarity, knowledge and understanding from one group, the creators, to another, the users, their intermediaries and assistants. The process resists acceleration or shortcutting, for these endanger the entire structure.

Challenger and Chernobyl belong in maturing but relatively young technologies—space and nuclear energy. The first is hardly one generation old, the second only two. Both require the safe handling of very large amounts of concentrated energy. They require supplemental technologies for control, safety and reliability. The chemical energy of a *Challenger* does not present the radioactive hazard of a Chernobyl (unless such material is part of the cargo), but the energy content of the fuel is large. At takeoff the space shuttle's propellant tanks contain the equivalent of about one-third of the energy released by the Hiroshima bomb. Fortunately, the laws of nature prevent a truly explosive release of this energy, but shuttles and nuclear reactors can be enormous hazards to many people.

In addition, *Challenger* and Chernobyl possessed known design flaws. This is not necessarily bad if the technology includes built-in safety and control mechanisms that protect against such flaws. The Chernobyl reactor had these features, but they were deliberately rendered inoperative. *Challenger* depended only on a human safety system, a decision to scrub the launch, and it was not used. In both instances people were at fault, people who undoubtedly were too far removed from the creators and mechanisms of the technology. Similar accidents are likely to happen when any technology reaches greater maturity. Technology in the hands of its creators is rarely ahead of its time; in the hands of the less able and more ignorant it is not technology that is ahead of its time, but humans who are behind.

Challenger-type operations undoubtedly will continue as will the use of Chernobyl-type reactors. What remains to be seen is if the inherent paradox of complex technologies receives adequate attention. The only apparent amelioration of this problem lies in reserving enough time to produce the experts who will bridge the gap between creators and users, as auto mechanics do, and to accumulate the experience necessary to obtain product reliability.

SAMUEL C. FLORMAN, 1987

"Doing" science implies a belief in science, and I think it is fair to say that this belief lies at the heart of engineering. The engineer does not believe in black magic, voodoo, or rain dances. The engineer believes in scientific truth, that is, truth that can be verified by experiment.

The search for scientific truth requires that we disregard, so far as possible, our personal value systems; yet, paradoxically, this approach creates its own values. As Jacob Bronowski has written: "Independence and originality, dissent and freedom and tolerance; such are the first needs of science, and these are the values which, of itself, it demands and forms." In the same vein, Bertrand Russell asserted that "those who forget good and evil and seek only to know the facts are more likely to achieve good than those who view the world through the distorting medium of their own desires."

So this is the beginning of the engineering view: a commitment to scientific truth and to the values that the search for this truth entails.

Occasionally our critics view this commitment as evidence of a lack of "soul." When one deals in hard facts it is easy for an observer to conclude that one has a hard heart. Using an even uglier metaphor, Theodore Roszak has said that scientists and technologists look at the world with "a dead man's eyes." I find such criticism to be particularly irritating. There is a place in the world for poetry and sermons and for visions of "what might be" rather than "what is." But there is also a need for facts and plain-speaking. As engineers we are pledged not to engage in merely wishful thinking. We are not the grasshopper; we are the ant who knows that winter is coming. We are the grumpy little pig who builds his house out of brick while his friends play and sing "Who's afraid of the big bad wolf?" We do this because we know what the big bad wolf can do when he huffs and puffs. This does not mean that, as individuals, we cannot love poetry and approve of sermons—or even preach sermons. But, because we are committed to scientific truth, we believe that poetry and sermons alone are not an adequate foundation on which to build human society. The evidence convinces us that God helps those who help themselves. It does not follow that we are selfish. The evidence also shows that God helps those who work together cooperatively and provide for one another.

There is a fringe benefit that comes along with our familiarity with science—and with the technological applications of science—and this is that it helps make us feel at home in the world. To the extent that the forces of nature have been comprehended, and the structure of the universe revealed, we share in the understanding and this gives us some measure of contentment. This comfort—this inner peace, if you will—is a basic ingredient of the engineering view.

This does not mean that engineers are, or have any reason to be, smug. We are humble before the unknown and stand in awe of the unknowable. But we do not feel alienated, as some people say they do, by the scientific advances of the age. And our message to our non-technical fellows is that to a certain extent this understanding—and the peace that goes with it—is available to all. Anyone who is willing to explore, however superficially, the findings of science, can share in this feeling of at-homeness in the world.

FOUR GENERALIZATIONS

Eugene B. Skolnikoff, 1989

Four generalizations about technology and its interaction with policy are worth noting . . . since they are so often misunderstood or not recognized in consideration of policy. . . .

351

• Technological change does not lead to immediate and dramatic social change, as often portrayed in "futures" literature. Social change evolves through the impact of incremental developments in which technology is but one factor. Most new technologies, even seemingly radical ones, lead to social effects through an evolutionary process of social learning rather than sudden quantum shifts. It may be arguable that the advent of atomic energy as a usable technology is one exception. Even if it is, there are not likely to be any others in the near future.

• There is no such thing as a pure technological fix. That is, technology alone cannot be expected to solve an important societal problem without creating new social, economic and political problems along the way, or in its wake. Technology can change the weight of the relevant factors, bring in new actors, and alter the costs and benefits. That is substantial; however, the societal consequences of technology—often unanticipated—mean that to turn to technology as an unencumbered way to solve a serious political problem is a delusive goal.

• As a corollary, all important technological issues in international politics must ultimately be dealt with in political terms. Technological factors are relevant, sometimes crucial, to understanding and dealing with many issues; but the policy choices of which technology is a part are not predetermined by technology. They will always finally turn on the political aspects.

• Technological development does not happen independently of human direction. Though there is a sense in which technological change has a certain independent momentum, for policy purposes it is important to recognize that allocation choices in R&D (research and development) do affect the characteristics of the resulting technology. At the same time, the cumulative effects and dispersed sources of technology continually create new and usually unforeseen requirements for policy. In effect, technology is both a dependent and an independent variable in the policy arena.

SURROUNDSOUND

GRAHAM NASH, 1991

Graham Nash, one member of the popular singing group Crosby, Stills & Nash, tells this story about singer Neil Young.

I once went down to Neil's ranch, and he rowed me out to the middle of the lake. Then he waved at someone invisible, and music started to play from the countryside. I realized Neil had his house wired as the left speaker and his barn wired as the right speaker. And from the shore Elliot Mazer, his engineer, shouted, "How is it?" And Neil shouted back, "More barn!"

A MODERN BABY

NANCY SMITHERS, 1991

A thirty-six-year-old American lawyer responds to the new technologies of reproduction, as reported by anthropologist Rayna Rapp.

I was hoping I'd never have to make this choice, to become responsible for choosing the kind of baby I'd get, the kind of baby we'd accept. But everyone—my doctor, my parents, my friends—everyone urged me to come for genetic counseling and have amniocentesis. Now, I guess I'm having a modern baby. And they all told me I'd feel more in control. But in some ways, I feel less in control. Oh, it's still my baby, but only if it's good enough to be our baby, if you see what I mean.

BIOSPHERICS

JOEL E. COHEN AND DAVID TILMAN

For two years, from 1991 to 1993, eight men and women lived inside a three-acre closed ecosystem in Oracle, Arizona, called Biosphere 2 (Earth=Biosphere 1). The immense glassed-in miniature world, containing a sea, a savanna, a mangrove swamp, a rain forest, a desert and a farm, had been under construction since 1984, paid for by Texas oil billionaire Edward P. Bass to evidence British scientist James Lovelock's Gaia hypothesis that the earth and its inhabitants form a self-regulating system. According to the project's chief ideologist, John Allen, it was also intended to serve as "the only prototype to date for research and development of extraterrestrial space colonies." It failed miserably, as this account attests, taught valuable lessons and has been converted to a testbed for studies of global warming.

NO existing closed-environment facilities for ecological research approach the size and sophistication of Biosphere 2: the original airtight footprint covered 13,000 [square meters] and enclosed 204,000 [cubic meters]. Despite the enormous resources invested in the original design and construction (estimated at roughly $200 million from 1984 to 1991) and despite a multimillion-dollar operating budget, it proved impossible to create a materially closed system that could support eight human beings with adequate food, water and air for two years. The management of Biosphere 2 encountered numerous unexpected problems and surprises, even though almost unlimited energy and technology were available to support Biosphere 2 from the outside. Isolating small pieces of large biomes and juxtaposing them in an artificial enclosure changed their functioning and interactions rather than creating a small working Earth, as originally intended.

The staff of Biosphere 2 and several reports revealed . . . numerous examples of surprises that had been encountered since the facility began its first "mission," the widely publicized enclosure of eight Biospherians from 1991 to 1993. By January 1993, 1.4 years after material closure of Biosphere 2, the oxygen concentration in the closed atmosphere fell from 21 percent to about 14 percent. This oxygen level, ordinarily found at an elevation of 17,500 feet, was barely sufficient to keep the Biospherians functioning. Carbon dioxide levels skyrocketed, with large daily and seasonal oscillations. Subsequent analyses discovered that microbial degradation of carbon in the highly fertile soils (needed for food production) consumed the atmospheric oxygen, producing carbon dioxide. Although no one knew it at the time, some of the carbon dioxide combined with the calcium in the concrete used to construct Biosphere 2 to produce calcium carbonate. The original atmos-

VISIONS OF TECHNOLOGY

353

pheric oxygen, in effect, became locked up in the walls of the structure. In early 1993, before the end of the first 24-month "mission," oxygen was added to Biosphere 2's atmosphere from outside. Another atmospheric problem was also unanticipated. The N_2O concentration in the air rose to 79 parts per million after three years of closure. At that level, N_2O may reduce vitamin B12 synthesis to a level that can impair or damage the brain. These and other such unforeseen problems made the biogeochemical regulation of a closed atmosphere a delicate problem.

Vines originally introduced as a carbon dioxide sink (such as morning glory . . .) proved to be exceptionally aggressive. The vines required a great deal of hand weeding, which was not entirely successful, to prevent them from overrunning other plants, including food plants. The trunks and branches of large trees became brittle and prone to catastrophic and dangerous collapses. Although some species were expected to go extinct, particularly among the plants, the extremely high fraction of species extinctions (for example, 19 of 25 vertebrate species) was unanticipated. All pollinators went extinct. Consequently, the majority of the plant species, which depend on insect or vertebrate pollinators for reproduction, had no future beyond the lifetime of the individuals already present. The majority of the introduced insects went extinct, leaving crazy ants (*Paratrechina longicornus*) running everywhere, together with scattered cockroaches and katydids. Despite the relatively small size of the Biosphere 2 ocean compared to the land areas, extinction rates in the ocean appeared to be lower than those on land. Air temperatures in the upper reaches of the glass structure were far higher than anticipated, while light levels were significantly lower. Areas designed to be deserts initially became chaparral or grasslands because of a failure to adjust the rainfall to reduced evaporative demand. Water systems became loaded with nutrients, polluting aquatic habitats. Nutrients had to be removed from the water by passage over plates on which algal mats grew. The algal mats were then harvested manually, dried, and stored within the enclosure. Water chemistry management made it necessary to separate a planned brackish estuary from the ocean.

These surprises left . . . the impression that Biospherians, despite annual energy inputs costing about $1 million, had to make enormous, often heroic, personal efforts to maintain ecosystem services that most people take for granted in natural ecosystems. Even these efforts did not suffice to keep the closed system safe for humans or viable for many nonhuman species. . . .

The major retrospective conclusion that can be drawn is simple. At present there is no demonstrated alternative to maintaining the viability of Earth. No one yet knows how to engineer systems that provide humans with the life-supporting services that natural ecosystems produce for free.

Dismembering major biomes into small pieces, a consequence of wide-spread human activities, must be regarded with caution. Despite its mysteries and hazards, Earth remains the only known home that can sustain life.

MULTIPLES

HOWARD REINGOLD, 1994

Similar to the way previous media dissolved social boundaries related to time and space, the latest computer-mediated communications media seem to dissolve boundaries of *identity* as well. One of the things that we "McLuhan's children" around the world who grew up with television and direct-dialing seem to be doing with our time, via Minitel in Paris and commercial computer chat services in Japan, England and the United States, as well as intercontinental Internet zones like [Multi-User Dungeons], is *pretending to be somebody else*, or even pretending to be several different people at the same time.

GETTING THE LEAD OUT

NORMAN BALABANIAN, 1994

[A] common concept describing the impersonal way that technology develops and is controlled is the idea of "the market." The interactions of producers and consumers in the market, it is claimed, will lead automatically to the optimum allocation of resources (material, financial and human) and the optimum development of technology. The sole measure of optimality in this picture is economics.

Dire consequences are predicted if all is not left to the market. If existing automobile engines have poor fuel economy, leave it to the market. If too much CFC's [i.e., chlorofluorocarbons] are produced in industrial processes and appliances, thus leading to the depletion of the ozone layer, leave it to the market. If too much carbon dioxide is being produced by our technological systems, leading to global warming, leave it to the market. Indeed, the judgment "too much" would be incomprehensible under this regime; if this is the level of carbon dioxide or any poisons released to the environment under market control, it must be just right since the market optimizes.

But economics cannot be the sole measure of the social good, and the market has failed "big time" in preventing the catastrophic ill effects of the wide-scale deployment of technology. Perhaps a study of lead can shed some light on the matter.

It has been known for at least several decades that lead and the human body are incompatible. When the body absorbs lead, it is damaged. Most susceptible are children, infants even more so. Yet, in spite of this knowledge, lead was widely introduced into the U.S. environment in several forms. Lead pipes continued to be used for carrying water in the construction of homes and other structures. Homes and other buildings continued to be painted by lead-based paint, long after manufacturers knew that the presence of this lead in the environment damaged human life. And leaded gasoline, whose fumes were particularly noxious in congested cities, continued to be manufactured.

Lead water pipes are gradually being replaced by copper pipes and lead paint is no longer being sold. But there are millions of homes and apartments that still carry lead-based paint, peeling and chipping through it may be. Children ingesting the dust and chips of such paint are at heavy risk. Many studies have shown the harmful effects of exposing children to lead-based paint. . . . Many people have paid, and will continue to pay, a high personal price for the advances in technology that gave us lead pipes, lead-based paint and leaded gasoline.

How was a reduction of lead use brought about? Was it free-market forces? Was lead removed from gasoline because people were lured into unleaded gasoline by lower prices? Did auto makers design engines to use unleaded gas because market forces drove them to do it? Was it the cheaper price of unleaded paint, brought about by the market, that caused people to switch to unleaded paint? No; none of these. It was strictly government regulation.

For a period, both leaded and unleaded gasoline were available at gas stations. Unleaded gas cost more. Unfettered market forces would have driven unleaded gasoline out of the market. However, simple technological means were introduced (different-sized nozzles) to prevent people whose car engines used unleaded gas from pumping the cheaper, leaded gas into it. (All new cars sold in the U.S. after a certain date were required to use just unleaded gasoline by U.S. federal government regulation.) People did not have a choice. The lead content of gasoline and the design of engines to use lead-free gas was regulated. If reliance on the market had prevailed, the people of the U.S. would still be inhaling exhaust fumes laced with lead, with the resulting damage to the health and lives of millions of people. Regulation (this time by states) is also the reason why lead-based paint is no longer available and strict lead-paint disclosure rules apply to the sale and rental of homes. Finally, regulations imposed by local housing codes have done in lead pipes.

The claim that the control of technology can be accomplished through the mechanism of the market fails to be verified empirically in case after case. I can see only two explanations why such a claim might still be maintained by some: a) the claim is based on ideology, as an article of unquestioning faith; or b) those who make and promote this claim have a vested interest in maintaining things as they are. In view of the overwhelming damage to human life, would a term like genocide be too harsh to describe the continuing reliance on the market to control technology?

One approach for controlling technology has not gotten the attention it deserves. Very often those who are most intimately acquainted with the properties of technological devices and systems, and their potential effects, are the engineers who design and build them. Mechanisms are needed in the education of engineers as professionals to "vaccinate" them with the ideal of service, and to protect them from reprisal when they act to control technology in the service of the public.

DIGGING DEEP

DAVID E. NYE, 1995

In the past the Grand Canyon invited reflection on human insignificance, but today much of the public sees it through a cultural lens shaped by advanced technology. The characteristic questions about the canyon reported by Park Service employees assume that humans dug the canyon or that they could improve it so that it might be viewed quickly and easily. Rangers reported repeated queries for directions to the road, the elevator, the train, the bus, or the trolley to the bottom. Other visitors request that the canyon be lighted at night. Many assume that the canyon was produced either by one of the New Deal dam-building programs or by the Indians—"What tools did they use?" is a common question.

Downloading

Gregg Easterbrook, 1995

Once musical sound, electrocardiograms and other time-dependent analog processes came to be digitized for computer processing, it was only a matter of time before electronic dreamers envisioned storing consciousness itself.

Suppose as biological life draws toward its inevitable conclusion a person's patterns of consciousness could be transferred to an electronic support apparatus. The part that matters about you might then exist a very long time, possibly an infinite time. You might possess full spiritual and mental awareness but no significant physical being—exactly the sort of existence religions have long supposed women and men should progress to after biological death. And the physical structures necessary to sustain consciousness

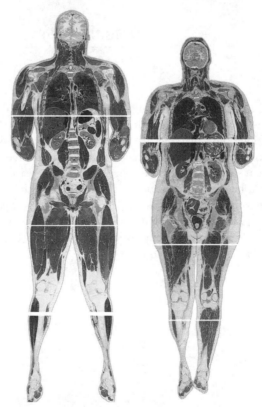

▶ Virtual Adam and Eve: male and female donors, frozen, sliced and scanned, made detailed human anatomy in three dimensions available on the Internet.

might be surprisingly small, an important ecological consideration given that with the passage of time, fantastic numbers of minds might end up requiring support. Those minds would not just be on hold but still be alive—continuing to learn, to have new emotions and new experiences, form new friendships and so on, only on a mental rather than physical basis.

Of course people might not want to have their consciousness go on after the body expires. Incorporeal mental existence might turn out to be boring or worse. Life as pure consciousness might have psychological consequences that would make the whole idea a bad dream. Weird paradoxes might result: for instances if whatever gizmo reads the pattern of your consciousness in order to preserve it made two copies, would you perceive yourself as alive in two places?

. . . Yes, once we have defined pure mental patterns as living consciousness, this means that there may someday be something approximately like electronic life. Unless there is an ineffable soul and hence a door to self-awareness that only a divinity may open, the development of forms of life that have no biological origin seems close to inevitable.

INFLUENCE

Hans Bethe, 1995

On the occasion of the fiftieth anniversary of the atomic bombing of Hiroshima, a Nobel laureate physicist whose gifts helped inaugurate the Nuclear Age publicly called for a strike, resolving an issue—whether a scientist should work on weapons of mass destruction—he had debated within himself for fifty years.

As the director of the Theoretical Division of Los Alamos, I participated at the most senior level in the World War II project that produced the first atomic weapons.

Now, at age 88, I am one of the few remaining such senior persons alive. Looking back at the half century since that time, I feel the most intense relief that these weapons have not been used since World War II, mixed with the horror that tens of thousands of such weapons have been built since that time—one hundred times more than any of us at Los Alamos could ever have imagined.

Today we are rightly in an era of disarmament and dismantlement of nuclear weapons. But in some countries nuclear weapons development continues. Whether and when the various nations of the world can agree to stop this is uncertain. But individual scientists can still influence this process by withholding their skills.

Accordingly, I call on all scientists in all countries to cease and desist

from work creating, developing, improving and manufacturing further nuclear weapons—and, for that matter, other weapons of potential mass destruction such as chemical and biological weapons.

137 MILLION LIVES

KEVIN M. WHITE AND SAMUEL H.
PRESTON, 1996

*Two American demographers assess the impact of public health on the twentieth
century by estimating how many Americans are alive because of its protections.*

Mortality reduction throughout the world has been more rapid in the
twentieth century than in any previous period. The expansion in
longevity ranks among the great social achievements of our time. Life expectancy at birth in the United States has increased from 47.3 years in 1900
to 75.7 years in 1994. . . . Life also has a certainty that did not exist before.
Fully 95 percent of women can expect to live from birth to age 50 under
current health conditions, while in 1900 fewer than 60 percent reached that
age. . . . Identification of the relative contribution of various factors to this
advance is controversial. . . . Among the causes are a better understanding
of disease etiology and the translation of this understanding into preventive
health practices; improvements in nutrition, housing and general living
standards; and the deployment of new therapeutic interventions. . . .

Most individuals recognize that health improvements have enhanced
their own survival prospects. But they may not be aware that their existence
itself is often a result of these improvements. To the extent that they are un-
witting beneficiaries, the advantages of gains in health will be underesti-
mated. In this article, we estimate the number of Americans currently alive
who literally owe their lives to health progress. This calculation involves
comparing the actual population to a hypothetical population in which
mortality improvements did not occur. In particular, we ask how many
Americans are alive today by virtue of twentieth-century mortality reductions and identify some of their characteristics. In addition to the conventional features of age and sex, we estimate the number of persons who would
have been born and then died; the number whose parents would have been
born but died before having them; the number whose grandparents would
have been born but died before bearing their parents; and so on. . . .

From 1900 to 1930, the largest gains occurred in infancy and early
childhood. Major improvements up to age 45 were achieved through 1960,
by which time young adult mortality had fallen to around 0.2 percent a year,

RICHARD RHODES

360

a level too low to leave much room for further gains. From 1930 on, women began experiencing large mortality improvements above age 50, while male improvements were less pronounced. Since 1960, improvements in survival have accelerated for men at ages 50 and above, and the largest gains have occurred for individuals of both sexes at ages above 75. . . .

If mortality had remained at 1900 levels throughout the century, holding everything else constant, the population in the year 2000 would be almost exactly half its actual size: 139 million people instead of 276 million. Half of Americans today can attribute their being alive to mortality improvements in the twentieth century: 51 percent of females and 49 percent of males. . . .

The addition to the population of 2000 produced by mortality decline during the century is approximately double that produced by twentieth-century immigrants to the United States and their descendants. . . .

Half (49.9 percent) of all the hypothetically dead in the year 2000 [i.e., those who are alive because of mortality improvements] represent direct deaths, those who would have been born but would subsequently have died without mortality improvements. . . . But most of the additional people below age 30 would never have been born. They are the indirect beneficiaries of mortality reductions among their mothers, grandmothers and great-grandmothers. . . .

Women are a majority of the population because of mortality declines. . . .

Over 68 million people alive in 2000, or one-quarter of the population, would have been born but died without the mortality improvements since 1900. Another quarter of the population would not have been born because their parents, grandparents or earlier ancestors would have died before giving birth to them or to their progenitors. Half of the population owes its existence to twentieth-century mortality improvements. Exactly who is in which category is impossible to say, since these population averages cannot be assigned to individuals. Because this 50 percent estimate shows little variation by age or sex, any individual could sensibly estimate whether he or she is alive due to survival improvements by merely flipping a coin. Fortunately, the decline in mortality in this century has made this experiment purely hypothetical.

DEEP BLUES

FRANK RICH, 1997

On May 11, 1997, an IBM computer named Deep Blue made history by winning a six-game exhibition match against the reigning world chess champion, Garry Kasparov. Kasparov's loss to a computer generated a range of comment. New York Times cultural critic Frank Rich thought the computer won.

In the end it didn't make any difference whether Deep Blue beat Garry Kasparov at chess or not, because the press was going to send the same message no matter who won. Had Mr. Kasparov triumphed, the headlines would have read: "Man Beats Machine!" When he lost, commentators worked hard to recast defeat as a moral victory. Deep Blue is not human or intelligent, we were reassured, but just a big, fat, soulless number cruncher. Deep Blue can't write a sonnet. Deep Blue can't hug a baby. Deep Blue has no sense of humor. (As if Garry Kasparov were George Carlin.)

All true, and all beside the point. The desperate propping up of humanity that followed IBM's stock-stoking publicity stunt is a backhanded testament to just how much ground man is losing to computer. Print and TV, directly threatened by the Internet, are in a particular panic and especially defensive. "Computers do what we make them do, period," wrote David Gelernter in *Time*, summing up the circle-the-wagons defense of mankind. Yes, but the real story is that we increasingly do what computers make *us* do. Just think of the friends you once regularly saw or talked to on the phone but now encounter only in the virtual relationship of email.

Such computer-driven changes in our culture, good and bad, big and small, are happening faster than we can calculate them, brought about not by 1.4-ton sideshow freaks like Deep Blue but by the ubiquitous PC's at home, work and school. The great engine of our economy (and the bull market on which so many Americans or their pension funds have staked their futures) is the computer; Intel, *The Economist* reports, should surpass General Electric as the US's most profitable business by 2000. The Internet will also be a major purveyor of entertainment and journalism, needing only two inevitable improvements (faster connections, TV-quality video) to re-make the entire information industry, from network news to movie rentals.

David Shenk, whose new book *Data Smog* is an indispensable guide to the big picture of technology's cultural impact, says that the press overcovered Deep Blue, "a story that doesn't affect mankind," because it "can be encapsulated into a man vs. machine drama." The subtler story of the computer's real effects on life, not chess competition, is "one of a slow creep, with the change being imperceptible from day to day."

Data Smog reports on some of the more negative changes, starting with a computer's ability to isolate its user, and increase stress by glutting the brain with data and images. Mr. Shenk also questions—as our Geeks in Chief, Al Gore and Newt Gingrich, never do—the fallout of wiring every classroom. Though the computer can serve as a great library, it is hardly a foolproof instructional tool. Used improperly, it will decrease kids' attention spans and pump them full of information rather than teach essential skills.

Old media often miss the computer's cultural impact because we see stories in, well, linear terms, as morality plays with clear plots. As Deep Blue was overcovered, so are the size of Bill Gates's house and fortune, the spread of Web pornography (and the doomed legal efforts to censor it) and the technological travails of America Online (another man-bests-computer fable).

Among the important non-linear stories being missed is the coming impact of the computer on politics—a story that has nothing to do with politicians' much-documented Web sites. As Mr. Shenk points out, cyberspace (and its corporate culture) is intrinsically libertarian Republican, a model of "highly decentralized, deregulated society." This is bad news for big-government Democrats, but not necessarily a bonanza of a GOP keen to restrict immigration (a huge source of digital-industry talent) and regulate morality, on the Web and beyond. Nor is it clear who on the digital political landscape will help the swelling underclass of information have-nots.

We can address such questions faster if we recognize that the battle between man and computer is over, and the computer has won. Our task now is not to deny that victory—or stomp off childishly in defeat like Mr. Kasparov—but to seize the countless creative opportunities before us to bend the machine's power to our own ends as we negotiate the most humane truce possible.

BEYOND SOCIAL CONSTRUCTION

FREDRICO CAPASSO, 1997

Science and sociology found themselves at intellectual war in the last decades of the century as sociologists explored the extent to which science is a social construct—an exploration which many scientists took to imply that science had no more claim to authority as a description of reality than religions do. Defending science in a letter to Nature, *a Lucent Technologies scientist invoked its application to technology as evidence of its firm grip on reality.*

Science constrains possible technology. When the latter (a new device, instrument or system) becomes reality, it validates the specific theories on which it is founded. If scientific knowledge was not an objective description of reality, but merely a social construct, we would not have technological realities such as transistors, microprocessors, personal computers and the Internet, or semiconductor lasers, fiber optics and compact disc players in addition to radios, cars and jets.

It is instructive to examine a few of the thousands of "food chains" connecting scientific knowledge (the product of hypotheses and experiments converging into a theory) to technology.

Consider lasers and in particular semiconductor lasers, an essential element of long-distance fiber-optic communications. Without the concepts of energy levels and energy bands, created by quantum mechanics, we would not have these lasers, modern telecommunications, CD players, CD-ROMs and much more. Similarly, integrated circuits, microprocessors and personal computers (used by sociologists in writing papers on science as a social construct!) would not have become a reality without quantum mechanics and its quantitative description of chemical bonds and energy bands in silicon and of the statistical behavior of electrons in doped semiconductors (Fermi levels and Fermi statistics), transistors and the subsequent elements of the "food chain." Of course much more (electronics, chemistry, optics and so on) than quantum mechanics, a necessary but not sufficient element, was needed for all of this to happen.

Another recent and wonderful but less known example is the Global Positioning System (GPS). This system, based on 24 orbiting satellites, allows anyone equipped with a suitable portable receiver to find almost instantaneously his/her position anywhere on the Earth with a precision of roughly 100 feet. Its applications also include navigation systems (from ships to jet aircraft and cars), surveys of the Earth and earthquake monitoring, synchronization of telecommunication networks, tracking of satellites and so on. This technology is based on atomic clocks, accurate to within one second in 100,000 years, which would not have been possible without the understanding of hyperfine transitions in caesium and rubidium atoms, made possible by quantum mechanics and atomic theory. In addition, general relativity and its effects on the timing of these clocks must be taken into account in the design of the GPS!

. . . In summary, the reality of successful technologies helps to validate science as an objective system of knowledge, with a firm grip on reality, as much as reproducible experiments and predictions do. In addition, through a fascinating feedback loop, technologies often open up new frontiers of scientific knowledge.

Autism on the Net

Harvey Blume, 1997

Despite the diverse accounts of autism in books like Oliver Sacks's *An Anthropologist on Mars* . . . the prevailing image of the autistic today is probably still that of the rocking child, prone to tantrums and averse to touch, or of an adult like the character Dustin Hoffman played in the movie *Rain Man* who can instantly multiply large numbers in his head, but cannot connect to other people or take care of himself.

Yet anyone who explores the subject on the Internet quickly discovers an altogether different side of autism. In cyberspace, many of the nation's autistics are doing the very thing the syndrome supposedly deters them from doing—communicating—often in celebration of the medium that enables them to do so.

"Long live the Internet," one autistic recently exulted in an on-line discussion, where "people can see the real me, not just how I interact superficially with other people."

Another explained why she prefers on-line to face-to-face interaction: "Ordinarily," she wrote to other members of her electronic mail forum, "the giving of support involves being with someone, and that's always draining for me. If someone does give me support in person, I will have to spend some time recovering from the experience of receiving that support."

Both writers subscribe to Independent Living, a suite of email forums created by and almost exclusively for autistics. Topics addressed by Independent Living . . . include jobs, hobbies, "sexuality and being different," and the recurrent question of how to relate to what the autistics refer to as neurologically typical people—or "NT's" in the community's parlance.

In a sense, autistics are constituting themselves as a new immigrant group on line, sailing to strange neurological shores on the Internet, and exchanging information about how to behave on arrival. They want to be able to blend in, to pass, and are intently studying the ways of the natives in order to do so. . . .

The impact of the Internet on autistics may one day be compared in magnitude to the spread of sign language among the deaf. By filtering out the sensory overload that impedes communication among autistics, the Internet opens vast new opportunities for exchange.

▶ New weapons evolve in a Darwinian struggle between offense and defense: the F-117 Stealth fighter-bomber, configured to minimize its radar signature.

BEAUTY AND TRUTH

Louis Brown, 1997

In his magisterial history of radar, Technical and Military Imperatives, *Carnegie Institution of Washington physicist Louis Brown proposes a surprising distinction between science and engineering.*

Within most physicists there lies an engineer eager to design something. The two branches of learning have always had a thin boundary separating them, one frequently crossed or dismissed. In principle the distinction is simple enough: physicists discover nature's laws, engineers synthesize these laws into apparatus for good or evil. But in the day to day performance of their trades the two are at times indistinguishable. An experimentalist designing equipment with which he plans to measure some atomic quantity looks for all the world like an engineer designing some part of an engine or radio; a theorist striving to describe a nuclear scattering process with quantum mechanics looks like an engineer analyzing the stress patterns of a bridge design. Names often confuse more than enlighten. Designers of lasers manipulate atomic properties so as to produce some form of radiation, operations no different from an electrical engineer designing yet another form of high-frequency oscillator, but the former seems invariably referred to as a laser scientist whereas the latter is an engineer.

Beauty is a common attribute for which both strive. There is never a more satisfying achievement for a physicist than the reduction of some complicated phenomenon to description by a simple, closed mathematical statement. It is one of the misfortunes of our civilization that the beauty of an equation cannot be shared with so many who specialize in the appreciation of beauty. It is in beauty that engineer and scientist differ: for the scientist it is nature's beauty, for the engineer it is the beauty of creation. Engineering is a form of art and has filled the world with things of obvious visual beauty but also with subtle forms, much like a theorist's equations. A poet may enjoy the grace of the Brooklyn Bridge but will never appreciate the charm of a well-designed electronic circuit, but the beauty is there for all that, and an electronics man will recognize grace, symmetry and style in a design without ever having had a course in electronic art appreciation.

OMEGA POINT

JARON LANIER, 1997

The computer scientist who coined the term "virtual reality" ponders the apocalypse.

I n the computer-science community, there's a perspective, which is difficult to communicate to the outside world, that things are going to continue to change in our field at such a rapid rate that at some point something very dramatic will change about the fundamental situation of people in the universe. I don't know if I share that belief, but it's a widespread belief. In the mythology of computer science, the limits for the speed and capacity of computers are so distant that they effectively don't exist. And it is believed that as we hurtle toward more and more powerful computers, eventually there'll be some sort of very dramatic Omega Point at which everything changes—not just in terms of our technology but in terms of our basic nature. This is something you run across again and again in the fantasy writings of computer scientists: this notion that we're about to zoom into a transformative moment of progress that we cannot even comprehend.

> Anita Thacher was at a Shelter Island yard sale recently when she noticed a boy, about ten years old, standing in front of, and scrutinizing, an old manual typewriter.
> Suddenly he shouted out excitedly, "Dad, look at this!" and when his father walked over to see what merited such enthusiasm, added, "Dad, look, it has the exact same keyboard as our computer."
> *New York Times* METROPOLITAN DIARY

THE END OF THE BEGINNING

FREEMAN DYSON, 1997

In the summer of 1995 I took part in a technical study of the future of the United States nuclear [weapons] stockpile. The study was done by a group of academic scientists together with a group of professional bomb designers from the weapons laboratories. The purpose of the study was to answer a question. Would it be technically feasible to maintain forever a stockpile of reliable nuclear weapons of existing designs without further nuclear tests? The study did not address the underlying political questions, whether reliable nuclear weapons would always be needed and whether further nuclear tests would always be undesirable. Each of us had private opinions about the political questions, but politics was not the business of our study. We assumed as the ground rule for the study that the weapons in the permanent stockpile must be repaired and remanufactured without change in design as the components deteriorate and decay. We assumed that the new components would differ from the old ones when replacements were made, because the factories making the old components would no longer exist. We looked in detail at each type of weapon and checked that its functioning was sufficiently robust so that minor changes in the components would not cause it to fail. We concluded our study with a unanimous report, saying that a permanently reliable nuclear stockpile without nuclear testing is feasible. . . .

The conclusion of our study was a historical landmark, commemorating the fact that the nuclear arms race is finally over. The nuclear arms race raged with full fury for only twenty years, the 1940s and 1950s. Then it petered out slowly for the next thirty years, in three stages. The science race petered out in the 1960s, after the development of highly efficient hydrogen bombs. Nuclear weapons then ceased to be a scientific challenge. The military race petered out in the 1970s, after the development of reliable and invulnerable missiles and submarines. Nuclear weapons then ceased to give a military advantage to their owners in real-world conflicts. The political race petered out in the 1980s, after it became clear to all concerned that massive nuclear weapons industries were environmentally and economically disastrous. The size of the nuclear stockpile then ceased to be a political status symbol. Arms control treaties were concluded at each stage, to ratify with legal solemnity the gradual petering out of the race. The atmospheric test ban ratified the end of the science race, the ABM and SALT treaties ratified the end of the military race, and the START treaties ratified the end of the political race.

R.U.R. REVISITED

MARVIN MINSKY, 1997

Journalist David Remnick elicited this prediction from the artificial-intelligence pioneer.

Look, the world is a rather dumb place. There's nothing special about it. It's accidental. The world was *terrible* before people came along and changed it. So we don't have much to lose by technology. The future of technology is about shifting to what people like to do, and that's entertainment. Eventually, robots will make everything. The trend is over time. When Henry Ford was around, a large percentage of the population was involved in manufacturing. Now it's much smaller. I'm telling you: all the money and the energy in this country will eventually be devoted to doing things with your mind and your time.

STUFF

IVAN AMATO, 1997

Robots may eventually make everything, as Marvin Minsky concludes; but there will continue to be vast amounts of things to make, this chronicler of "stuff" observes.

Human beings extract about 15 billion tons of raw material—that's 30 trillion pounds—from the earth each year, and from that they make every kind of stuff that you can find in every kind of thing. Mined ore becomes metal becomes wire becomes part of a motor becomes a cooling fan in a computer. Harvested wood becomes lumber becomes a home. Drilled petroleum becomes chemical feedstock becomes synthetic rubber becomes automobile tires. Natural gas becomes polyethylene becomes milk jugs and oversize, multicolored yard toys. Mined silica sand becomes silicon crystal becomes the base of microelectronic chips. Each kind of stuff is a link to enormous industrial trains whose workers process the world's raw materials into usable forms that constitute the items of our constructed landscape.

Each kind of stuff is also a palimpsest of innovations in the use of materials, some going back to prehistoric times. The wood-pulp paper from which books are made today comes from a pedigree of cotton and linen rags, animal-skin parchment, Nile-reed papyrus and Sumerian clay tablets. The ink, a black pigment made from the ground waste of some carbon-bearing fuel and then suspended in a rapidly evaporating solvent, has its roots in crushed ore and charcoal mixed with spit or animal grease for use on cave walls and faces. The materials in every book tell a tale that rivals the one conveyed in the words.

It is the same for every other material thing that you encounter. Train your attention on the stuff of things rather than on their function. What you see is a rich medley of materials: the liquid crystal display of your laptop computer; the gritty concrete sidewalk on which you are strolling; the nylon of your raincoat's zipper; the carbon-fiber-reinforced epoxy polymer of your tennis racket; the Kevlar polymer in your police force's bulletproof vests; the oak of your dresser; the diamond in your engagement ring; the nickel-based superalloy in the turbine blades in the engines of an airliner you are flying in; the warm, supple skin of your newborn; the cool, transparent glass of your office window; the combination of slick, high-density polyethylene and stainless steel that makes up the artificial hip which a surgeon may have implanted into you; the cotton of the shirt you are wearing; the aluminum of the can you just drank from. . . .

That books and buildings and the things in the world are supposed to be made of materials suited for their functions is so obvious that it almost goes without saying. That is why the materials that make up the world are most often not on people's minds. . . .

Simply noticing the diversity of stuff in the material world can evoke an openmouthed sense of wonder akin to visiting a zoo filled with animals that you had never really seen up close. That sense of wonder might even grow into awe when you consider this: the constructed world brims with metals, ceramics, plastics, fabrics, glass and thousands of other specific materials that are found nowhere in the world.

SOCIETY EVOLVING

BRAN FERREN, 1997

The chief Disney Imagineer assesses the Internet.

Trying to assess the true importance and function of the Net now is like asking the Wright brothers at Kitty Hawk if they were aware of the potential of American Airlines Advantage Miles. We're always very bad at predicting how a given technology will be used and for what reasons. Originally, cable television was just meant to improve your picture reception. Theater people used to think that the idea of motion pictures was ridiculous. But the Net, I guarantee you, really is fire. I think it's more im-

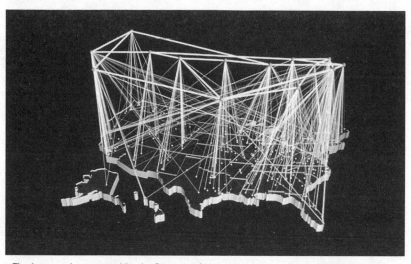

▸ The Internet, inaugurated by the Pentagon for communication during nuclear war, expanded into a vast interconnected library of the world.

portant than the invention of movable type. But you have to wait and see. Society evolves like the species. It's not smooth and linear. You'll have something like the industrial revolution—it comes like a jolt, and then you kind of dick around for the next fifty years getting used to it.

PREDICTIONS: ROBOTIC SEX

JOEL SNELL, 1997

Robots that provide sexual companionship are likely to become common in the future. Reportedly, prototype models have already been developed in Japan. The future "sexbots" will have humanlike features and will be soft and pliant. Sexbots will contain vibrators to provide tactile stimulation and sound systems to provide love talk. They could certainly alter human relations in any or all of the following ways:

• Marriages may be destroyed by sexbots when husbands choose sex with the sexbots rather than with their wives. Jealous wives may destroy sexbot rivals and sue the manufacturer.

• Heterosexual people may use same-sex sexbots to experiment with homosexual relations. Or gay people might use other-sex sexbots to experiment with heterosexuality.

- Robotic sex may become addictive. Sexbots would always be available and never say no. People may become obsessed by their ever faithful, ever pleasing sexbot lovers and rearrange their lives to accommodate their addictions. Eventually, support groups will likely form.
- Technovirgins will emerge. An entire generation of humans may grow up never having had sex with other humans.
- Robotic sex may become "better" than human sex. Like many other technologies that have replaced human endeavors, robots may surpass human technique; because they would be programmable, sexbots would meet each individual user's needs.

Will electronic and robotic sex reduce teen pregnancy, sexually transmitted diseases, abortions, pedophilia and prostitution? The jury is still out on these implications. However, boundaries, barriers and beliefs will be challenged.

THE FUTURE AS A STORY

DAVID REMNICK, 1997

The future is itself a story, and predictions are stories we tell to amaze ourselves, to give hope to the desperate, to jolt the complacent. The trouble comes when the storyteller tries to predict the future, as [Walt] Disney sometimes did, in a voice of certainty. H. G. Wells, for one, was counted as a visionary, able to foresee skyscrapers and superfast trains, yet he was also convinced of the coming of a socialist World State. When Wells laid out his vision in *A Modern Utopia*, G. K. Chesterton remarked that unfortunately the creators of visionary futures "first assume that no man will want more than his share, and then are very ingenious in explaining whether his share will be delivered by motor-car or balloon."

Speculative storytelling, from Thomas More's *Utopia* to William Gibson's *Virtual Light*, is always about the present: what confuses us, what we desire, what we fear. In *Paris in the Twentieth Century*, a novel abandoned in a safe and published only a couple of years ago, Jules Verne imagined easily enough a world of electric elevators and modern facsimile machines, but he also predicted a culture whose bookstores would have room only for scientific tracts and not the classics of French literature. Verne, of course, was worried about the plight of Verne.

Disney's attempts to tell the story of an American future is part of a larger tradition in both high and mass culture. Of all the American stories set in the future, probably the most influential was Edward Bellamy's *Look-*

ing Backward, published in 1888. Bellamy was born in 1850, in the mill town of Chicopee Falls, Massachusetts. The town was a typical product of the Gilded Age—crowded tenements, illnesses, strikes—and Bellamy, who in growing up absorbed an ethic of self-denial and charity, quickly became a crusading journalist. In the late eighteen-seventies, he also turned to fiction and, as a kind of antidote to his times, wrote *Looking Backward*. Bellamy's surrogate in the novel is young Julian West, who goes to sleep in Boston in "the present" and magically wakes up in the year 2000. As he wanders the city, he discovers that the corruptions of his time have disappeared. Technology has come along to meet everyone's needs: there are shopping malls, credit cards, and electric lighting, though no Internet. Most important of all—at least to Bellamy—is that a religion of solidarity has driven out the ills of the Gilded Age. People are no longer selfish. Bellamy, like the Puritans before him, restated the dream of an American golden age, a new Jerusalem, and his book not only sold tens of thousands of copies a week but also spawned more than a hundred Bellamy clubs across the country.

The problem was, as Alfred Kazin suggests in *On Native Grounds*, that Bellamy, with his industrial conscript army and collectivism, had also managed to propose a kind of "beneficent Socialist totalitarianism." A lord of high capitalism answered in 1894, when John Jacob Astor IV, the dilettante scion of the Astor fortune, published *A Journey in Other Worlds*, another utopia set in the year 2000. Like Bellamy and other futurists of the time, Astor took for granted an acceleration of technological progress to burnish his utopia. The Terrestrial Axis Straightening Company, for example, tries to pump water from one pole to the other so that the earth's axis will stay perpendicular to its rotation, thus creating eternal spring everywhere. But, while Bellamy was surrounded by misery and corruption and sought to correct it, Astor was swaddled in privilege and sought to protect it. In his novel, English is fast becoming the language of the entire world, the poorer, duskier peoples retreat, and Anglo-Saxons take over from one end of the earth to the other. "If much of Astor's literary and scientific work had a certain Buck Rogers flavor, reflecting dreams and unearthly speculations, it was because he himself lived in a world as unreal as any he ever imagined," Harvey O'Connor, a biographer of the Astor dynasty, wrote.

The utopian novel faded out decades ago, with products like B. F. Skinner's *Walden Two* and Ayn Rand's dreams of a modern Atlantis, but in an atmosphere quickened by the millennium (and all its marketing possibilities) a certain kind of nonfiction fantasy has proliferated. If you fly fairly often and sneak a look at what the passengers have brought along to pass the time, it's striking that anyone who is not reading Grisham or Crichton is immersed in a kind of self-help book that promises the reader to keep him from drowning come the new age: *The 500 Year Delta. The Third Wave. Vi-*

sions: How Science Will Revolutionize the 21st Century. Being Digital. What Will Be. The Road Ahead.

These are books read more out of anxiety than out of pleasure. Part of that anxiety, no doubt, is the approach of the millennium—a thrill both dramatic and meaningless, rather like that instant so long ago when the family watched the dashboard as the Chevy's odometer finally turned from 9999 to 10000. (Those zeros!) But the anxiety also stems from something more meaningful—an undeniable sense that, as all those books insist, we are entering a technological era in which change itself has accelerated beyond all precedent in human history. Alvin Toffler's books may be loud with the bark of the future-hustler, but he was correct about that: change has changed—it has speeded up, and leads to a sickening anxiety (think of the hapless passenger stuck in the back of a taxi driven by an unblinking maniac). To fall behind is to risk humiliation, loss of status, poverty.

The hemline of the moment is technological bonhomie. A recent article by Peter Schwartz and Peter Leyden in *Wired* entitled "The Long Boom," which claims that we are just reaching the midpoint of forty years of millennial peace, prosperity and interconnectedness, has had enormous resonance in futurist circles. Occasionally, there appears an alarmingly dark vision—science-fiction writers warning of a techno dystopia, various environmentalists forecasting scenarios of apocalypse, writers like Sven Birkerts and Alvin Kernan mourning the burial of print culture under the great blizzard of digital bytes and streams—but in the main the chorus of giddiness overwhelms them.

Optimism is a sound economic decision. Optimism brings in the corporate consultancies, the two-million-dollar book advances, the forty-thousand-dollar lecture fees. Optimism sells. (This is a principle that Disney understands better than anyone.) A corporate audience wants to know that it can master the future, and it wants to know how. Watts Wacker, for one, had been on the road giving speeches to the Pacific Coast Gas Association and a convention of confectioners when I reached him. From talking to Wacker and reading his book *The 500 Year Delta*, I got the impression that he had no idea what he was talking about, but, all the same, he was having a wonderful time pitching happy days. "I think things are pretty cool," he said cheerfully. "I think it was Emerson who said, 'This time like all times is a great time as long as we know what the fuck to do with it.' Emerson's my hero, by the way, though he wasn't much of a poet." Wacker laughed. And then he said, "I'm a futurist, and I make all these things up, of course. The only thing I know is that an optimist is going to have a good future and a pessimist is going to have a bad one."

Wacker is no scientist, you will be surprised to learn, but he told me with utter confidence that babies born in the year 2000 will probably live to be a hundred and fifty, and babies born five hundred years from now will have a chance to live to be eight hundred. This notion of eternal life, or close to it, is not entirely alien to futurist circles. An electrical-engineering professor at MIT, Gerald Sussman, has said, "I don't think the time is quite right, but it's close. I'm afraid, unfortunately, that I'm in the last generation to die." All this brings to mind Woody Allen's remark "I don't want to achieve immortality through my words. . . . I want to achieve it through not dying."

SUREFIRE PREDICTIONS

JULIAN L. SIMON, 1995

A confidently optimistic economist who was the chief nemesis of doomsayers in and out of the environmental movement predicts continuing improvement in the human condition.

These are my most important long-run predictions, contingent on there being no global war or political upheaval: (1) People will live longer lives than now; fewer will die young. (2) Families all over the world will have higher incomes and better standards of living than now. (3) The costs of natural resources will be lower than at present. (4) Agricultural land will continue to become less and less important as an economic asset, relative to the total value of all other economic assets. These four predictions are quite certain because the very same predictions, made at all earlier times in history, would have turned out to be right. . . .

Almost as certain is that (5) the environment will be healthier than now—that is, the air and water people consume will be cleaner—because as nations continue to get richer, they will increasingly buy more cleanliness as one of the good things that wealth can purchase. People will probably continue to be worried about pollution nevertheless, both because new sorts of pollutions will occur as new kinds of economic activities develop, and because ability to detect pollutions increases. But the danger of the pollutions that catch our attention will diminish, because we address the worst pollutions first and leave the lesser ones for later, and because our capacity to foresee newly created pollutions in advance will increase. And (6) not only will accidents such as fires continue to diminish in number, but losses to natural disasters such as hurricanes and earthquakes will get smaller, as our buildings become stronger and our methods of mitigating disasters improve.

Perhaps the easiest and surest prediction is that (7) nuclear power from

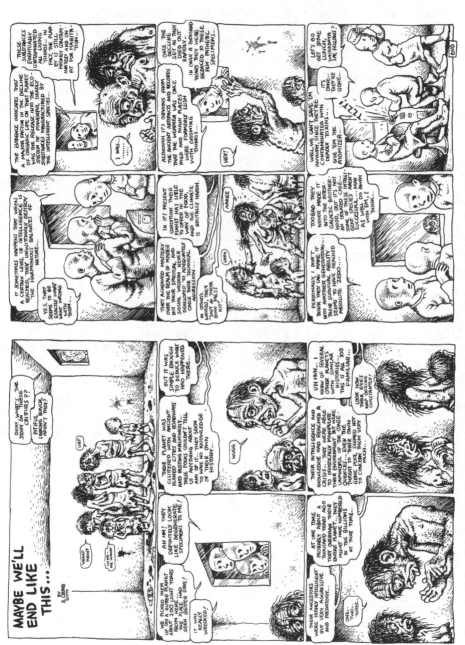

▲ Alternative Futures I: Robert Crumb's 1979 vision of humanity degenerated by environmental apocalypse.

fission will account for a growing proportion of our electricity supply, and probably our total energy supply as well, until it is displaced by some other cheaper source of energy (perhaps fusion). (8) Nuclear power will never be displaced by solar energy using the kinds of technology that are currently available, or by any ordinary development of those kinds of technology.

Can one really make almost surefire long-run predictions? Check for yourself: although the stock market gyrates from day to day and week to week, its course from decade to decade has almost always been upward. The story is the same in reverse with natural resource prices. Copper, iron, wheat, rice, sugar, and every other natural resource have fallen in price, and therefore risen in availability, throughout the two hundred years of US history, and over the thousands of years of human history wherever records exist. Indeed, the history of civilization is a history of increased knowledge to produce goods more efficiently and cheaply. This goes hand in hand with liberty becoming more widespread, and with increased mobility and communication. All this progress is reflected in the long-run trends of human welfare. . . .

Some wonder: How can we be sure that scientific and technical progress will continue indefinitely? Related to this question is another: What will be the rate of economic growth in the future? . . . No sensible answer about future rates of advance is possible in principle, in considerable part because of the inherent impossibility of comparative measurement of the value of progress in science from one period to the next.

But uncertainty about the rate of advance in technology does not imply uncertainty about the direction of the future of humankind. Our future material welfare is already assured by our knowledge of how to obtain energy from nuclear fission in unlimited quantities at constant or declining cost, even if no other source of energy is discovered until fissionable material runs out at some almost infinitely distant time. And energy is the only strong constraint on the supplies of all other raw materials.

Assurance about our raw-material future does not imply absence of need for more technology in the short run or the long run, however. There are, and always will be, endless ways to improve human life, and plenty of pressing problems to challenge us. But we can confidently face the future without worrying about threats to the end of civilization from "overconsumption" and raw-material shortages that technology is unable to deal with. . . .

People alive now are living in the midst of what may well be the most extraordinary three or four centuries in human history. The "industrial revolution" and its technical aftermath—even including the spectacular rise in living standards for most human beings from near subsistence to the level of today's modern nations—is only part of the upheaval. The process has already been completed for perhaps a third of humanity . . . and within a century or two (unless there is a holocaust) the rest of humanity is almost sure to attain the amenities of modern living standards; the worst holocaust imaginable could only delay the process by a century or two.

The most spectacular development, and by far the most meaningful in both human and economic terms, is the revolution in health that we are witnessing in the second part of the twentieth century. Barring catastrophic surprises in the first half of the twenty-first century, most of humanity will come to share the long healthy life that is now enjoyed by the middle-class contemporary residents of the advanced countries.

The technical developments of the past two centuries certainly depended on earlier discoveries. But the knowledge that emerged before the last two centuries was only infrastructure. Until the most recent generations, most people could not observe the effects or gain the benefits of this progress in their own lives. Now we are reaping the full fruits of that earlier investment.

The spreading of a high level of living will be speeded by another phenomenon that can be predicted for sure: increased migration from poor to rich countries. The lure of a higher standard of living pushes migrants from their native countries and pulls them to richer places. And the felt need in the richer countries for youthful persons in the labor force to balance the ever greater concentrations in the older age cohorts fuels the demand for them. One can see the drama of this process in the youthful medical and custodial staffs in big-city hospitals who came from abroad to tend to the aged natives and the veterans of earlier migrations.

Might there not be even more and faster and more radical change in the future? In human terms, I doubt it. Life expectancy and child mortality cannot fall much faster unless we change genetically. Concorde-like supersonic speed of travel demonstrably does not matter very much (and the trip to and from airports will take a long time to speed up). Communications cannot become much faster, many being at the speed of light now. The distances one can travel—to the planets and beyond—will eventually increase greatly, and this may alter life significantly, though I cannot imagine how.

The only impending shortage is a shortage of economic shortages. . . . As to the non-material aspects of human existence—good luck to us.

ENVOY: QUO VADIS?

EDWARD O. WILSON, 1978

The eminent Harvard entomologist asks and tentatively answers the ultimate question about technology: where do we go when it's done?

N o species, ours included, possesses a purpose beyond the imperatives created by its genetic history. Species may have vast potential for material and mental progress but they lack any immanent purpose or guidance from agents beyond their immediate environment or even an evolutionary goal toward which their molecular architecture automatically steers them. I believe that the human mind is constructed in a way that locks it inside this fundamental constraint and forces it to make choices with a purely biological instrument. If the brain evolved by natural selection, even the capacities to select particular esthetic judgments and religious beliefs must have arisen by the same mechanistic process. They are either direct adaptations to past environments in which the ancestral human populations evolved or at most constructions thrown up secondarily by deeper, less visible activities that were once adaptive in this stricter, biological sense.

The essence of the argument, then, is that the brain exists because it promises the survival and multiplication of the genes that direct its assembly. The human mind is a device for survival and reproduction, and reason is just one of its various techniques. Steven Weinberg has pointed out that physical reality remains so mysterious even to physicists because of the extreme improbability that it was constructed to be understood by the human mind. We can reverse that insight to note with still greater force that the intellect was not constructed to understand atoms or even to understand itself but to promote the survival of human genes. The reflective person knows that his life is in some incomprehensible manner guided through a biological ontogeny, a more or less fixed order of life stages. He senses that with all the drive, wit, love, pride, anger, hope, and anxiety that characterize the species he will in the end be sure only of helping to perpetuate the same cycle. Poets have defined this truth as tragedy. Yeats called it the coming of wisdom:

> Though leaves are many, the root is one;
> Through all the lying days of my youth
> I swayed my leaves and flowers in the sun;
> Now I may wither into the truth.

The . . . dilemma, in a word, is that we have no particular place to go. The species lacks any goal external to its own biological nature. It could be

VISIONS OF TECHNOLOGY

379

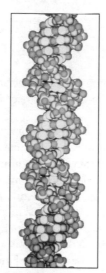

► Alternative Futures II: A computer simulation of a DNA molecule, symbolizing technology applied to improving human life.

that in the next hundred years humankind will thread the needles of technology and politics, solve the energy and materials crises, avert nuclear war, and control reproduction. The world can at least hope for a stable ecosystem and a well-nourished population. But what then?

> For every expert, there is an equal and opposite expert.
> ARTHUR C. CLARKE, 1998

Bibliography

Abbey, Edward, *Desert Solitaire*. Simon & Schuster, 1968.

Adams, Henry, *The Education of Henry Adams*. Modern Library, 1918.

Agee, James, and Walker Evans, *Let Us Now Praise Famous Men*. Houghton Mifflin, 1941, 1969.

Allen, John, *Biosphere 2*. Viking, 1991.

Allport, Floyd H., "This Coming Era of Leisure," *Harper's*, November 1931.

Amato, Ivan, *Stuff: The Materials the World Is Made Of*. Basic Books, 1997.

Asbell, Bernard, *The Pill*. Random House, 1995.

Asimov, Isaac, *I, Robot*. Doubleday, 1950.

Balabanian, Norman, "Controlling Technology," *IEEE Technology and Society Magazine* 13(1), Spring 1994.

Balsley, Gene, "The Hot-Rod Culture," *American Quarterly* 2, 1950.

Beard, Charles A., ed., *Whither Mankind*. Longmans, Green, 1928.

Bell, Daniel, "Technology, Nature and Society: The Vicissitudes of Three World Views and the Confusion of Realms," *The American Scholar* 42(3), Summer 1973.

Berlin, Isaiah, "Virtue and Practicality," in Melvin Kranzberg, ed., *Ethics in an Age of Pervasive Technology*. Westview Press, 1974.

Bloch, Arthur, *Murphy's Law*. Price/Stern/Sloan, 1977.

Blume, Harvey, "Technology on the Net," *New York Times*, 30 June 1997.

Board of Consultants to the Secretary of State's Committee on Atomic Energy, *A Report on the International Control of Atomic Energy*. Department of State Publication 2498, USGPO, 1946.

Boas, George, "In Defense of Machines," *Harper's*, June 1932.

Bogue, Donald J., "The End of the Population Explosion," *The Public Interest*, Spring 1967.

Boorstin, Daniel J., *The Image*. Atheneum, 1962.

381

Boorstin, Daniel J., *The Americans: The Democratic Experience*. Random House, 1973.

Bowden, B. V., ed., *Faster Than Thought: A Symposium on Digital Computing Machines*. Sir Isaac Pitman & Sons, Ltd., 1953.

Bright, James R., *Research, Development and Technological Organization*. Richard D. Irwin, 1964.

Brooks, Harvey, "Technology: Hope or Catastrophe?" *Technology in Society* 1(1), Spring 1979.

Brown, Louis, *Technical and Military Imperatives*. Naval Institute Press, 1998.

Burlingame, Roger, *Engines of Democracy*. Charles Scribner's Sons, 1940.

Burton, Charles Pierce, "The Cost of Progress," *Harper's*, September 1923.

Bush, Vannevar, "As We May Think," *Atlantic Monthly*, July 1945.

————, *Modern Arms and Free Men*. Simon & Schuster, 1949.

————, *Science Is Not Enough*. William Morrow, 1967.

————, *Pieces of the Action*. William Morrow, 1970.

Calder, Nigel, *Technopolis*. Simon & Schuster, 1970.

Capasso, Fredrico, "Correspondence," *Nature* 387, 5 June 1997.

Capek, Karel, *R.U.R.* Doubleday, Page, 1923.

Carey, John, ed., *Eyewitness to History*. Harvard University Press, 1987.

Carleton, William G., "The Century of Technocracy," *Antioch Review* 25(4), Winter 1965–1966.

Caro, Robert, *The Power Broker*. Knopf, 1974.

Carpenter, M. Scott, et al., *We Seven*. Simon & Schuster, 1962.

Carson, Rachel, *Silent Spring*. Houghton Mifflin, 1962.

Clark, Allan H., "Victims of Misplaced Values," *Vital Speeches* 51(17), 15 June 1985.

Clark, Ronald W., *Tizard*. MIT Press, 1965.

Clarke, Arthur C., *Ascent to Orbit: A Scientific Autobiography*. John Wiley & Sons, 1984.

Clarke, Arthur C., "The Righter Stuff," *Nature* 370, 14 July 1994.

Cohen, Joel E., and David Tilman, "Biosphere 2 and Biodiversity—The Lessons So Far," *Science* 274(5290):1150.

Collingridge, David, *The Social Control of Technology*. St. Martin's Press, 1980.

Commoner, Barry, *Science and Survival*. Viking, 1966.

Conant, James Bryant, "A Skeptical Chemist Looks into the Crystal Ball," *Chemical and Engineering News* 29(38), 17 September 1951.

Cooke, Morris Llewellyn, "Some Factors in Municipal Engineering," *Mechanical Engineering* XXXVII, February 1915.

Corn, Joseph J., *Imagining Tomorrow*. MIT Press, 1986.

Cranwell, John Philips, "Air Power and Future History," *Harper's*, December 1940.

Daniels, George H., *Science in American Society*. Knopf, 1971.

Didion, Joan, *The White Album*. Farrar, Straus and Giroux, 1979.

Drucker, Daniel C., "The Engineer and the Establishment," *Bulletin of the Atomic Scientists* 27(10), December 1971.

Dyson, Freeman, *Imagined Worlds*. Harvard University Press, 1997.

Dyson, George B., *Darwin Among the Machines*. Addison-Wesley, 1997.

Easterbrook, Gregg, *A Moment on the Earth*. Viking, 1995.

Eckman, Philip K., ed., *Technology and Social Progress—Synergism or Conflict?* AAS Science and Technology Series, Vol. 18, AAS Publications Office, 1968.

Elliot, Gil, *Twentieth Century Book of the Dead*. Charles Scribner's Sons, 1972.

Ellul, Jacques, *The Technological Society*. Knopf, 1964.

Etzioni, Amitai, "Humane Technology," *Science* 179(4077), 9 March 1973.

Fisher, Allan G. B., "The Clash Between Progress and Security," *Harper's*, July 1944.

Florman, Samuel C., *The Civilized Engineer*. St. Martin's Press, 1987.

Ford, Henry, *My Life and Work*. Doubleday, Page, 1922.

———, *Today and Tomorrow*. Doubleday, Page, 1926.

Forum, "Our Machines, Ourselves," *Harper's*, May 1997.

Frost, Robert, *The Poetry of Robert Frost*. Henry Holt, 1969.

Gilfillan, S. C., *The Sociology of Invention*. MIT Press, 1935.

Goldberg, Steven E., and Charles R. Strain, eds., *Technological Change and the Transformation of America*. Southern Illinois University, 1987.

Goodman, Paul, "The Morality of Scientific Technology," *Dissent*, January–February 1967.

Green, Harold P., "The New Technological Era: A View From the Law," *Bulletin of the Atomic Scientists* 23, November 1967.

Green, Venus, "Goodbye Central: Automation and the Decline of 'Personal Service' in the Bell System, 1878–1921," *Technology and Culture* 36(4), October 1995.

Groves, Leslie R., *Now It Can Be Told*. Harper & Bros., 1962.

Harrison, R. H., "Out of the Clouds" (Letter), *American Heritage of Invention & Technology*, Winter 1994.

Hecht, Jeff, *Laser Pioneer Interviews*. High Tech Publications, 1985.

Helton, Roy, "The Anti-Industrial Revolution," *Harper's*, December 1941.

Hendrick, Burton J., "The Safe and Useful Aeroplane," *Harper's*, April 1917.

Hills, L. Rust, *How to Do Things Right*. Warner, 1972.

Hoover, Herbert, "The Nation and Science," *Science* LXV, 14 January 1927.

———, *Memoirs*. Macmillan, 1951.

Howe, Frederic C., "The Remaking of the American City," *Harper's*, July 1913.

Hughes, Thomas P., *Networks of Power*. Johns Hopkins University Press, 1983.

Huxley, Aldous, *Proper Studies*. Harper & Brothers, 1927, 1955.

———, *Brave New World*. HarperPerennial, 1932.

Jewkes, John, David Sawers and Richard Stillerman, *The Sources of Invention*. St. Martin's Press, 1959.

Jordan, Michael H., "Humanism in a Technological Age," *Vital Speeches* 51(7), 15 January 1985.

Kaempffert, Waldemar, "Why Flying-Machines Fly," *Harper's*, April 1911.

Kelly, Fred C., ed., *Miracle at Kitty Hawk: The Letters of Wilbur and Orville Wright*. Farrar, Straus and Young, 1951.

Koon, Sidney G., "10,000 Automobile Frames a Day," *The Iron Age*, 5 June 1930.

Koppes, Clayton R., "The Social Destiny of the Radio: Hope and Disillusionment in the 1920's," *South Atlantic Quarterly* 68, Summer 1969.

Kouwenhoven, John A., *The Beer Can by the Highway*. Johns Hopkins University Press, 1961.

Kreinberg, Nancy, and Elizabeth K. Stage, "EQUALS in Computer Technology," in Jan Zimmerman, ed., *The Technological Woman*. Prager, 1983.

Kubrick, Stanley, and Terry Southern, "Dr. Strangelove, or How I Learned to Stop Worrying and Love the Bomb." Hawk Films Ltd., Shepperton Studios, 1963.

Land, Edwin H., "Research by the Business Itself," in *Selected Papers on Industry*, Polaroid Corporation, 1983.

LaPorte, Todd R., "Technology Observed," *Science* 188(4184).

Layton, Edwin T., Jr., "Through the Looking Glass, or News From Lake Mirror Image," *Technology and Culture* 28(3), July 1987.

Lerner, Max, ed., *The Portable Veblen*. Viking, 1948.

Licklider, J. C. R., "Man-Computer Symbiosis," *IRE Transactions* HFE-1(1), March 1960.

———, "The Computer as a Communication Device," *Science and Technology*, April 1968.

Lindbergh, Charles A., *The Spirit of St. Louis*. Charles Scribner's Sons, 1953.

Lundberg, George A., "Can Science Save Us?" *Harper's*, December 1945.

Mackaye, Benton, and Lewis Mumford, "Townless Highways for the Motorist," *Harper's*, August 1931.

MacLeish, Archibald, "Machines and the Future," *Nation* 136(3527), 8 February 1933.

Maclaurin, Richard C., "Mr. Edison's Service for Science," *Science* XLI, 4 June 1915.

Maddocks, Melvin, *The Great Liners*. Time-Life Books, 1979.

Manchester, Harland, "Inventors and Fighters," *Atlantic Monthly*, July 1943.

Manley, John, "Return Technology to Human Hands," *Bulletin of the Atomic Scientists* 43(10), December 1987.

Matthews, Robert A. J., "The Science of Murphy's Law," *Scientific American*, April 1997.

Mauskopf, Seymour H., ed., *Chemical Sciences in the Modern World*. University of Pennsylvania Press, 1993.

Mazlish, Bruce, "The Fourth Discontinuity," *Technology and Culture* 8(1), January 1967.

McCullough, David, *The Path Between the Seas*. Simon & Schuster, 1978.

McFarland, Marvin W., ed., *The Papers of Wilbur and Orville Wright*. McGraw-Hill, 1953.

Meikle, Jeffrey L., *American Plastic: A Cultural History*. Rutgers University Press, 1995.

Merwin, W. S., *The Lice*. Macmillan, 1967.

Mesthene, Emmanuel G., ed., *Technology and Social Change*. Bobbs-Merrill, 1967.

Mitchell, William, *Memoirs of World War I*. Random House, 1960.

Moore, Gordon E., "Progress in Digital Integrated Electronics," *Electronics*, 1965.
———, "Lithography and the Future of Moore's Law," *SPIE* 2438, 1995.
Morison, Elting E., *From Know-How to Nowhere*. Basic Books, 1974.
Mumford, Lewis, *Technics and Civilization*. Harcourt, Brace, 1934.
———, *The Pentagon of Power*. Harcourt Brace Jovanovich, 1970.
Nakicenovic, Nebojsa, "The Automobile Road to Technological Change," *Technological Forecasting and Social Change* 29, 1986.
Nicholson, Joseph L., and George R. Leighton, "Plastics Come of Age," *Harper's*, August 1942.
Nisbet, Robert A., "The Year 2000 and All That," *Commentary* 45(6), June 1968.
Olesen, Douglas E., "Technology," *Vital Speeches* 53(3), 15 November 1986.
Oppenheimer, J. R., "Physics and the Contemporary World," *Bulletin of the Atomic Scientists* 4(3), March 1948.
Orwell, George, *1984*. Harcourt Brace, 1949, 1977.
Patterson, James T., *America in the Twentieth Century*. Harcourt Brace Jovanovich, 1983.
Piel, Gerard, *The Acceleration of History*. Knopf, 1972.
Printers' Ink 122(6), 8 February 1923.
Pursell, Carroll W., Jr., *Readings in Technology and American Life*. Oxford University Press, 1969.
Pursell, Carroll W., Jr., ed., *Technology in America*. MIT Press, 1981.
Rapp, Rayna, "Communicating About the New Reproductive Technologies: Cultural, Interpersonal and Linguistic Determinants of Understanding," in Judith Rodin and Aila Collins, eds., *Women and New Reproductive Technologies*. Lawrence Erlbaum Associates, 1991.
Reed, Henry, *A Map of Verona and Other Poems*. Harcourt, Brace, 1947.
Reingold, Howard, *The Virtual Community*. Addison-Wesley, 1994.
Remnick, David, "Future Perfect," *New Yorker*, 20 and 27 October 1997.
Rhodes, Richard. *The Inland Ground* (revised edition). University Press of Kansas, 1991.
———, *Nuclear Renewal*. Whittle Books/Viking, 1993.
Rich, Frank, "Computer Bites Man," *New York Times*, 18 May 1997.
Ridgway, Robert, "The Modern City and the Engineer's Relation to It," *Transactions of the American Society of Civil Engineers*, LXXXVII, 1925.
Riesman, David, *The Lonely Crowd*. Yale University Press, 1950.
Romanyshyn, Robert D., *Technology as Symptom and Dream*. Routledge, 1989.
Roszak, Theodore, *Where the Wasteland Ends*. Doubleday, 1972.
Sanger, Margaret, *Woman and the New Race*. Maxwell Reprint Co., 1969.
Scarry, Elaine, *The Body in Pain*. Oxford University Press, 1985.
Schumacher, E. F., *Small Is Beautiful*. Harper & Row, 1973.
Sears, Paul B., "Science and the New Landscape." *Harper's*, July 1939.
Simon, Herbert A., "A Computer for Everyman," *The American Scholar* 35(2), Spring 1966.

Simon, Julian L., *The State of Humanity*. Blackwell, 1995.

Skinner, B. F., "Freedom and the Control of Man," *The American Scholar* 25, Winter 1955–1956.

Skinner, Charles M., "The Electric Car." *Atlantic Monthly*, June 1902.

Skolnikoff, Eugene B., "Technology and the World Tomorrow," *Current History* 88(534), January 1989.

Sloan, Alfred P., Jr., "The Forward View," *Atlantic Monthly*, September 1934.

———, *Adventures of a White-Collar Man*. Doubleday, Doran, 1941.

Smith, Alice Kimball, and Charles Weiner, eds., *Robert Oppenheimer: Letters and Recollections*. Harvard University Press, 1980.

Smith, Bernard B., "The Radio Boom and the Public Interest," *Harper's*, March 1945.

Smith, Cyril Stanley, *A Search for Structure*. MIT Press, 1981.

Snell, Joel, "Impacts of Robotic Sex," *The Futurist*, July–August 1997.

Snow, C. P., *The Two Cultures*. New American Library, 1959, 1963.

Starr, Chauncey, "Social Benefit Versus Technological Risk," *Science* 165(3899), 19 September 1969.

Steinbeck, John, *Travels with Charley*. Penguin, 1962.

Sullivan, Mark, *Our Times*. 6 vols. Charles Scribner's Sons, 1926–1935.

Susskind, Charles, *Understanding Technology*. Johns Hopkins University Press, 1973.

Technological Trends and National Policy, USGPO, 1937.

Temporary National Economic Committee, *Investigation of Concentration of Economic Power* (Hearings before the Committee, 76th U.S. Congress, 3rd Session), USGPO, 1940.

———, Investigation of Concentration of Economic Power, *Final Report of the Executive Secretary*, 77th Congress, 1st Session, USGPO, 1941.

Thomas, Lewis, *Late Night Thoughts on Listening to Mahler's Ninth Symphony*. Viking, 1983.

Thompson, Roy A. H., "What's Happening to the Timber?" *Harper's*, August 1945.

Train, Arthur, Jr., "Catching Up with the Inventors," *Harper's*, March 1938.

Tufte, Edward, *Visual Explanations*. Graphics Press, 1997.

U.S. House of Representatives, "Investigation of Taylor System of Shop Management." Hearings before the Committee on Labor of the House of Representatives, 62nd Congress, 1st Session, on House Resolution 90, Washington, D.C., 1911.

Veblen, Thorstein, *The Engineers and the Price System*. Viking, 1921.

von Neumann, John, "Can We Survive Technology?" *Fortune*, June 1955.

Vonnegut, Kurt, Jr., *Slaughterhouse Five*. Dell, 1968.

Wainwright, Loudon, "A Dark Night to Remember," *Life*, 19 November 1965.

Weingart, Jerome Martin, "The Helios Strategy," *Technological Forecasting and Social Change* 12(4), 1978.

Wells, H. G., *The World Set Free*. Dutton, 1914.

White, E. B., "The World of Tomorrow," in *One Man's Meat*. Tilbury House, 1997.

White, Kevin M., and Samuel H. Preston, "How Many Americans Are Alive Because of Twentieth-Century Improvements in Mortality?" *Population and Development Review* 22(3): 415–428, September 1996.

Whitehead, Alfred North, *Science and the Modern World.* Macmillan, 1948.

Wiener, Norbert, "Some Moral and Technical Consequences of Automation," *Science* 131(3410), 6 May 1960.

———, *God and Golem, Inc.* MIT Press, 1964.

Wigglesworth, V. B., "DDT and the Balance of Nature," *Atlantic Monthly*, December 1945.

———, "The Contribution of Pure Science to Applied Biology," *Annals of Applied Biology* 42, 1955.

Wilson, Edward O., *On Human Nature.* Harvard University Press, 1978.

Winner, Langdon, *Autonomous Technology.* MIT Press, 1977.

Index

THE SLOAN TECHNOLOGY SERIES

Dark Sun: The Making of the Hydrogen Bomb by Richard Rhodes

Dream Reaper: The Story of an Old-Fashioned Inventor in the High-Stakes World of Modern Agriculture by Craig Canine

Turbulent Skies: The History of Commercial Aviation by Thomas A. Heppenheimer

Tube: The Invention of Television by David E. Fisher and Marshall Jon Fisher

The Invention that Changed the World: How a Small Group of Radar Pioneers Won the Second World War and Launched a Technological Revolution by Robert Buderi

Computer: A History of the Information Machine by Martin Campbell-Kelly and William Aspray

Naked to the Bone: Medical Imaging in the Twentieth Century by Bettyann Kevles

A Commotion in the Blood: A Century of Using the Immune System to Battle Cancer and Other Diseases by Stephen S. Hall

Beyond Engineering: How Society Shapes Technology by Robert Pool

The One Best Way: Frederick Winslow Taylor and the Enigma of Efficiency by Robert Kanigel

Crystal Fire: The Birth of the Information Age by Michael Riordan and Lillian Hoddesen

Edwin Land: American Innovator by Victor McElheny

City of Light by Jeff Hecht (forthcoming)

Visions of Technology: A Century of Vital Debate About Machines, Systems and the Human World edited by Richard Rhodes

Silicon Sky by Gary Dorsey

Permissions, continued from page 8.

Addison Wesley Longman for an excerpt from *Darwin Among the Machines* by George B. Dyson © 1997 by George B. Dyson.

American Astronautical Society for an excerpt from "National Programs and the Progress of Technological Societies" by T. J. Gordon and A. L. Shef, originally presented at the Sixth AAS Goddard Memorial Symposium, March 1968, Washington, D.C., and originally published in *Technology and Social Progress—Synergism or Conflict?*, American Astronautical Society *Science and Technology Series*, Vol. 18, edited by Philip K. Eckman © 1969 by American Astronautical Society Publications Office.

The American Scholar, Spring 1966, for an excerpt from "A Computer for Everyman" by Herbert A. Simon © 1966 by the Phi Beta Kappa Society.

American Society of Civil Engineers for an excerpt from "The Modern City and the Engineer's Relation to It" by Robert Ridgway, published in *Transactions of the American Society of Civil Engineers,* LXXXVII © 1925 by ASCE.

Antioch Review 25(4), Winter 1965/66, for an excerpt from "The Century of Technocracy" by William Carleton © 1966 by the Antioch Review, Inc.

Bantam Doubleday Dell Publishing Group, Inc., for excerpts from "Technology, Nature and Society" by Daniel Bell, from *Technology and the Frontiers of Knowledge* by Saul Bellow, et al., © 1975 and *Today and Tomorrow* by Henry Ford and Samuel Crowther © 1926 by Doubleday, a division of Bantam Doubleday Dell Publishing Group, Inc.

Basic Books for an excerpt from *Stuff: The Materials the World is Made Of* by Ivan Amato © 1997 by Basics Books, a subsidiary of Perseus Books Group LLC.

Blackwell Publishers for an excerpt from *The State of Humanity* by Julian L. Simon, copyright © 1995 by Julian L. Simon.

Bobbs-Merrill for an excerpt from "Technology and Wisdom" by Emmanuel G. Mesthene in *Technology and Social Change*, edited by Emmanuel G. Mesthene © 1967 by Emmanuel G. Mesthene.

Donald J. Bogue, Community and Family Study Center at University of Chicago, for an excerpt from "The End of the Population Explosion" by Donald J. Bogue in *The Public Interest* © 1967 by National Affairs, Inc.

Georges Borchardt, Inc., for reprint of "For a Coming Extinction" from *The Lice* by W. S. Merwin (New York: Atheneum, 1969) © 1963, 1964, 1965, 1966, 1967 by W. S. Merwin.

The Bulletin of the Atomics Scientists (December 1971) for excerpts from "The Engineer and the Establishment" by Daniel C. Drucker (December 1971); "The New Technological Era: A View from the Law" by Harold P. Green (November 1967); "Return Technology to Human Hands" by John Manley (December 1987); all © 1997 by the Educational Foundation for Nuclear Science.

Vannevar Bush estate for an excerpt from "As We May Think" © July 1945 by Vannevar Bush.

Cambridge University Press for an excerpt from *The Two Cultures* by C. P. Snow © 1959 by C. P. Snow.

Frederico Capasso, Lucent Technologies, for reprint of a letter published in *Nature* 187, June 5, 1997.

Copyright Management, Inc., for reprint of "The Pill" by Loretta Lynn © 1973 by Coal Miners Music and Guaranty Music. All Rights Reserved. International Copyright Secured.

Commentary (June 1968) for an excerpt from "The Year 2000 and All That" by Robert A. Nisbet © 1968 by Commentary.

Bessie Sparkes Eagles and Dorothy Sparkes Primrose for an excerpt from *Adventures of a White-Collar Worker* by Alfred P. Sloan, Jr. (New York: Doubleday, Doran, 1941) © 1941 by Alfred P. Sloan, Jr.

Gil Elliot for an excerpt from *Twentieth Century Book of the Dead* © 1972 by Gil Elliot.

David R. Godine, Publisher, Inc., for an excerpt from *How to Do Things Right* by L. Rust Hills © 1993 by L. Rust Hills.

Elsevier Science for an excerpt from "The Helios Strategy" by Jerome Martin Weingart, in *Technological Forecasting and Social Change* 12(4) © 1978 by Elsevier Science.

Farrar, Straus & Giroux, Inc., for an excerpt from "At the Dam" from *The White Album* by Joan Didion © 1979 by Joan Didion.

Fortune for an excerpt from "Can We Survive Technology?" by John von Neumann © 1955 by Time, Inc.

Sally Goodman for an excerpt from "The Morality of Scientific Technology" by Paul Goodman, from *People or Personnel and Like a Conquered Province*, Vintage Books © 1968 by Paul Goodman.

Harcourt Brace & Company for excerpts from *The Myth of the Machine: The Pentagon of Power*, Vol. II by Lewis Mumford © 1967 by Lewis Mumford; *Technics and Civilization* by Lewis Mumford © 1934 by Harcourt Brace & Company, renewal © 1961 by Lewis Mumford; and *Nineteen Eighty-Four* by George Orwell © 1949 by Harcourt Brace & Company, renewal © 1977 by Sonia Brownell Orwell.

HarperCollins Publishers, Inc., for excerpts from *Brave New World* by Aldous Huxley © 1932, 1960 by Aldous Huxley and *Small Is Beautiful: Economics As If People Mattered* by E. F. Schumacher © 1973 by E. F. Schumacher.

Harper's Magazine for excerpts from "This Coming Era of Leisure" by Floyd H. Allport © 1931; "In Defense of Machines" by George Boas © 1932; "Air Power and Future History" by John Philips Cranwell © 1940; "The Clash Between Progress and Security" by Allan G. B. Fisher © 1944; "The Anti-Industrial Revolution" by Roy Helton © 1941; "Can Science Save Us" by George A. Lundberg © 1945; "Townless Highways for the Motorist" by Benton Mackaye and Lewis Mumford © 1931; "Plastics Come of Age" by Joseph L. Nicholson and George R. Leighton © 1942; "Science and the New Landscape" by Paul B. Sears © 1939; "The Radio Boom and the Public Interest" by Bernard B. Smith © 1945; "What's Happening to the Timber" by Roy A. H. Thompson © 1945; "Catching Up with the Inventors" by Arthur Train, Jr. © 1938; all by *Harper's Magazine*.

Harvard University Press for excerpts from *Imagined Worlds* by Freeman Dyson © 1997 and *On Human Nature* by E. O. Wilson © 1978 by the President and Fellows of Harvard College.

Henry Holt & Company, Inc., for an excerpt from *The Poetry of Robert Frost* edited by Edward Connery Lathem © 1944 by Robert Frost; © 1969 by Henry Holt & Company, Inc.

The Johns Hopkins University Press for excerpts from "The Hot-Rod Culture" by Gene Balsley in *American Quarterly* © 1983 and *Networks of Power* by Thomas P. Hughes © 1983 by Johns Hopkins University Press.

Houghton Mifflin Co. for an excerpt from *Silent Spring* by Rachel Carson © 1962 by Rachel L. Carson, renewal © 1990 by Roger Christie.

Huxley Estate for an excerpt from *Proper Studies* by Aldous Huxley © 1955 by Aldous Huxley.

Institute of Electrical and Electronics Engineers, Inc., for an excerpt from "Controlling Technology" by Norman Balabanian in *IEEE Technology and Society Magazine* (13)1, Spring 1994 © 1994 by IEEE.

Institute of Electrical and Electronics Engineers, Inc., and Tracy Licklider for an excerpt from "Man-Computer Symbiosis" by J. C. R. Licklider in *IRE Transactions* HFE-1(1), March 1960 © 1960 by IEEE.

Alfred A. Knopf for excerpts from *The Technological Society* by Jacques Ellul, trans. John Wilkinson © 1964 by Alfred A. Knopf, Inc., and *The Acceleration of History* by Gerard Piel © 1972 by Gerard Piel.

Stanley Kubrick for an excerpt from *Dr. Strangelove, or How I Learned to Stop Worrying and Love the Bomb* by Stanley Kubrick and Terry Southern (Hawk Films, Ltd.) © 1963 by Stanley Kubrick.

Tracy Licklider for an excerpt from "The Computer as a Communication Device" by J. C. R. Licklider in *Science and Technology* (April 1968) © 1968 by J. C. R. Licklider.

David McCullough for an excerpt from *The Path Between the Seas* © 1977 by David McCullough.

The MIT Press for an excerpt from *A Search for Structure* by Cyril Stanley Smith © 1981 by The MIT Press.

Elizabeth F. Morison for an excerpt from *From Know-How to Nowhere* by Elting E. Morison © 1945 by Elting E. Morison.

William Morrow & Company, Inc. for an excerpt from *Pieces of the Action* by Vannevar Bush © 1970 by Vannevar Bush.

The Nation for an excerpt from "Machines and the Future" by Archibald MacLeish © 1933 by *The Nation*.

The New Yorker for an excerpt from "Future Perfect" by David Remnick © 1997 by The New Yorker Magazine, Inc.

The New York Times Company for an excerpt from "Computer Bites Man" by Frank Rich © 1997 by The New York Times Company.

Oxford University Press, Inc., for excerpts from *The Body in Pain* by Elaine Scarry © 1985 and "Naming of Parts" by Henry Reed from *Collected Poems*, edited by Jon Stallworthy © 1991 by Oxford University Press, Inc.

Polaroid Corporation for an excerpt from "Research by the Business Itself" by Edwin H. Land, from *Selected Papers on Industry* © 1983 by the Polaroid Corporation.

The Population Council for an excerpt from "How Many Americans Are Alive Because of Twentieth-Century Improvements in Mortality?" by Kevin M. White and Samuel H. Preston, from *Population and Development Review* 22(3) © 1996 by The Population Council.

Random House, Inc., for an excerpt from *Memoirs of World War I* by Brigadier General William Mitchell © 1928 and renewal by Lorraine Lester and Associates © 1960 by Lucy M. Gilpin.

Theodore Roszak for an excerpt from *Where the Wasteland Ends* © 1972 by Theodore Roszak.

Russell & Volkening as agents for the author for an excerpt from "Inventors and Fighters" by Harland Manchester, published in the *Atlantic Monthly* © 1943 by Harland Manchester, renewal © 1971 by Harland Manchester.

St. Martin's Press, Inc., for an excerpt from *The Civilized Engineer* by Samuel C. Florman © 1987 by Samuel C. Florman.

PICTURE CREDITS